Biohydrogen Production

Fundamentals and Technology Advances

Debabrata Das

Namita Khanna

Chitralekha Nag Dasgupta

CRC Press
Taylor & Francis Group
Boca Raton London New York

CRC Press is an imprint of the
Taylor & Francis Group, an **informa** business

CRC Press
Taylor & Francis Group
6000 Broken Sound Parkway NW, Suite 300
Boca Raton, FL 33487-2742

First issued in paperback 2017

© 2014 by Taylor & Francis Group, LLC
CRC Press is an imprint of Taylor & Francis Group, an Informa business

No claim to original U.S. Government works

ISBN-13: 978-1-4665-1799-8 (hbk)
ISBN-13: 978-1-138-07320-3 (pbk)

Library of Congress Cataloging-in-Publication Data

Das, Debabrata, 1953-
  Biohydrogen production : fundamentals and technology advances / Debabrata Das, Namita Khanna, and Chitralekha Nag Dasgupta.
       pages cm
  Includes bibliographical references and index.
  ISBN 978-1-4665-1799-8 (alk. paper)
  1. Hydrogen--Biotechnology. 2. Biomass energy. I. Khanna, Namita, 1982- II. Dasgupta, Chitralekha Nag, 1975- III. Title.

TP248.65.H9D37 2013
665.8'1--dc23   2013034177

Visit the Taylor & Francis Web site at
http://www.taylorandfrancis.com

and the CRC Press Web site at
http://www.crcpress.com

Biohydrogen Production

Fundamentals and Technology Advances

Contents

Foreword

Hydrogen energy was proposed some four decades ago as a permanent solution to the interrelated global problems of the depletion of fossil fuels and the environmental problems caused by their utilization. It was formally presented at the landmark The Hydrogen Economy Miami Energy (THEME) Conference, March 18–20, 1974, Miami Beach, Florida, the United States. Hydrogen is the most efficient and the cleanest fuel. Its combustion will produce no greenhouse gases, no ozone layer–depleting chemicals, little or no acid rain ingredients, no oxygen depletion, and no pollution.

Of course, hydrogen is a synthetic fuel and it must be manufactured. There are various hydrogen-manufacturing methods such as direct thermal, thermochemical, electrochemical, biological, and so on. Among the hydrogen production methods, the biological method has the potential of resulting in the most cost-effective hydrogen. Because of this, many research groups around the world are working on biological hydrogen production. In several cases, bench-scale production systems have come up with encouraging results.

Clearly, the time has arrived for a book on biohydrogen production technologies. I congratulate the authors Debabrata Das, Namita Khanna, and Chitralekha Nag Dasgupta, for seeing the need for such a book and producing it. This book entitled *Biohydrogen Production: Fundamentals and Technology Advances* covers biological hydrogen production authoritatively from A to Z, including microbiology, hydrogen production processes, biohydrogen feedstocks, molecular biology of hydrogenases, genetic and metabolic engineering for enhanced hydrogen production, influence of physicochemical parameters on biohydrogen production, photobioreactors, mathematical modeling and simulation of biohydrogen production processes, scale-up and energy analysis of biohydrogen production processes, and biohydrogen production process economics, policy, and environmental impact.

I strongly recommend this excellent book to energy scientists, engineers, and students who are interested in hydrogen production in general and biological hydrogen production in particular, as well as to industrial concerns that are looking for inexpensive hydrogen production technologies.

T. Nejat Veziroğlu
President, International Association for Hydrogen Energy

Preface

The future is green energy, sustainability, renewable energy.

Arnold Schwarzenegger
Former Governor of California

Today, the world consumes 85 million barrels of oil per day, and the demand is only growing exponentially. Oil production in 38 out of 44 countries including Kuwait, Russia, and Mexico has already peaked. It is widely acknowledged that 95% of all the recoverable oil has been extracted and to date we have consumed 2.5 trillion barrels of oil. The discovery of oil wells peaked in the 1960s and has followed a steady decline ever since. There have been no significant discoveries of oil wells since 2002. In 2001, there were eight large-scale discoveries, and in 2002 there were three such discoveries. In 2003, there were no large-scale discoveries of oil. Since 1981 we have consumed oil at a much faster rate than its discovery. This intensive energy fuel took around 50–300 million years to develop; however, we have managed to consume it in 125 years or so!

Along with oil, natural gas reservoirs are also diminishing. However, coal is still an abundant resource but its energy profile (15–19 MJ/kg) is less than half that of oil (40–45 MJ/kg). Moreover, it is distributed unevenly with countries such as the United States, Russia, and China having the largest reserves. Developed countries and a booming economy would require more reserves. In such a scenario, its export would be much dearer as compared to oil.

In view of this, the hype about alternative sources of fuel appears to be the need of the hour. Governments all around the world are heavily investing in this research. Sustainable alternatives based on high-energy content and a low emission rate of pollutants are desired. In this respect, renewable energy resources are critical in the search for alternatives for fossil-based raw materials. Biomass-based renewable resources along with solar, wind, tidal geothermal, and nuclear together may offer a promising solution. Among all the alternatives, biohydrogen is touted as the most promising by virtue of the fact that it is renewable, does not evolve into greenhouse gas, has a high energy content per unit mass of any known fuel (143 GJ/t), and on combustion gives water as the only by-product. Presently, only about 1% is produced from biomass while 40% H_2 is produced from natural gas, 30% from heavy oils and naphtha, 18% from coal, and 4% from electrolysis. However, today, biological H_2 production processes are becoming important mainly due to two reasons: (i) they can utilize renewable energy resources and (ii) they are usually operated at ambient temperature and atmospheric pressure.

Hydrogen production using biological processes is new, innovative, and potentially more efficient in the direct conversion of solar energy and biomass to hydrogen. However, in order to realize the complete potential of this technology, practical hydrogen processes, advanced low-cost technologies, bioreactors, and systems with oxygen-tolerant hydrogenase with high-efficiency need to be developed and engineered. The work is challenging and there is an immediate need to develop global cooperation, understanding, and concerted R&D efforts in this direction.

This book is an attempt to present the fundamentals and the state-of-the-art biohydrogen production technology to the research community, entrepreneurs, academicians, and industrialists. It is a comprehensive collection of chapters related to microbiology, biochemistry, feedstock requirements, and molecular biology of the biological hydrogen production processes. Additionally, the book gives the readers an deep insight into the scale-up of the processes and the engineering perspective of this technology. Besides, the beauty of any innovation is its applicability, socioeconomic concern, and cost of energy analysis. The book comprehensively covers each of these points to give the reader a holistic picture about this technology. Further, the book summarizes the recent research advances that have been made in this field and also discusses the bottlenecks of the various processes that currently limit the commercialization of this technology.

This book is aimed at a wide audience, mainly undergraduates, postgraduates, energy researchers, scientists in industries and organizations, energy specialists, policy makers, research faculty, and others who wish to know the fundamentals of the biohydrogen technology, and also the authors of this book wished to keep abreast with the latest developments. Each chapter begins with a fundamental explanation for general readers and ends with in-depth scientific details suitable for expert readers. Various bioengineering and biohydrogen process laboratories may find this book a ready reference for their routine use.

We hope this book will be useful to our readers!

Authors

Dr. Debabrata Das earned his doctoral studies from the Indian Institute of Technology Delhi. He is an associate professor at IIT Kharagpur. He has pioneered promising R&D of bioenergy production processes by applying fermentation technology. He has been actively involved in research of hydrogen, biotechnology for the last 13 years. His commendable contributions toward development of a commercially competitive and environmentally benign bioprocess began with the isolation and characterization of the high-yielding bacterial strain *Enterobacter cloacae* IIT-BT 08, which, as of today, is known to be the highest producer of hydrogen by fermentation. Dr. Das has conducted basic scientific research on the standardization of physicochemical parameters in terms of maximum productivity of hydrogen by fermentation and made significant contribution toward enhancement of hydrogen yield by redirection of biochemical pathways. Dr. Das has also conducted a modeling and simulation study of a continuous immobilized whole cell hydrogen production system using lignocellulosic materials as solid matrix. Apart from pure substrates, the utilizations of several other industrial wastewaters such as distillery effluent, starchy wastewater, de-oiled cake of several agricultural seeds such as groundnut, coconut, and cheese whey were also explored successfully as feedstock for hydrogen fermentation. The aim was to synchronize the bioremediation of wastewater with clean energy generation.

He has also been associated as MNRE Renewable Energy Chair Professor at IIT Kharagpur. *Thomson Reuters ISI h-index of his published research papers is 26.* Dr. Das has about 100 research publications in the peer-reviewed journals and contributed more than 14 chapters in books published by international publishers and has two Indian patents. He is the editor-in-chief of the *American Journal of Biomass and Bioenergy* and is also a member of the editorial board of the *International Journal of Hydrogen Energy; Biotechnology for Biofuels;* and the *Indian Journal of Biotechnology*. He has successfully completed six pilot plant studies in different locations in India and is involved in several national and international sponsored research projects such as NSF, USA, and DAAD, Germany etc. He has been leading a technology mission project of the Ministry of New and Renewable Energy (MNRE), Government of India, for the installation of several pilot plants on biohydrogen production processes in different locations in India for its commercial exploitations. Dr. Das has organized several international Conferences/Workshops in India and has been awarded the IAHE Akira Mitsue Award 2008 for his contribution to hydrogen research.

Dr. Namita Khanna earned her M.Sc. degree in biotechnology from the University of Bhubaneswar, Orissa, India in 2007. She then moved to the Indian Institute of Technology (IIT) Kharagpur, West Bengal, India where she earned her PhD under the guidance of Dr. Debabrata Das. Currently, she works as a postdoc at the Department of Photochemistry and Molecular Science, Uppsala University, Sweden. Her research interests focus on hydrogen production from mesophilic and photosynthetic bacteria.

Dr. Chitralekha Nag Dasgupta earned her PhD in plant biotechnology from the Department of Botany, University of Calcutta, India and moved to Faculté de Médecine Necker, INSERM U1001, Paris, France for her postdoctoral research. Her work was on mutagenesis and prokaryotic DNA repair mechanisms. She then joined as a research associate the Department of Biotechnology, IIT Kharagpur, India to work on biohydrogen production processes. She has published several articles in different journals of international repute. Currently, Dr. Dasgupta is a scientist (DST WOS-A) at CSIR-National Botanical Research Institute, Lucknow, India, where she is involved in screening different microorganism for efficient biohydrogen production and trying to improve the process through genetic and metabolic engineering.

1

Introduction

1.1 Introduction

After the March 2012 earthquake and tsunami incident, though Japan soon rebuilt its highway, however, the villages and towns that had been swept away were harder to reestablish. The disaster killed nearly 19,000 people, devastated towns, and caused a meltdown of the Fukushima Daiichi nuclear power plant that kept the island nation on tenterhooks. This incident has made its people wary of relying on nuclear energy in the future. Japan was well prepared to face tough times but probably not the worst. The nuclear power plants were built to withstand quakes and tsunamis that Japan has suffered many times, however, the recent tsunami was bigger in magnitude than the country has witnessed during the past 1000 years. Before the 2012 disaster, roughly 30% of Japan's electricity came from nuclear power, however, post disaster, the Japanese government is rethinking over the continuation and expansion of its nuclear technology program. The disaster has not only left the Japanese government thinking of renewable energy as alternatives, but it also has reaffirmed the faith of other nations on sustainable renewable alternatives to replace the diminishing fossil fuels.

On the basis of such concerns, the governments across the world are looking for safer and greener alternatives to fossil fuels. Biohydrogen appears to be one of the more promising alternatives to the rapidly depleting and polluting fossil fuels. However, before we begin on the merits and demerits of hydrogen, let us first broadly survey our main environmental and energy concerns.

1.1.1 Global Environmental Issues

As early as 1896, the Swedish scientist Svante Arrhenius had predicted that anthropogenic activities would eventually destroy the natural balance resulting in growing environmental concerns. Unfortunately, his predictions have come true. Today, the earth is riddled with major environmental issues such as

- Climate change
- Ozone layer depletion

- Global warming
- Loss of biodiversity
- Desertification

These issues have been discussed in detail elsewhere (Solomon, 1999; Walther et al., 2002; Dirzo and Raven, 2003; Veron et al., 2006; Jacobson, 2009). In view of the severe environmental problems affecting society, the last few decades have seen many treaties, conventions, and protocols for the cause of global environmental protection. However, this is not enough. The problem has to be dealt with at the fundamental level.

Most of the energy we consume today is derived from fossil fuel such as coal, oil shales, tar sands, petroleum, bitumen, and natural gas. These are carbon-rich fossilized remnants of the prehistoric plants and animals, buried between the layers of earth and converted into high-energy molecules by high pressure and temperature inside the core of the earth (Tissot and Welte, 1978). The formation of fossil fuels is a continuous process, albeit an extremely slow one. The oil utilized today was formed due to the process of fossilization almost a billion years ago. In the world of modern day technology, the consumption of fossil fuels is at a much higher rate compared to its production. For this reason, fossil fuels are considered to be nonrenewable. With the current consumption and demand, we may soon run out of sustainable oil. Moreover, burning of the fossil fuels has led to major environmental concerns as discussed above. Thus, there is an emerging need to focus on nonrenewable energy sources that are inexhaustible and also environmentally benign.

1.2 Nonconventional Energy Resources

Today, the need for renewable energy technologies is the foremost challenge to humanity due to the continuing explosion of the human population and anthropogenic climate change. It is estimated that by the year 2050 the Earth will carry 8.9 billion people (Cohen, 2001). Our oil reserves would not be sufficient to support the oil requirements of the overwhelming population. In view of this, Steven Chu, the current United States secretary of energy, addressed the need for renewable energy, as "Necessity is the mother of invention and this is the mother of all necessities."

The renewable energy sources such as biomass, hydropower, wind, solar (thermal and photovoltaic [PV]), geothermal, marine, and hydrogen will play important roles in the world's future energy crisis. The European Union energy policy predicts that by 2050 approximately half of the global energy supply will be derived from renewable resources. Further, it also postulates

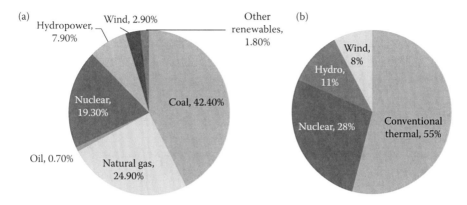

FIGURE 1.1
(a) The United States. (Adapted from the U.S. Environmental Protection Agency (EPA.) Inventory of U.S. Greenhouse Gas Emissions and Sinks: 1990–2010, 2012. http://www.epa.gov/ climatechange/ghgemissions/usinventoryreport.html.) (b) Europe: electricity generation by source, 2011 (in %). (Adapted from Annual report on European Supply, transformation, consumption of electricity (Eurostat) 2013. Retrieved from http://appsso.eurostat.ec.europa.eu/ nui/show.do?dataset=nrg_105a&lang=en on 01 March, 2013.)

that electricity generation from renewables will contribute more than 80% to the total global electricity supply by the year 2040 (Wilkes et al., 2011). This appears to be a significant leap as compared to the current scenario where only 12.6% and 17% electricity is produced by renewable energy sources in the United States and Europe, respectively (Figure 1.1).

Unlike the problem with petroleum fuels regarding their uneven distribution in the world, renewable energy resources are more evenly distributed than fossil or nuclear resources. Also, the energy flow from renewable resources is more than three orders of magnitude higher than current global energy needs. Considering this, it is predicted that sustainable renewable energy sources such as biomass, hydropower, wind, solar (both thermal and PV), geothermal, and marine will play an important role in the world's future energy supply. Some of the major advantages of the use of renewable fuels include

1. Renewable energy is as freely available as the sun and wind and can have the potential to considerably influence the nations' economic parameters.

2. The end use of renewable energy or its production entails no greenhouse gas emission.

3. The cost of construction and operation of power plants is economical. Moreover, the power plants do not consume conventional fossil fuels and hence operational costs are restricted.

4. Accession to remote locations with renewable energy sources is comparably easier as high transmission costs can be curtailed.

5. Low requirement of capital investments.

6. May help generate local employment if designed, constructed, man-ufactured, and operated locally.

7. May be economically more attractive for end users, especially if gov-ernment provides subsidy benefits.

In view of the above advantages, it is predicted that there will be a boom in the scientific and technological advancement of these techniques in the next 8–10 years making it imperative for the international decision makers and planners to keep abreast of these developments.

The various alternative sources of energy and their advantages are briefly discussed below. The advantages and disadvantages of the technology are discussed in Table 1.1.

1.2.1 Solar Energy

Solar energy is the primary energy received from the sun that sustains life on the earth. For many decades, solar energy has been considered as a huge source of energy and also an economical source because it is freely available. However, it is only now after years of research that technology has made it possible to harness this unlimited supply of free energy. On the surface of the earth's orbit, normal to the sun, solar radiation hits at the rate of 1.366 kW/m^2 (Chen, 2011). This is known as solar constant. While 19% of this energy is absorbed in the atmosphere, 35% gets reflected by clouds. Thus, the solar energy that reaches the sea level is much reduced. However, it has been estimated that even if we can harness 5% of this energy, it is more than sufficient to meet the energy requirements of an individual on this earth.

Electricity can be produced from solar energy by PV solar cells and/or by solar panels, which convert the solar energy directly into electricity (Figure 1.2). However, these solar panels can be installed only in places that get optimal sunlight. The mode of operation of solar panels is simple. Each panel consists of several PV cells. When sunlight (photons) strikes the PV cell, some photons pass right through, some are reflected, and the other photons are absorbed. The absorbed photons hit electrons and make them lose their place around the nucleus of the atom. The electrons then cross a barrier that is located inside the panels. The only way the electrons can get back is by connecting the positive side to the negative side of the panel with a wire. When this happens, it creates a direct current (DC). The DC has to go through a certain machine called an inverter that changes DC into alternating current (AC). AC instead of going in straight line goes up and down (periodically reverses direction). This is required as most of the residential appliances work on AC current because it is safer to use as compared to DC current. The most sig-nificant applications of the PV cell include local power (residential), local

TABLE 1.1

Advantages and Disadvantages of the Various Renewable Energy Technologies

Type of Energy	Suitable Location for Harnessing	Benefits	Lacuna
Solar energy	Countries which get high amount of sunlight including equatorial and tropical countries	• Renewable and clean • No greenhouse gas emission • Once installed no utility bills • High government subsidy • Technology is becoming economy driven and affordable	• The initial cost of the installation and equipment is high • Costly and polluting chemicals are used in its manufacture • Requires space for installing solar panels • Can be installed only at places with optimal sunlight • Does not provide energy at night. This requires storing energy in batteries which are expensive
Wind energy	Coastal areas having high wind velocity	• Free and technology to harness it is well developed • Requires small area of land • No greenhouse gas emission • Remote areas that are not connected to the electricity power grid can use wind turbines to produce their own supply	• The strength of the wind is not constant, thus electricity produced by the turbine is not constant • Wind turbines are noisy. Each one can generate the same level of noise as a family car traveling at 70 mph • Setting up a wind turbine is not pollution free • Several windmills are required to power whole village or communities, thereby requiring large areas of free unobstructed land
Hydropower	Anywhere near a small canal to a large river	• Requires more readily available locations • Can be manually controlled by engineers by controlling the flow of water • Can be transported over long distances • Is clean • Reservoirs may offer recreational opportunities, such as swimming and boating	• Damming rivers may destroy or disrupt wildlife and other natural resources • Hydropower plants can also cause low dissolved oxygen levels in the water, which is harmful to river habitats

continued

TABLE 1.1 (continued)
Advantages and Disadvantages of the Various Renewable Energy Technologies

Type of Energy	Suitable Location for Harnessing	Benefits	Lacuna
Tidal power	Suitable only near coastal regions	• As 71% of the earth's surface is covered by water, there is scope to generate this energy on large scale • We can predict the rise and fall of tides as they follow cyclic fashion • Efficiency of tidal power is far greater as compared to coal, solar, or wind energy • Low maintenance costs • Long life of tidal power plant • Energy density of tidal energy is relatively higher than other renewable energy sources	• High cost of construction • End consumers located far away • Few suitable locations for installation • Velocity of incoming waves is unpredictable • Can only generate when the tide is flowing in or out in other words, only for ~10 h each day • Effects aquatic life
Geothermal	Areas which have rich resources for hot water springs	• Clean technology • Can be directly used • Facilitates job creation • Economical	• Finding geothermal hot spots to extract energy at a suitable location is difficult • Sometimes along with steam harmful gases may also escape from the earth's interior
Biomass	Freely available waste material of plants and animals, wood, agricultural wastes, dead parts of plants and animals	• Provides manure as by-product which can be used for agriculture • Relatively cheap and reliable technology • Effective recycling of waste	• Cost of construction of biogas plant is high • Continuous supply of biomass is required to fuel the plant • Difficult to store biogas in cylinders • Long-distance transport of biogas is difficult • Crops to produce biogas are seasonal and not available throughout the year

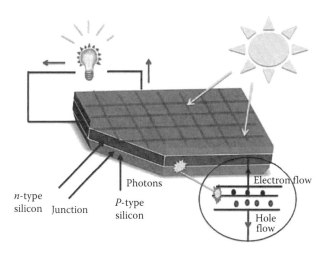

FIGURE 1.2
Functional operation of a solar panel.

industry/agriculture (irrigation pumps, etc.) and local community (street lamps, public parks, etc.).

1.2.2 Wind Energy

Wind is nature's gift and when harnessed optimally can provide great advantages economically. Primarily, the sun heats the earth's surface unevenly. As heat is transferred to the air, we get regions of warm and cool air that can turn into regions of low and high pressure. This difference in pressure makes a force that causes the wind to blow. Thus, in the true sense, wind is the indirect source of solar conversion. Wind power can be used to drive wind turbines such as windmills, which, in turn, drive a generator to produce electricity. A wind energy system usually requires an average annual wind speed of at least 15 km/h. Table 1.2 represents a guideline of different wind speeds and their potential in producing electricity.

To harness maximum energy, the wind turbines are mounted approximately 30 m or more above the ground. Turbines catch the wind's energy with their propeller-like blades. Usually, two or three blades are mounted on a shaft to form a rotor. A blade acts much like an airplane wing. When the wind blows, a pocket of low-pressure air forms on the downwind side of the blade. The low-pressure air pocket then pulls the blade toward it, causing the rotor to turn. This is called lift. The force of the lift is actually much stronger than the wind's force against the front side of the blade, which is called drag. The combination of lift and drag causes the rotor to spin like a propeller, and the turning shaft spins a generator to make electricity.

TABLE 1.2

Guidelines of Different Wind Speeds and Their Potential in Producing
the Electricity

Average Wind Velocity (km/h)	Suitability
<15	Not suitable to operate wind turbines
15–18	Wind turbines can be operated with low output
22	Wind turbines can be operated with moderate output
>25	Locations which have wind velocity above 25 km/h are the most suitable for operating wind turbines

Source: http://www1.agric.gov.ab.ca/$department/deptdocs.nsf/all/eng4445.

Among the different renewable energy sources, wind energy is currently
making a significant contribution to the installed capacity of power generation, and is emerging as a competitive option. India with an installed capacity of 3000 MW ranks fifth in the world after Germany, the United States,
Spain, and Denmark in wind power generation.

Among other advantages, wind turbines are available in a range of sizes,
which means that a vast range of people, and businesses can use them.
Single households to small towns and villages can make good use of a range
of wind turbines available today.

1.2.3 Hydropower

Hydropower is electricity generated using the energy of moving water.
Rain or melted snow, usually originating in the hills and mountains, create
streams and rivers that eventually flow into the ocean. Therefore, the major
advantage of this technology is that it does not have a location constrain.
Hydroelectric power provides almost one-fifth of the world's electricity.
China, Canada, Brazil, the United States, and Russia are currently the five
largest producers of hydropower. In fact, one of the world's largest hydroplant is located at Three Gorges dam on China's Yangtze River.

A typical hydroplant is a system with three parts: an electric plant where
the electricity is produced; a dam that can be opened or closed to control
water flow; and a reservoir where water can be stored. The water behind
the dam flows through an intake and pushes against blades in a turbine,
causing them to turn. The turbine spins a generator to produce electricity.
The amount of electricity that can be generated depends on how far the
water drops and how much water moves through the system. The electricity can be transported over long-distance electric lines to homes and
factories.

1.2.4 Tidal Energy

Tides are the waves caused due to the gravitational pull of the heavenly bodies such as the moon and also the sun (though its pull is very low). The rise is called high tide and fall is called low tide. This building up and receding of waves happens twice a day and causes enormous movement of water. Thus, tidal energy forms a large source of energy and can be harnessed in some of the coastal areas of the world. Tidal dams are built near shores for this purpose. During high tide, the water flows into the dam and during low tide, water flows out which results in turning the turbine.

A major drawback of the tidal power is that it can be operated only for around 10 h including high and low tides. Moreover, their maximal power output does not naturally combine with the maximum human activity/consumption. Further, they need to be transported long distances for optimal use. Therefore, Gorlov (2001) suggested the use of tidal power *in situ* for powering the hydrogen fuel cell. The hydrogen, thus generated, can be stored and transported over long distances when required.

1.2.5 Geothermal Energy

The terms geo means the earth and thermal means heat. Therefore, geothermal energy is the heat from the earth. It is clean and sustainable. Resources of geothermal energy range from the shallow ground to hot water and hot rock found a few miles beneath the earth's surface, and down even deeper to the extremely high temperatures of molten rock called magma. It is known that for every 100 m you go below ground, the temperature of the rock increases by about 3°C.

Geothermal power plants are like regular power plants except that no fuel is burned to heat water into steam. The steam or hot water in a geothermal power plant is heated by the earth. Holes are drilled and tubes are lowered to channelize the steam into the tubes. The tubes are connected to a special turbine. The turbine blades spin, and the shaft from the turbine is connected to a generator to make electricity. The steam then gets cooled off in a cooling tower. The cooled water can then be pumped back below ground to be reheated by the earth.

The technology is relatively cheap and easy to use. However, the limitation of the process is mostly the identification of a suitable location for building the power plant. Predominantly, these areas are located near volcanoes or along the fault lines, which are not ideal locations for the construction of these power plants.

1.2.6 Biomass Energy

The use of biomass as energy dates back to prehistoric times when man first learnt to use wood to fuel his needs. Till date, wood continues to be our largest biomass and energy source. However, today, besides wood other sources can be used including plant residues from agriculture or forestry and the organic component of municipal and industrial wastes. Even the fumes from landfills can be used as a biomass energy source. The use of biomass energy has the potential to greatly reduce our greenhouse gas emissions. Biomass generates the same amount of CO_2 as fossil fuels; however, the net emission will be zero if a sapling is planted for every plant removed. The biggest attraction of the technology is its capability to use municipal waste to generate energy. Moreover, unlike other processes, it can directly be used to produce liquid and gaseous fuel. The three major biomass energy applications include (i) production of biofuels such as alcohol and hydrogen to be used for transportation; (ii) green biopower production to generate electricity; and (iii) production of bioproducts. Biomass can be treated in

DIE ENERGIEWENDE (THE ENERGY CHANGE)

In the wake of the Fukushima disaster in Japan in March 2011, Germany has accelerated its proposed phase out of the nuclear power plants. Germany has already suspended eight of its power plants—at least temporarily. With renewed efforts, the Germans are depicting admirable courage in turning the country into a laboratory for energy policies. In its renewable quest for energy, Germany has ambitious plans. *Energiewende* or energy transition is a long-term plan to clean up the country's energy system. The primary motivation of the *Energiewende* is to combat climate change. Germany hopes to generate 35% of its electricity from green sources by 2020; by 2050 the share is expected to surpass 80%. With this, by 2020, Germany aims to cut its greenhouse gas emissions by 40% below 1990 production levels, and further it hopes to achieve a reduction of 80% by 2050 (Schiermeier, 2013). *Energiewende* is the world's most extensive embrace of wind and solar power as well as other forms of renewable energy and enjoys extensive government and public support. To reach its target, Germany is currently investing more than €1.5 billion per year in energy research. The *Energiewende* is already visible in the German countryside where expensive solar panels adorn more than one million houses, farms, and warehouses, thanks to generous subsidies. Germany's Renewable Energy Act (EEG) allows owners of solar panels and wind turbines to sell their electricity to the grid at a fixed, elevated price. The excess electricity is stored by using pumps to push water uphill into reservoirs. Germany with its population of 80 million people, a stable economy, and socially and geographically diverse region serves as the ideal ground to carry out this experiment while the world looks on....

Ein hoch auf die energiewende (Long live the energy change.)

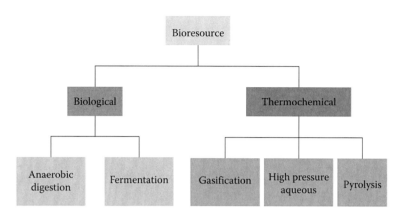

FIGURE 1.3
Pathways showing conversión of biomass into energy.

different ways: biological treatment or thermochemical treatment to extract energy out of the same (Figure 1.3).

1.2.7 Hydrogen Energy and Fuel Cell

Today hydrogen energy technology and fuel cells are also considered as renewable sources of energy. In recent years, hydrogen has been receiving worldwide attention as a clean and efficient energy carrier with a potential to replace liquid fossil fuels. Hydrogen produces only water as a by-product and is, therefore, environmentally benign. Moreover, it has the highest energy per unit mass and lowest CO_2 emission of any known fuel (Table 1.3).

The current total annual worldwide hydrogen consumption is in the range of 400–500 billion Nm^3 (Demirbas, 2009). However, the bulk of the hydrogen produced is utilized by various industries such as food, electronics, petrochemical, and metallurgical processing industries while only a small fraction is utilized by the energy sector. The current utilization of hydrogen as energy is equivalent to 3% of the total energy consumption and with a growth rate estimated at 5–10% per year (Mohan et al., 2006). The global market for hydrogen is already greater than U.S. $40 billion per year; including hydrogen used in ammonia production (49%), petroleum refining (37%), methanol production (8%), and miscellaneous smaller-volume uses (6%) (Konieczny et al., 2008).

Statistics shows that hydrogen has been produced for a long time. However, till date, it is produced using the conventional energy technology, which utilizes fossil fuel. Thus, though the end product is carbon free, the path to obtain it involves heavy emissions of greenhouse gases. In view of the same, it is necessary to identify the limitations of the conventional hydrogen technologies and adopt the biological routes to fulfill the requirements for the same.

TABLE 1.3

Comparison of Different Fuel Types with Respect to Energy Per Unit Mass and
Specific Carbon Dioxide Emission

Fuel Type	Chemical Structure	Carbon Content (%)	Fuel Material (Feedstock)	Energy Per Unit Mass (J/kg)	Specific Carbon Emission (kg C/kg Fuel)
Hydrogen	H_2	0	Natural gas, methanol, and electrolysis of water, biomass	141.90	0.00
Ethanol	C_2H_5OH	52	Corn, grains, or agricultural waste (cellulose)	29.90	0.50
Biodiesel	Methyl esters of C_{12} to C_{22} fatty acids	77	Fats and oils from sources such as soy beans, waste cooking oil, animal fats, and rapeseed	37.00	0.50
Methanol	CH_3OH	37.5	Natural gas, coal, woody biomass	22.30	0.50
Natural gas	CH_4	75	Underground reserves	50.00	0.46
Gasoline	C_4 to C_{12}	74	Crude oil	47.40	0.86

1.3 Conventional Hydrogen Technologies and Limitations

Hydrogen is a carrier of energy, not a source (Box 1.1). Despite being the most
common and abundant element in the universe, molecular hydrogen must
be produced from hydrogen-rich feedstock such as water, biomass, or fossil
fuel. The technologies for producing pure hydrogen from these feedstocks
also require energy to power the production process.

Out of the total global production of hydrogen, 48% is produced from steam
methane reforming (SMR), about 30% from oil/naphtha reforming from
refinery/chemical industrial off-gases, 18% from coal gasification, and 3.9%
from water electrolysis (Baghchehsaree et al., 2010). These figures imply that
globally 96% of the hydrogen production is derived from fossils (Abánades,
2012). This industrial output is expected to increase around 3.5% annually
through 2013 (Freedonia Group, Inc., 2010). The figures suggest that hydrogen has immense potential, however, the challenge is to tap it economically
and in an environment-friendly manner.

Further, the key point to be noted here is that hydrogen is not present like
oil or petroleum but rather has to be made like electricity. The current major

BOX 1.1   ENERGY SOURCES AND CARRIERS

Primary sources of energy include coal, oil, and natural gas that can be drilled from the earth and directly used to fuel our needs. They are rich in kinetic or potential energy. As a rule of physics, energy can neither be destroyed nor created; it can only be converted from one form to the other. This transformation of energy allows it to be converted into more useable secondary forms of energy such as electricity. Such secondary sources of energy are also known as energy carriers. Hydrogen is also a secondary source, as it must be produced using a hydrogen-rich source. It can be converted to energy (heat) either through combustion or through an electrochemical reaction to generate heat and electricity.

limitation involves the greenhouse gas emissions and the requirement of fossil fuel for its production by conventional methods. The various conventional methods to produce hydrogen from feedstock can be grouped into three major categories. These include

- Thermal and thermochemical processes
- Electrolytic processes
- Photolytic processes

Thermochemical processes, such as SMR, partial oxidation (POx), or gasification, involve the use of heat and pressure to break molecular, usually hydrocarbon bonds. Electrolytic processes, such as simple water electrolysis, involve running water through electricity to separate water into its constituent oxygen and H_2. Photolytic processes involve extracting H_2 from the waste gases of biological organisms, such as algae (Padro and Putsche, 1999). Table 1.4 compares the three different processes. The vast majority (99%) of H_2 used for industrial purposes is produced using thermochemical processes to extract H_2 from fossil fuels. Approximately 95% of this H_2 production involves SMR of natural gas (USDoE, 2003). SMR is a well-established commercial process and is the most common and least expensive method to produce large quantities of H_2. Nickel is used as a cheap catalyst. In most cases, natural gas (methane) is the raw material. Initially methane reacts with steam to produce carbon monoxide and hydrogen. The carbon monoxide, passed over a hot iron oxide or a cobalt oxide catalyst, then reacts with the steam to produce carbon dioxide and additional amounts of hydrogen.

$$CH_4 + H_2O \text{ (steam)} \rightarrow CO + 3H_2 \tag{1.1}$$

$$CO + H_2O \rightarrow CO_2 + H_2 \tag{1.2}$$

TABLE 1.4

Comparison of the Current Three Major Conventional Hydrogen Production Technologies

Process/Parameters	Thermal	Electrolytic	Photolytic
Definition	Uses the energy stored in resources such as coal or biomass and/or chemicals to simply release the hydrogen contained within their molecular structures	Water electrolysis uses electricity to split water into hydrogen and oxygen	Photolytic processes use light energy to split water into hydrogen and oxygen
Associated processes	• Natural gas Reforming • Bio-derived liquids reforming • Coal and biomass gasification • Thermochemical production	• Water electrolysis	• Photo-electrochemical hydrogen production • Biological hydrogen production
Challenges	• High capital costs • High operation and maintenance costs • Plant design	• High capital costs • High operation and maintenance costs • Plant design • Feedstock quality and quantity	• High reactor costs • Feedstock • Carbon capture and storage
Key benefits	• Currently the most economical • Well-established infrastructure	• Well-established infrastructure • Operates at low temperatures • Clean and sustainable—using only water and solar energy	• Sustainable

POx of methane is also used to produce hydrogen. The process involves reacting methane with oxygen to produce hydrogen and carbon monoxide, which is then reacted with water to produce more hydrogen and carbon dioxide. Overall conversion efficiency is generally lower than for steam reforming, which is why the latter technique dominates commercial production today and is expected to continue to do so.

$$CH_4 + O_2 \rightarrow CO + 3H_2 \tag{1.3}$$

$$CO + H_2O \rightarrow CO_2 + H_2 \tag{1.4}$$

Gasification of coal is the oldest technique for making hydrogen, and is still used in some parts of the world. Coal is heated until it turns into a gaseous state, and is then mixed with steam in the presence of a catalyst to produce

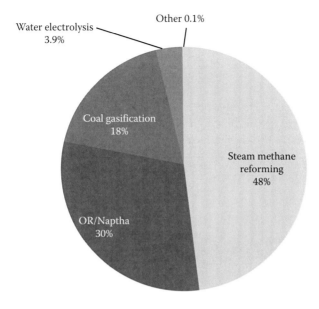

FIGURE 1.4
Production of hydrogen from various resources.

a mixture of hydrogen (around 60%), carbon monoxide, carbon dioxide, and oxides of sulfur and nitrogen. This synthesis gas may then be steam-reformed to extract the hydrogen, or simply burned to generate electricity.

Hydrogen production using water electrolysis is normally used only to produce hydrogen of very high purity, required in some industrial processes, or other products, such as chloro-alkali, where hydrogen is produced as a by-product.

However, all conventional processes are energy intensive, therefore expensive. Considering long-time goals, these conventional means have several limitations. For a sustainable solution, hydrogen production must be carbon neutral. In the early 1970s, hydrogen was first detected as a by-product of microbial metabolism. Since then, considerable efforts have been made in the genetic and physiological levels to establish microbial hydrogen at commercial scale. However, out of the total amount of hydrogen produced till date <0.1% is of microbial origin (Figure 1.4).

1.4 Biological Hydrogen Production Technology

In nature, hydrogen economy was seen 3 billion years ago when the process of photosynthesis converted CO_2, water, and sunlight into hydrogen and

TABLE 1.5

Advantages of Biohydrogen Production

Advantages of Biohydrogen	Rationale
Naturality	Can be produced by microorganisms under ambient conditions of temperature and pressure.
Neutrality	Microbial hydrogen production produces no greenhouse gas emissions.
Versatility	Can be produced from a wide variety of substrate including municipal solid waste and waste water.
Reliability	It is reliable as waste is always available in large quantities.
Sustainability	As compared to the fossil fuels, it is more sustainable.
Energy security	It reduces the dependence of nations on politically unstable countries like the OPEC nations from which the oil is imported to sustain energy needs.
Affordability	Much more cost effective as compared to fossil fuels.
Efficiency	The combustion of hydrogen in automobiles is 50% more efficient than gasoline. Hydrogen battery is deemed as future supply for automobiles (Balat and Balat, 2009).

oxygen. Biological hydrogen production processes are not only environment friendly but also inexhaustible (Benemann, 1997; Greenbaum, 1990). Biohydrogen production has several advantages over fossil fuels (Table 1.5). Moreover, they have the highest energy per unit mass as compared to any other fuel (Table 1.3). Studies on hydrogen production have focused on direct and indirect biophotolysis mediated by cyanobacteria and green algae, photofermentative hydrogen production by photofermentative bacteria, and dark hydrogen production by fermentative bacteria. Figure 1.5 depicts typical biohydrogen production processes. However, each process has its own advantage and disadvantage (Table 1.6). Many say that "hydrogen is a futuristic fuel"; some add "and it shall always be!" Nevertheless, opportunists still believe that in the future the share of hydrogen will increase in meeting the final energy needs (Table 1.7).

POWER OF MICROBIAL HYDROGEN

Estimates are that approximately 150 million tons of hydrogen is annually formed by microorganisms used to fuel methanogens. The combustion of 150 million tons of hydrogen yields 18×10^{18} J of energy. This is equivalent to 3.75% of the primary energy consumed in 2006 by the world population (455×10^{18} J) (Thauer et al., 2010)!

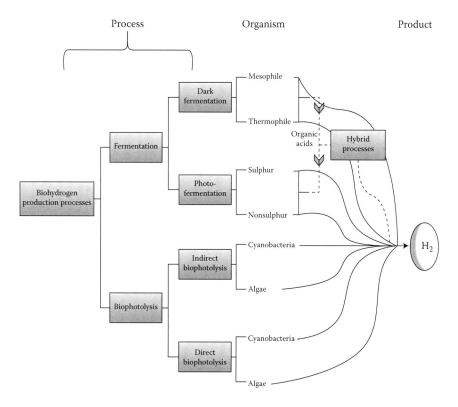

FIGURE 1.5
Overview of biohydrogen production processes. (Reprinted from Khanna, N. and Das, D. 2012. In *Biohydrogen Production by Dark Fermentation, Energy and Environment*, eds. P. Lund and S. Basu. USA: Wiley-Blackwell Publication. With permission.)

However, to obtain this goal, significant research efforts are needed to overcome the major bottlenecks toward commercialization of the process.

- The yield of H_2 from any of the processes defined above is low for commercial application. The pathways of H_2 production have not been identified and the reaction remains energetically unfavorable.
- Processing of some biomass feed stock is too costly. There is a need to develop low-cost methods for growing, harvesting, transporting, and pretreating energy crops and/or biomass waste products.
- There is no clear contender for a robust, industrially capable microorganism that can be metabolically engineered to produce more than 4 mol H_2/mol of glucose.
- Several engineering issues need to be addressed which include the appropriate bioreactor design for H_2 production, difficult to sustain steady continuous H_2 production rate in the long-term, scale-up,

TABLE 1.6

Scientific and Technical Challenges for Microbial Hydrogen Production

Bottlenecks	Challenges	Plausible Solution
Organism	• Lack of a robust organism • Insufficient knowledge of metabolic pathways • Thermodynamic limitation • Cannot naturally produce more than 4 mol H_2/mol glucose • Lack of industrially suitable strain	• Isolate more novel microbes • Establish metabolic pathways of known promising organisms • Reverse electron transport to drive H_2 production past barrier • Implementation of low hydrogen partial pressure • Genetic engineering for strain development
Enzyme (hydrogenase)	• Lack of sufficient knowledge of their structure and function • Oxygen sensitivity	• More studies on enzyme crystallization should be encouraged • Isolation of naturally less oxygen-sensitive enzymes or development of oxygen-insensitive enzyme by genetic engineering
Engineering aspects	• Lack of suitable reactor • Kinetic and theoretical considerations • Scale-up challenges	• Implementation of process engineering concepts to develop a suitable reactor for basic studies and scale up of the process
Economics	• Fuel is not cost effective	• Integration of different processes based on system economy

preventing interspecies H_2 transfer in nonsterile conditions and separation/purification of H_2.

- Sensitivity of hydrogenase to O_2 and H_2 partial pressure severely disrupts the efficiency of the processes and adds to the problems of lower yields.
- Insufficient knowledge on the metabolism of H_2-producing bacteria and the levels of H_2 concentration tolerance of these bacteria.
- A lack of understanding on the improvement of economics of the process by integration of H_2 production with other processes.

TABLE 1.7

Share of Individual Primary Energy Sources in Meeting the Final Energy Needs (%)

Source of Energy	2025	2050
Fossil fuels	62	29
Nuclear energy	2	2
Solar energy	25	35
Hydrogen (from different sources)	11	34

Source: Balat, M. and Balat, M. 2009. *International Journal of Hydrogen Energy*, 34, 3589–3603.

This book attempts to describe in detail all the different processes for production of hydrogen, the limitations, and the advancements that have been made. As discussed, in the book, there have been considerable improvements in both yield and volumetric production rate of hydrogen. However, to be practical, yields must considerably extend past the current metabolic limitation of 4 H_2/glucose. Thus, the major outstanding question is "Can a practical biological process be created to extract nearly all the H_2 from the substrate (12 H_2/glucose)?" Attempting to address this challenge should be the focus of future biohydrogen research.

1.5 Properties of Hydrogen

At normal atmospheric conditions, hydrogen is a colorless and odorless gas. It is stable and coexists harmlessly with free oxygen until an input of energy drives the exothermic (heat releasing) reaction that forms water. Hydrogen is the lightest element occurring in nature and contains a large amount of energy in its chemical bond. Boiling point of hydrogen is −250.87°C (22.28 K). The solubility of gas in liquid, known as the Bunsen coefficient (α) is 0.018 at 20°C. The diffusion coefficient (D_w) in water at 20°C is nearly 4×10^{-9} m²/s. The homolytic cleavage of hydrogen in the gas phase is endergonic by +436 kJ/mol and the heterolytic cleavage in water at 20°C is endergonic by about +200 kJ/mol (pK_a near 35) (Kubas, 1988). The combustion energy of hydrogen is 120 MJ/kg.

1.5.1 Fuel Properties of Hydrogen

Hydrogen's physical and chemical properties make it a good candidate to be used as fuel. Further, it must be noted that hydrogen may be directly used as a transportation fuel. However, nuclear and solar fuels do not have this advantage. The properties of hydrogen make it a promising candidate as an alternate source of fuel for internal combustion engines. Hydrogen can be used as a fuel directly in an internal combustion engine not much different from the engines used with gasoline. Hydrogen has key properties as a transportation fuel, including a rapid burning speed, a high effective octane number, and no toxicity or ozone-forming potential (Table 1.8). It has much wider limits of flammability in air (4–75% by volume) than methane (5.3–15% by volume), and gasoline (1–7.6% by volume). Moreover, hydrogen-fueled ICEs (internal combustion engines) and gas turbine engines have negligible emissions of air pollutants. Hydrogen-powered-fuel-cell vehicles have zero emissions. However, platforms powered by petroleum-based fuels emit significant amounts of air pollutants (hydrocarbons, carbon monoxide, nitrogen oxides, sulfur oxides, and particulate matter), air toxics (either confirmed or

TABLE 1.8

Fuel Properties of Hydrogen

Fuel Property	Value
Lower-heating value (MJ/kg)	120.7
Higher-heating value (MJ/kg)	141.9
Density under ambient conditions (kg/m^3)	0.08
Auto ignition temperature in air[a] (°C)	566–582
Phase under ambient conditions	Gas
Ignition limit in air[b] (vol.%)	4.1–74
Diffusion coefficient (cm^2/s)	0.61

[a] Auto ignition: Lowest temperature at which a fuel will ignite when an external source of ignition is present.
[b] Ignition limit: Range of concentration within which the fuel will ignite.

suspected human carcinogens, including benzene, formaldehyde, 1,3-buta-diene, and acetaldehyde), and carbon dioxide. The health effects of these pollutants range from headaches to serious respiratory damage such as lung cancer.

Thus, hydrogen holds the potential to provide energy service to all sectors of the economy: transportation, residential, and industries. It can complement or even replace network-based electricity. It can also provide storage options for intermittent renewable-based electricity technology such as solar and wind. It can be used as an input for fuel cells to generate green electricity. Moreover, since it can be produced from a variety of sources, it impactfully may reduce our dependence on fossil fuel and hence may help resolve energy security issues.

1.6 Book Overview

This book is an attempt to explain, in lucid details, all the biological processes leading to the production of biohydrogen. It step by step details the fundamentals of this technology. Chapters 2 and 3 review the microbiology and the basic biochemistry involved in the biological process of hydrogen production. Chapter 4 reviews the availability of different types of feedstocks used for biohydrogen production and also assesses their cost benefits. Chapter 5 details in depth the key enzymes involved in the hydrogen production processes. Chapter 6 reviews the advancements made in this field using genetic and metabolic engineering principles. Chapter 7 outlines all the different process parameter considerations for optimization of the process. Chapters 8 through 10 deal with the engineering aspects of photobiorectors, theoretical considerations of the process, and scale up of the technology.

Finally, Chapter 11 focuses on the current state of process economy, policy, and environmental impacts of such a carbon neutral technology.

The question remains to be answered as to what extent the renewables will be able to replace the fossil fuel-based industries. However, with the dwindling fossil fuel reserves and the mounting environmental issues, one wonders till what time the supremacy of the fossil fuel-based industry can last! By adopting the biorefinery model with sustainable use of biomass to generate green fuel technologies, we may be able to accelerate the shift from heavy polluted to carbon neutral technologies. This book is a first attempt to describe the process technologies and the relative merits and challenges toward the building up of a biohydrogen-based economy.

References

Abánades, A. 2012. The challenge of hydrogen production for the transition to CO_2-free economy. *Agronomy Research, International Scientific Conference Biosystems Engineering*, Special Issue, *1*, 11–16.

Annual report on European Supply, transformation, consumption of electricity (Eurostat) 2013. Retrieved from http://appsso.eurostat.ec.europa.eu/nui/show.do?dataset=nrg_105a&lang=en on 01 March, 2013.

Baghchehsaree, B., Nakhla, G., Karamanev, D., and Argyrios, M. 2010. Fermentative hydrogen production by diverse microflora. *International Journal of Hydrogen Energy*, *35*, 5021–5027.

Balat, M. and Balat, M. 2009. Political, economic and environmental impacts of biomass-based hydrogen. *International Journal of Hydrogen Energy*, *34*, 3589–3603.

Benemann, J. R. 1997. Feasibility analysis of photobiological hydrogen production. *International Journal of Hydrogen Energy*, *22*, 979–987.

Chen, F. F. 2011. The future of energy II: Renewable energy. In: *An Indispensable Truth: How Fusion Power Can Save the Planet*. New York: Springer, pp. 450.

Cohen, J. E. 2001. World population in 2050: Assessing the projections. In *Conference Series*, Federal Reserve Bank of Boston, *46*, 83–113.

Demirbas A. 2009. *Biohydrogen for Future Engine Fuel Demands*. London: Springer.

Dirzo, R. and Raven, P. H. 2003. Global state of biodiversity and loss. *Annual Review of Environment and Resources*, *28*, 137–167.

Freedonia Group, Inc. 2010. World hydrogen demand and 2013 forecast. *Journal of Arid Environments*, *6*, 751–763.

Gorlov A. M. 2001. Tidal energy. In: *Encyclopedia of Ocean Science*. London: Academic Press, 2955–2960.

Greenbaum, E. 1990. Hydrogen production by photosynthetic water splitting. In *Hydrogen Energy Progress VIII, Proceedings 8th WHEC*, Hawaii, eds. T. N. Veziroglu and P. K. Takashashi, pp. 743–754. New York: New York Pergamon Press.

Jacobson, M. Z. 2009. Review of solutions to global warming, air pollution, and energy security. *Energy Environmental Sciences*, *2*, 148–173.

Khanna, N. and Das, D. 2012. *Biohydrogen Production by Dark Fermentation, Energy and Environment*, eds. P. Lund and S. Basu. The USA: Wiley-Blackwell Publication.

Konieczny, A., Mondal, K., Wiltowski, T., and Dydo, P. 2008. Catalyst development for thermocatalytic decomposition of methane to hydrogen. *International Journal of Hydrogen Energy, 33,* 264–272.

Kubas, G.J. 1988. Molecular hydrogen complexes—coordination of a sigma bond to transition metals. *Accounts of Chemical Research,* 21, 120–128.

Mohan, D., Pittman, C. U., and Steele, P. H. 2006. Pyrolysis of wood/biomass for bio-oil: A critical review. *Energy Fuels, 20,* 848–889.

Padro, C. and Putsche, V. 1999. *Survey of the Economics of Hydrogen Technologies.* Golden, CO: National Renewable Energy Laboratory, The U.S. Department of Energy.

Schiermeier, Q. 2013. Renewable power: Germany's energy gamble. *Nature, 496,* 156–158.

Solomon, S. 1999. Stratospheric ozone depletion: A review of concepts and history. *Review of Geophysics, 37,* 275–316.

Thauer, R. K., Kaster, A.K., Goenrich, M., Schick, M., Hiromoto, T., and Shima, S. 2010. Hydrogenases from methanogenic archaea, nickel, a novel cofactor, and H_2 storage. *Annual review of Biochemistry, 79,* 507–536.

Tissot, B. P. and Welte, D. H. 1978. *Petroleum Formation and Occurrence: A New Approach to Oil and Gas Exploration,* pp. 554. New York, NY: Springer.

U.S. Department of Energy (USDoE), Office of Energy Efficiency and Renewable Energy. 2003. *Opportunities for Hydrogen Production and Use in the Industrial Sector.* Washington, DC: U.S. Department of Energy.

U.S. Environmental Protection Agency (EPA), Inventory of U.S. Greenhouse Gas Emissions and Sinks: 1990–2010, 2012. Retrieved from http://www.epa.gov/climatechange/ghgemissions/usinventoryreport.html on 12th November, 2012.

Veron, S. R., Paruelo, J. M., and Oesterheld, M. 2006. Assessing desertification. An interim report on world agriculture: Towards 2030/2050. Global Perspective Studies Unit Food and Agriculture Organization of the United Nations, Rome.

Walther, G. R., Post, E., Convey, P. et al., 2002. Ecological responses to recent climate change. *Nature, 416,* 389–395.

Wilkes, J., Moccia, J., Wilczek, P., Gruet, R., and Radvilaitė, V. 2011. *Report on EU Energy Policy to 2050. Achieving 80–95% Emissions Reduction.* 1–68.

2

Microbiology

2.1 Introduction

In the course of evolution there emerges different genetic diversity of microorganisms with new functions to use different energy sources. This co-operation results in a stable, self-regulatory, and sustainable system that converts complex organic matter into a wide range of products often including hydrogen gas (H_2).

The microbes having hydrolytic properties solublize the insoluble complex components of the organic matter (anaerobic digestion) with the help of some other microbes (Box 2.1). These solublized intermediates then act as a feed for H_2-producing microorganisms. Photosynthetic microorganisms use light as an energy source. A wide variety of microorganisms are hydrogen producers including archaea, strictly anaerobic and facultative aerobic bacteria, cyanobacteria, and lower eukaryotes such as green algae and protists (Tamagnini et al., 2002; Boichenko et al., 2004), which may be operated singly or as consortia of similar types or as mixed cultures (Figure 2.1).

Versatility in habitats and nutritional types of microorganism help them withstand extreme conditions such as complete darkness, high temperature, and absence of oxygen. Archaea are phototrophs, lithotrophs or

---

BOX 2.1   ANAEROBIC DIGESTION

During the process of anaerobic digestion, microorganisms hydrolyze biodegradable material in its simplest form. During this process, organic matter such as carbohydrate is hydrolyzed and is available for other bacteria. Different group of bacteria perform different functions, one group of bacteria hydrolyze easily accessible polysaccharides such as starch, pectin, glucans, cellulose, hemicelluloses, and gums; others, which are cellulytic in nature actively, work on carboxymethyl cellulose and hemicelluloses. As a result of this process, the gases, mainly methane and carbon dioxide are produced. Methane is used as cooking fuel and helps to replace fossil fuel.

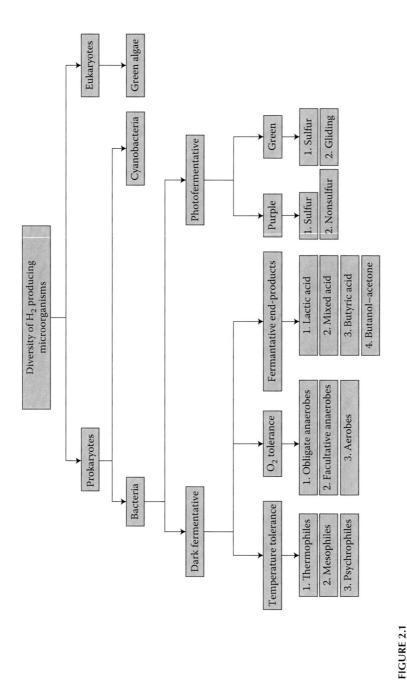

FIGURE 2.1
Schematic diagram represents the diversity of hydrogen-producing microorganisms.

organotrophs found in a broad range of habitats. They are mostly extremophiles, lived in harsh environments, such as hot springs and salt lakes. Bacteria have their own hydrogen fuel cycle. In nature, low partial pressure of H_2 is maintained by the presence of H_2 consumers with syntrophic association with H_2 producers. Hydrogen-producing bacteria are mostly heterotrophic, and they produce H_2 during fermentative metabolism. Some of them do not require light energy and can withstand oxygen deficiency designated as dark fermentative bacteria and some of them require light energy and produce H_2 in oxygen-deficient condition called photo-fermentative bacteria. Cyanobacteria, marine and fresh water algae are photoautotrophic, and they produce hydrogen by biophotolysis in the absence of oxygen. Certain *Trichomonas* and other anaerobic protozoa are also known to produce H_2 (Robson, 2001). Different hydrogen-producing microorganisms, their yield, and production rate with their optimal physicochemical condition have been given in Table 2.1.

2.2 Dark Fermentative Bacteria

Dark fermentative bacteria evolved with the appearance of organic material on earth. They are adapted to different temperatures and substrates. They are heterotrophs and produce hydrogen in anoxia. These bacteria grow on organic substrates (heterotrophic growth) and metabolized energy by partial oxidation of organic compounds using organic intermediates as electron donors and electron acceptors instead of oxygen. ATP in fermentative organisms is produced by substrate-level phosphorylation, where a phosphate group is transferred from a high-energy organic compound to ADP to form ATP. Fermentative organisms use NADH and other cofactors to produce many different reduced metabolic by-products, small organic acids, and alcohols, often including hydrogen gas (H_2). Dark fermentative bacteria can be distinguished in different groups according to their adaptation to different temperatures, tolerance to oxygen, and different fermentative end products.

2.2.1 Adaptation to Temperature

Microorganisms are distributed globally in different temperature niches. They can be classified according to their adaptation to different temperature ranges. Thermophiles are adapted to high temperature range (hyperthermophiles: 353–388 K, with optimum growth at >353 K) or little lower (313–383 K, optimum ~333 K). Some of them are mesophiles ranging from 283 to 323 K, optimum ~310 K among them some are psychrotrophs (273–303 K, optimum ~295–298 K). Those that are growing in very low temperature are known as psychrophiles (273–293 K, optimum ≤288 K) (Budiman et al., 2011).

TABLE 2.1

Different Diverse Group of Hydrogen-Producing Microorganisms are Listed in the Table along with Their Optimal Physicochemical Parameters for Achieving the Maximum Hydrogen Yield and Production Rate

Organism	Process Parameter	H$_2$ Yield	H$_2$ Production Rate	Reference
Dark-Fermentative Bacteria				
Thermophiles				
Thermotoga maritime	80°C	4.0 mol/mol-glucose	2.78 mmol/m^3 s	Schröder et al. (1994)
Caldicellulosiruptor saccharolyticus	Sweet sorghum juice	58% of the theoretical maximum	5.84 mmol/m^3 s	Claassen et al. (2004)
Caldicellulosiruptor saccharolyticus DSM 8903 and *C. kristjanssonii* DSM 12137	Continuous flow-stirred tank reactor, dilution rate 0.06/h	3.7 mol/mol-glucose		Zeidan et al. (2010)
Mesophiles				
Citrobacter amalonaticus Y19	37°C, pH 8.0		18.6 ± 1.7 × 10^2 μmol/ kg-cell s	Seol et al. (2008)
Escherichia coli K-12 MG1655	45°C, pH 8.0		23.6 ± 1.1 × 10^2 μmol/ kg-cell s	
Escherichia coli DJT135	45°C, pH 8.0		41.1 ± 2.2 × 10^2 μmol/ kg-cell s	
Enterobacter aerogenes	37°C, pH 8.0		47.3 ± 0.6 × 10^2 μmol/ kg-cell s	
Psychrophiles				
Geobacter psychrophilus dominant strain in mixed culture	Single-chamber MECs, 4°C/9°C	2.66 ± 0.22 to 2.94 ± 0.02 mol/ mol-acetate	0.64 ± 0.08 to 1.47 ± 0.1 × 10^{-4} m^3/H$_2$ m^3 s	Lu et al. (2011)
Obligate Anaerobes				
Clostridium beijerinckii AM21B	36°C, pH 6.5	6.2 × 10^{-3} mmol/s from glucose 5.1 mmol/s from starch		Taguchi et al. (1993)

Organism	Conditions	Yield	Rate	Reference
Clostridium sp. strain no. 2	36°C, pH 6.0	1.5×10^{-3} mmol/kg-arabinose 3.6×10^{-3} mmol/kg xylose		Taguchi et al. (1994)
Clostridium sp. strain X53	40°C, pH 5.0, Xylan		1.5×10^{-5} m³/m³ s	Taguchi et al. (1996)
Clostridium paraputrificum M-21 (chitinolytic bacterium)	pH 6.0	2.2 mol/mol-GlcNAc		Evvyernie et al. (2001)
Clostridium tyrobutyricum JM1	pH 6.7–6.8	0.2 m³/kg-glucose	20×10^{-4} m³/m³ s	Jo et al. (2008)
Ruminococcus albus		2.4 mol/mol-glucose		Iannotti et al. (1973)
Enterococcus gallinarum strain G1 and *Ethanoigenens harbinense* B49	37°C, pH 6.5	2.97×10^{-3} mmol/kg-cellulose		Wang et al. (2009)
Facultative Anaerobes				
Enterobacter aerogenes, strain HO-39	38°C	1.0 mol/mol-glucose		Yokoi et al.(1995)
Enterobacter aerogenes E.82005	40.5°C, pH 6.0		1.4×10^{-4} m³/m³ s	Tanisho et al. (1983)
Enterobacter cloacae IIT-BT 08	36°C, pH 6.0	2.2 mol/mol-glucose, 6.0 mol/mol sucrose and 5.4 mol/mol cellobiose	8.24 mmol/kg-cell s	Kumar and Das (2000)
Escherichia coli WDHL		0.30 mol/mol-glucose 1.12 mol/mol-galactose 1.02 mol/mol-lactose		Rosales-Colunga et al. (2012)
Citrobacter intermedius	pH 5.75 and 6.0	1.1 mol/mol-glucose		Brosseau et al. (1980)
Citrobacter sp. Y19	30–40°C, pH 5.5–7.5		7.53 mmol/kg-cell s	Jung et al. (2002)
Citrobacter freundii		1.286 mol/mol-glucose		Kumar and Vatsala (1989)
Aerobes				
Bacillus lichenformis	40°C, pH 7.0	1.3×10^{-2} m³/mol-glucose		Kumar et al. (1995)
Mixed culture	Immobilized	0.8×10^{-2} m³/mol-glucose		

continued

TABLE 2.1 (continued)

Different Diverse Group of Hydrogen-Producing Microorganisms Are Listed in the Table along with Their Optimal Physicochemical Parameters for Achieving the Maximum Hydrogen Yield and Production Rate

Organism	Process Parameter	H₂ Yield	H₂ Production Rate	Reference
Photosynthetic Bacteria				
Purple–Sulfur				
Thiocapsa roseopersicina			0.5×10^{-6} mmol/m³ s	Horvath et al. (2004)
Purple–Nonsulfur				
Rhodobacter sphaeroides	Immobilized on polyurethane foam, light 300 W/m²; dilution rate 0.023/h		0.58×10^{-4} m³/m³ s	Fedorov et al. (1998)
Green–Sulfur				
Chlorobium vibrioforme strain 1930 and *Desulfuromonas acetoxidans* strain 5071	pH 6.8, light intensity 110 µmol/m² s	3.1 mol/mol-acetate		Warthmann et al. (1992)
Cyanobacteria				
Anabaena cylindrical B 629	7000 lux, 5% CO₂		0.28×10^{2} µmol/kg-dry wt s	Lambert and Smith (1977)
Anaebaena variabilis ATCC 29413	150 µE/m² s· 1% CO₂		0.68×10^{2} µmol/kg-dry wt s	Berberoglu et al. (2008)
Synechocystis sp. PCC 6803	50 µE/m² s· NaHCO₃		2.25×10^{2} µmol/kg-chl s	Burrows et al. (2008)
Cyanothece sp. ATCC 51142	Glycerol		1.3×10^{5} µmol/kg-chl s	Bandyopadhyay et al. (2010)

Organism	Conditions			Reference
Nostoc linckia HA-46	31°C, pH 8.0, CO_2:Ar ratio 2:10.		2.53×10^4 µmol/kg-chl s	Mona et al. (2011)
Arthrospira maxima		0.036 m³/kg-dry wt		Ananyev et al. (2008)
Green Algae				
Chlamydomonas reinhardtii	pH 7.0, 100 µE/m² s, acetate, S-deprived		5.8×10^{-7} m³/m³ s	Laurinavichene et al. (2006)
Chlamydomonas MGA 161	pH 8.0		12.5×10^{-7} m³/m³ s	Ohta et al. (1987)
Chlorella sorokiniana Ce	120 µE/m² s, acetate, S-deprived		1.7×10^{-2} mmol/m³ s	Chader et al. (2009)
Chlorella vulgaris	120 µE/m² s	0.530 ± 5 m³/m³		Rashid et al. (2011)
Chlorella vulgaris	37°C	10.8×10^{-3} m³/kg-volatile solids	96.7×10^{-7} m³/m³ s	Lakaniemi et al. (2011)
Dunaliella tertiolecta (marine microalga)		12.6 ml/g-volatile solids		
Chlorella sp.	Immobilized, pH 8.0	1.3 m³/m³ cult with sucrose	66.7×10^{-7} m³/m³ s with fructose	Rashid et al. (2013)
Mixed Culture				
Mixed culture previously enriched from soya bean meal	35°C, pH 6.0, CSTR	0.85 mol/mol-glucose		Mizuno et al. (2000)

Bacteria's response to temperature depends on adaptive changes in cellular proteins and lipids. Changes in proteins are genotypic and are related to the properties of enzymes and translation systems, whereas changes in lipids are genotypic or phenotypic, important in regulating membrane fluidity and permeability. The hydrogen production from thermophiles, mesophiles, and psychrophiles is described in the following three subsections.

2.2.1.1 Thermophiles

Fermentative hydrogen production reaction is an endothermic reaction. It needs more energy in the form of heat to perform the reaction. Thermophiles are therefore considered to be a more promising candidate for hydrogen production than mesophiles as the fermentation occurs in high temperature. Furthermore, the added advantages are with high temperature contamination of other mesophilic hydrogen-consuming microorganisms that can be avoided; hot industrial effluent is useable directly as substrate. Thermophiles are mostly obligate anaerobes. They are naturally found in hot springs, in deep sea hydrothermal vents and in other geothermally heated regions of the earth. They derive metabolic energy from oxidation of external substrates. The substarte can be cellulose, hemicellulose, and pectin-containing biomass (Van de Werken et al., 2008).

The different species of cellulolytic and hyper-thermophilic bacteria, such as *Thermotoga, Thermoanaerobacter, Caldicellulosiruptor, Anaerocellum, Clostridium, Dictyoglomus, Fervidobacterium* and *Spirocheta*, have the capacity to produce hydrogen (Bergquist et al., 1999). They can produce hydrogen near the stochiometry that is 4 moles of hydrogen per mole of glucose (Schröder et al., 1994; de Vrije et al., 2007). Schröder et al. (1994) reported H_2 yield 4 mol/mol glucose from *Thermotoga maritima* on batch fermentations at 80°C. Similar stoichiometries were obtained for two moderate thermophiles, *Acetomicrobium flavidum* and *Acetothermus paucivorans* (Winter and Zellner, 1990). Extensive work has been done on *Caldicellulosiruptor saccharolyticus*, a strictly anaerobic, extreme-thermophilic, Gram-positive bacterium that belongs to the Clostridia class. It has used sweet sorghum juice as carbon and energy source and 58% of hydrogen yield has been achieved (Claassen et al., 2004). Sometimes, coexistence of more than one strain has been observed beneficial to be for hydrogen production (Zeidan et al., 2010). An interspecies interaction has been proposed as the reason behind this ability of the two strains to coexist in the system rather than only competing for the growth-limiting substrate (Zeidan et al., 2010). It has been found to be enriched with the bacteria from Clostridia family at around 60°C (Nissilä et al., 2011).

2.2.1.2 Mesophiles

Mixed microflora at mesophilic temperatures is a suitable approach for commercial development of hydrogen production. Bacteria of family

Enterobacteriaceae and Clostridia are mostly mesophilic and produce hydrogen from complex carbohydrates (Penfold et al., 2003). Clostridial-based cultures from natural sources have been widely used for hydrogen production (Hawkes et al., 2007). It has also been reported that the viable number of dominated *Clostridium* sp. could directly affect the hydrogen production rate. Some bacterial species are able to break down the starch into small molecules such as *Bifidobacterium* sp. These less complex compounds were then utilized by the *Clostridium* species for hydrogen production (Cheng et al., 2008). But the recent discoveries on bacterial community composition elaborate the role of other microorganisms that coexist with *Clostridium* and *Enterobacter* (Hung et al., 2011).

2.2.1.3 Psychrophiles

The ability of psychrophiles to survive and proliferate at low temperatures implies that they have evolved features, genotypic and/or phenotypic to overcome key barriers in low temperature. They have to overcome reduced enzyme activity; decreased membrane fluidity; altered transport of nutrients and waste products; decreased rates of transcription, translation and cell division; protein cold-denaturation; inappropriate protein folding; and intracellular ice formation. Hydrogen production is also depressed under psychrophilic conditions. Although Lu and co-researchers (Lu et al., 2011) have described a single-chamber microbial electrolysis cells (MECs), which were enriched successfully with *Geobacter psychrophilus,* it has been operated at 277 K or 282 K to inhibit the growth of hydrogenotrophic methanogens such as *Methanobrevibacter arboriphilus* for the production of hydrogen.

2.2.2 Tolerance to Oxygen

Hydrogen has been produced in absence of oxygen irrespective of the adaptation to temperature and light. The dark adapted microorganisms are mostly found in those places where oxygen is absent, as for example, under the soil layers, water, sewage, inside the vegetables, etc. According to the tolerance to oxygen, they can be distinguished as strict anaerobes, facultative anaerobes, and aerobes. The difference in oxygen tolerance among bacteria depends on the defensive enzymes possessed by them against reactive oxygen species (ROS).

2.2.2.1 Obligate Anaerobes

Instead of oxygen, obligate anaerobes use alternate electron acceptors for cellular respiration such as sulfate, nitrate, iron, manganese, mercury, and carbon monoxide. The aerobes possess enzymes scavenging the ROS, the super oxide dismutase (SOD), and catalase, whereas aerotolarent anaerobes are mostly devoid of catalase and have some SOD activity (McCord, 1971).

Previously, it has been widely accepted that obligate anaerobes are wholly intolerant of oxygen due to the absence of the enzymes superoxide dismutase and catalase. In further investigation, these enzyme activities have been identified in some obligate anaerobes, and genes for these enzymes and related proteins have been found in their genomes (Meinecke et al., 1989; Sawers and Watson, 1998; Pan and Imlay, 2001). In *Clostridium acetobutylicum*, Helmann and co-workers in the laboratory of Hubert Bahl discovered a homologue of PerR, which is the master repressor of peroxide stress responses in Gram-positive bacteria (Lee and Helmann, 2006). Perhaps, gradual entry of oxygen in anaerobes might be possible but the full-bore oxygenation generates severe ROS stress. Other works revealed that superoxide and hydrogen peroxide are not the sole effectors of obligate anaerobiosis. In the presence of oxygen, the central metabolic pathways of anaerobes can be affected irrespective of intracellular levels of superoxide and hydrogen peroxide (Meinecke et al., 1989; Sawers and Watson, 1998; Pan and Imlay, 2001). Other than Clostridiaceae family, rumen bacteria are mostly inside the obligate anaerobes, details of which are provided in the following sections.

2.2.2.1.1 Clostridium

The genus *Clostridium* is Gram-positive, rod-shaped bacterium under family Clostridiaceae. They are mostly found in soil, sewage, and animals' intestines. Most of them are obligate anaerobes. A few species, such as *C. butyricum* and *C. pasteurianum*, fix nitrogen (Chen et al., 2001). They are fast-growing and capable of forming spores during times of stress, and can persist in toxic environment where the other anaerobic bacteria cannot, which make the bacteria easy to handle in industrial application. Several species of clostridia are used industrially for the production of alcohols and commercial solvents. It has high hydrogen production rate naturally. Clostridia could be thermophiles or mesophiles, produce hydrogen within a pH range of 5.0–6.5 and sometimes accompanied by other bacteria which help to degrade and convert organic waste material into H_2 and organic acids. The clostridia produce H_2 via a reversible hydrogenase enzyme (Calusinska et al., 2010).

Clostridia could produce H_2 continuously from glucose (Andel et al., 1985; Heyndrickx et al., 1986; Kataoka et al., 1997). It also has capability to produce H_2 from a large number of other carbohydrates, such as glucose, xylose, arabinose, galactose, cellobiose, sucrose, and fructose and waste. It suggests that it can be very effective organism for hydrogen production from industrial waste water (Suzuki and Karube, 1981).

Taguchi and colleagues isolated various Clostridia strains from termites. *C. beijerinckii* AM21B could utilize a large number of other carbohydrates (Taguchi et al., 1992, 1993). *Clostridium* sp. strain no. 2 produced H_2 more efficiently from xylose and arabinose (13.7 and 14.6 mol/kg or 2.1 and 2.2 mol/mol) than from glucose (11.1 mmol/g or 2.0 mol/mol) (Taguchi et al., 1994). *Clostridium* sp. strain X53 from wild termites produced xylanase in a batch culture and converted xylan to hydrogen. In comparison to xylose, the kinetics of hydrogen production from xylan was not significantly different, but the total yield from xylan was lower than from xylose (Taguchi et al., 1996).

2.2.2.1.2 Rumen Bacteria

The rumen is a complex ecosystem. It forms the larger part of the reticulorumen, which is the first chamber in the alimentary canal of ruminant animals. It serves as the primary site for microbial fermentation of ingested feed. The smaller part of the reticulorumen is the reticulum.

Rumen bacteria are mostly fiber digesters, protein, starch and sugar digesters, lactate using, and hydrogen-using bacteria. There are few bacteria which have long been known to produce H_2 together with other products such as acetate, ethanol, formate, and CO_2 from carbohydrates as, for example, *Ruminococcus albus* in a continuous culture produce H_2 2.4 mol/mol glucose (Iannotti et al., 1973). Joyner et al. (1977) has studied hydrogen production in different rumen anaerobes using different substrates. *Enterococcus gallinarum* strain G1 isolated from rumen fluid has been cocultured with *Ethanoigenens harbinense* B49 and produced approximately 2.97 mol H_2/kg of cellulose (Wang et al., 2009).

2.2.2.2 Facultative Anaerobes

Facultative anaerobic organisms grow very quickly and undergo aerobic respiration and ATP generation when oxygen is present; they are capable of switching from aerobic to anaerobic metabolism progressively as oxygen becomes depleted. The concentrations of oxygen and fermentable material in the environment influence the organism's use of aerobic respiration vs. fermentation to derive energy. In industrial scale facultative microorganisms are more preferable for fermentative hydrogen production as they can survive in contamination of oxygen. They have the advantage of rapidly consuming oxygen thereby restoring anaerobic conditions immediately in reactors.

Bacteria of family Enterobacteriaceae is mostly hydrogen-producing facultative anaerobes. They produce hydrogen from complex carbohydrates in the absence of oxygen, for example *Enterobacter*, *Escherichia coli*, *Citrobacter*, and so on. The members of the Enterobacteriaceae family can have several beneficial properties favorable for H_2 production such as high growth rates and utilization of a wide range of carbon sources. Seol et al. (2008) has compared hydrogen production in different strains of family Enterobacteriaceae and found that final hydrogen yield was very similar to the four strains (1.7–1.8 mol H_2/mol glucose), *Citrobacter amalonaticus* Y19, *Escherichia coli* K-12 MG1655, *Escherichia coli* DJT135, and *Enterobacter aerogenes*.

2.2.2.2.1 Enterobacter

The genus *Enterobacter* is a pathogenic, Gram-negative, rod-shaped bacterium of the family Enterobacteriaceae. Enterobacter species are found in the natural environment in habitats such as water, sewage, vegetables, and soil. In *Enterobacter*, hydrogen production is catalyzed by the enzyme hydrogenase (Mishra et al., 2004).

Enterobacter produces hydrogen mostly around 303 K, and hydrogen production increases with increasing pH ranges from 4.6 to 6.6 (Sen and Das 2005). Only the initial pH was found to have a profound effect on hydrogen production potential without controlling pH throughout the fermentation (Yokoi et al., 1995). While regulating the pH throughout, the fermentation was found to increase the cumulative hydrogen production rate and yield significantly (Khanna et al., 2011). Flushing with argon enhanced the H_2 yield and suggested that the removal of CO_2 was responsible for the improvement (Tanisho et al., 1998). Compared to *Clostridia*, the H_2 yield on glucose is normally lower, however the mixed culture of clostridia and *Enterobacter* can be industrially advantageous as the total yield will increase and the trace amount of oxygen will be utilized by the facultative one. There are many reports about the successful hydrogen production from different strains of *Enterobacter* species isolated from different places, which are capable of using different substrates (Tanisho et al., 1983; Tanisho and Ishiwata 1994; Kumar and Das, 2000).

Mutants of *E. aerogenes* and *E. cloacae* were also developed for improved H_2 production by blocking the production of other metabolites, alcohols, and organic acids and redirect the reducing equivalents toward H_2 production. These are described in detail in Chapter 6.

2.2.2.2.2 Escherichia coli

Escherichia coli strains are mostly Gram-negative, rod-shaped, and non-sporulating pathogen, and they belong to the family Enterobacteriaceae. *Escherichia coli* are mostly found in lower intestine of warm-blooded animals. It has been shown to be capable of producing H_2 and CO_2 from formate in the absence of oxygen (Stickland, 1929; Yudkin, 1932). A membrane-bound multienzyme complex called formate hydrogen lyase, consisting of a formate dehydrogenase and a hydrogenase, catalyzes the reaction (Gray and Gest 1965). Sustained lysis of formate required blocking of other anaerobic reductases (Nandi and Sengupta, 1996).

E. coli produces hydrogen under mesophilic condition. The rate of hydrogen production increases around fourfold with increasing temperature from 291 K to 303 K while decreasing the initial pH from 7.3 to 5.5 decreased the volume of evolved H_2. It has been observed that pH lower than 5.5 and greater than 7 is not suitable for hydrogen production (Penfold et al., 2003).

E. coli could produce hydrogen from different carbohydrates (Stickland, 1929). The fermentation of lactose, glucose, and galactose was studied with *E. coli* WDHL, a strain in which *hycA* and *lacI* were deleted to improve hydrogen production (Rosales-Colunga et al., 2012). The *hycA* gene codes for the negative regulator of the hydrogen pathway. With glucose as substrate, pyruvate was mainly routed to the lactate pathway (H_2 yield of 1.04×10^{-3} m^3 and

0.30 mol H_2/mol of glucose) but when galactose was the substrate the pyruvate formate lyase pathway was the main route for pyruvate (H_2 yield of 2080 mL and 1.12 mol H_2/mol of galactose). The fermentation of lactose or glucose plus galactose showed a similar yield of 1.02 mol H_2/mol of hexose consumed. The kinetics of H_2 and metabolites production as well as the H_2 yield was affected by the type of sugar used as substrate (Rosales-Colunga et al., 2012).

2.2.2.2.3 Citrobacter

Citrobacter is a genus of Gram-negative coliform bacteria in the Enterobacteriaceae family. These bacteria can be found almost everywhere in soil, water, wastewater, and so on. They can also be found in the human intestine. Presence of both formate hydrogenlyase (FHL) activity and other hydrogenases activity has been observed in *Citrobacter* (Kim et al., 2008).

The optimum pH for H_2 production for *Citrobacter* was found to be 5.75 and 6.0 which gave a yield of 1.1 mol H_2/mol glucose with rate of 0.04 moles H_2/s (Brosseau et al., 1980). *Citrobacter* sp. Y19 isolated from sludge digesters has been shown to produce hydrogen from CO and H_2O by the water–gas shift reaction under anaerobic conditions (Jung et al., 1999, 2002). *C. freundii*, isolated from sewage produced 8.9×10^{-3} m³ in the volumetric ratio of 63% H_2 and 37% CO_2 in 3.96×10^4 s from 0.031 kg glucose (Kumar and Vatsala, 1989).

2.2.2.3 Aerobes

Hydrogen can be produced by some aerobic bacteria. Some of them grow lithoautotrophically and consume molecular hydrogen. They have the ability to produce hydrogen during heterotrophic metabolism. *Alcaligenes eutrophus* has been shown to grow heterotrophically on gluconate and fructose, and when exposed to anaerobic conditions produced hydrogen (Kuhn et al., 1984). It contains a soluble NAD-reducing hydrogenase (Pinchukova et al., 1979). A hydrogen-producing *Bacillus lichenformis* was isolated from cattle dung (Kalia et al., 1994). It produced 0.5 mol H_2/mol glucose. Immobilized cells had an average H_2 yield of 1.5 mol per mol glucose, and cells were stable during 60 days (Kumar et al., 1995). Methanogens are characterized by the presence of hydrogenase, which is usually involved in the oxidation of H_2 coupled to CH_4 production and CO_2 reduction. However, Bott et al. reported that production of H_2 and CO_2 in stoichiometric amounts from CO and H_2O by a strain of *Methanosarcina barkeri*, the so-called water–gas shift reaction under conditions of inhibition of CH_4 formation (Bott et al., 1986).

2.2.3 Fermentative End Products

The hydrogen-producing bacterial communities can be distinguished by their fermentative end products into the following groups.

2.2.3.1 Lactic Acid Fermentation

Bacteria (*Lactobacillus* and most Streptococci) mostly use dairy products as their substrate during lactic acid fermentation. Lactic acid fermentation is of two types, homolactic fermentation ($C_6H_{12}O_6 \rightarrow 2CH_3CHOHCOOH$) and heterolactic fermentation ($C_6H_{12}O_6 \rightarrow CH_3CHOHCOOH + C_2H_5OH + CO_2$). But there is no H_2 production (Salminen et al., 2004).

2.2.3.2 Mixed Acid Fermentation

Hydrogen is produced by mixed acid fermentation. End products are mostly a mixture of lactic acid, acetic acid, formic acid, succinate, butanediol, and ethanol, with the formation of CO_2 and H_2 (Rachman et al., 1997). This pathway is mostly exhibited by *Enterobacteriaceae*, such as *Klebsiella*, *Enterobacter*, *E. coli*, and so on.

2.2.3.3 Butyric Acid Fermentation

Mostly, the Clostridia produce butyric acid during fermentation of sugars along with acetic acid, CO_2, and H_2. Small amounts of ethanol and isopropanol may also be formed (Ren et al., 2008).

2.2.3.4 Butanol–Acetone Fermentation

Butanol and acetone were discovered as the main end products of fermentation by *Clostridium acetobutylicum*. Butanol production inhibits H_2 generation (Kim et al., 1984).

There are other unusual fermentation carried out by some bacteria, which produce some acids, alcohol, and gases but end products do not include hydrogen. As for example, propionic acid bacteria were used in the manufacturing of Swiss cheese (Beresford et al., 2001).

2.3 Photosynthetic Fermentative Bacteria

Metabolism of dark-adapted microorganism depends on the availability of substrate. Any time, they can starve with depletion of substrate. In substrate crisis, energy can be derived only by limited endogenous catabolism and dissipates energy mostly in a nonreusable form. To sustain life and fulfill all metabolic demands, life evolved to utilize the light energy from the sun. Sun is the unlimited source of energy. Microorganism has developed different light-harvesting complexes for capturing photons. Light energy is converted as chemical energy via photophosphorylation. Two principal classes of photosynthetic bacteria, the purple bacteria and the greenbacteria, carry out photosynthesis with single photosystem (Blankenship, 1992). However, there is no oxygen evolution during photosynthesis. Phylogenetic studies have shown that purple and green bacteria are far distantly related (Gibson et al., 1979; Fox et al., 1980). They are probably metabolically most versatile organisms on the earth. They can adapt to any probable situation that might arise. In anaerobic conditions, they can live photoheterotrophically by using different organic acids as substrate. This process is known as photofermentation. Light energy is used for generation of ATP via cyclic phosphorylation. Most of the photosynthetic bacteria are active diazotrophs (exception green gliding) that fix N_2 by nitrogenase under anoxic nitrogen-limiting conditions (Madigan et al., 1984; Kimble and Madigan, 1992; Warthmann et al., 1992) although some of them may contain hydrogenases (Wu and Mandrand, 1993). The hydrogen evolution is observed normally by nitrogenase when ATP and electrons are available (Sasikala et al., 1993). Some of them may also possess uptake hydrogenase (Gogotov, 1984).

2.3.1 Purple Bacteria

Purple bacteria are phototrophic proteobacteria. They are pigmented with bacteriochlorophyll a or b, together with various carotenoids, which give them colors ranging between purple, red, brown, and orange. Photosynthesis takes place at reaction centers on the cell membrane, which is folded into the cell to form sacs, tubes, or sheets, increasing the available surface area. Purple bacteria contain bacteriochlorophyll and PSII like reaction center. It generates ATP via cyclic electron flow. Purple sulfur bacteria use sulfide, elemental sulfur or organic substrate, and others, hence they are called purple nonsulfur bacteria which typically use hydrogen although some may use other compounds in small amounts. Light intensity is an important factor for hydrogen production by purple bacteria. Under limiting light intensity, light energy conversion is higher but hydrogen production rate is lower. However, in supersaturating light intensities hydrogen production was not stable. The light saturation growth lies in the range of 40–80 W/m^2 (Tsygankov et al., 1996).

2.3.1.1 Sulfur Bacteria

The purple sulfur bacteria are a group of Proteobacteria. They are anaerobic or microaerophilic, and are often found in hot springs or stagnant water. In dark, some species are capable of chemolithoautotrophic or chemoorganoheterotrophic growth in the presence of oxygen. Members of Chromatiaceae use sulfide or organic substrate as electron donors (Kampf and Pfennig, 1980). In these bacteria, H_2 evolution is coupled with inorganic (sulfur-containing compounds) or organic substrate-driven reserve electron flow. Hydrogen is evolved when electrons are transferred toward nitrogenase via ferredoxin by expense of ATP. The majority of purple sulfur bacteria are active diazotrophs. They are able to fix N_2 to ammonia (Madigan, 2004). Nitrogen fixation is linked with hydrogen production. Under nitrogen starvation, most of the purple sulfur bacteria are able to produce molecular hydrogen (Mitsui, 1975). No oxygen is evolved in this process. Some species contain membrane-bound hydrogenase, which has hydrogen uptake capability instead of production of hydrogen. This capability was first detected in *Allochromatium vinosum* strain D (Gaffron, 1935).

2.3.1.1.1 Thiocapsa roseopersicina

Thiocapsa roseopersicina is Gram-negative, nonmotile purple sulfur bacteria from the family of Chromatiaceae. It has spherical cells that can form tetrads. *T. roseopersicina* can use hydrogen, sulfide, thiosulfate, acetate, propionate, pyruvate, malate, succinate, fumarate, fructose, or glycerol as substrates. It can fix atmospheric N_2 usually accompanied by H_2 production. It has four distinct NiFe hydrogenases, which are particularly important as they are tolerant to oxygen (Kovács et al., 2004). They can produce H_2 when the uptake hydrogenase activity has been eliminated (Fodor et al., 2001).

2.3.1.2 Nonsulfur Bacteria

Most of the purple nonsulfur bacteria are well studied for H_2 production (Kumazawa and Mitsui, 1982). Purple nonsulfur bacteria exhibit aerobic respiration in photoautotrophic growth. But in the presence of oxygen there is no evolution of hydrogen as the oxygen is the electron acceptor instead of protons. Also, the activity of enzyme nitrogenase is suppressed in the presence of oxygen. Hydrogen evolution in purple nonsulfur bacteria has been observed in the absence of oxygen. They depend on external organic substrate for carbon and electron source. In nitrogen starvation, nitrogenase catalyzes the formation molecular hydrogen from protons. The final amount of hydrogen produced is influenced also by the activity of uptake hydrogenase (Koku et al., 2002). Photosynthetic nonsulfur purple bacteria are considered the best means of photobiological hydrogen production (Das and Veziroğlu, 2001). The advantage of this process relies on application of organic wastes as the source of organic carbon (photofermentation). This adds extra advantage in economy of hydrogen production. Some success has been found using industrial waste water as substrate (Yetis et al., 2000).

2.3.1.2.1 Rhodobacter sphaeroides

Rhodobacter sphaeroides is a rod-shaped, Gram-negative, photoheterotrophic bacterium belonging to Proteobacteria. It is found in soil, in anoxic zones of waters, mud, sludge, and in organic-rich water habitats. It is a metabolically diverse organism capable of aerobic respiration, anaerobic anoxygenic photosynthesis, fermentation, and diazotrophic growth. In the absence of oxygen, it produces hydrogen by nitrogenase. The inner membrane of the organism undergoes morphological changes forming the intra cytoplasmic membrane (ICM). Photosynthetic apparatus contains the reaction center

(RC), encircled by the light-harvesting complex I (LHI), which is surrounded by a variable number of light-harvesting complexes II (LHII).

2.3.2 Green Bacteria

2.3.2.1 Sulfur Bacteria

Green sulfur bacteria are obligate anaerobic photoautotrophic bacteria. In contrast to purple bacteria, they have the PSI type reaction center, which directly reduced ferredoxin. In green bacteria, ferredoxin is reduced by FeS protein and serves directly as electron donor for the H_2 production. Most of them can fix nitrogen by the enzyme nitrogenase and in N_2 limited condition nitrogenases catalyze the reduction of protons to hydrogen similar to purple bacteria. When H_2S donates electrons to green bacteria, sulfur globules remain outside of the cell. This is unlike purple bacteria, where the globules of sulfur remain inside of the bacterial cell (Imhoff, 2004; Friedrich et al., 2005). Reaction center donates electrons only to the nitrogenase for hydrogen production coupled with nitrogen fixation but not for CO_2 fixation. They lack RuBisCO, the key enzyme of Calvin cycle. Green sulfur bacteria fix CO_2 by reductive TCA cycle (Holo and Sirevåg, 1986; Hügler et al., 2003). Therefore, there is no competitive inhibition of hydrogen production by CO_2 fixation. CO_2-free environment within a reactor is not required, but N_2-free environment is required to inhibit NH_3 formation. Storage product in green sulfur bacteria is mainly glycogen. In dark, they can catabolize storage product to produce different organic acids similar to dark fermentation. It has been suggested that storage product serves as energy source in dark and reducing power in light (Sirevåg and Ormerod, 1977).

2.3.2.1.1 Chlorobium

Chlorobium is Gram-negative, mostly thermophilic green sulfur bacteria. They have special light-harvesting complexes called chlorosomes that contain bacteriochlorophylls and carotenoids. They grow in dense mats over hot springs as well as in other hot water bodies that contain sufficient hydrogen sulfide. It has been observed that the marine green sulfur bacterium *Chlorobium vibrioforme* strain 1930 produced H_2 and elemental sulfur from sulfide or thiosulfate under N_2 limitation in the light. H_2 production depended on nitrogenase and occurred only in the absence of ammonia. The coculture of *Chlorobium vibrioforme* strain 1930 and *Desulfuromonas acetoxidans* strain 5071 formed significant amount of H_2 from acetate (Warthmann et al., 1992).

2.3.2.2 Gliding Bacteria

Very few studies have been carried out for the hydrogen production from green-gliding bacteria. They have photosystem II type of reaction center like purple bacteria but having light-hervesting structure chlorosome like green sulfur bacteria. Most of them probably do not have nitrogenase (Heda and Madigan, 1986). Hydrogenase was isolated from green-gliding bacteria *Chloroflexus aurantiacus* and used as efficient biocatalyst for hydrogen production (Gogotov et al., 1991). This organism can survive in dark having orange color in the presence of oxygen, but in light it is green. *Chloroflexus* uses reduced sulfur compounds such as hydrogen sulfide, thiosulfate, or elemental sulfur, and also hydrogen. *Chloroflexus aurantiacus* is thought to grow photoheterotrophically in nature, but it has the capability of fixing inorganic carbon through photoautotrophic growth.

2.4 Cyanobacteria

Cyanobacteria form a large and diverse group of oxygenic photoautotrophic, unicellular, or filamentous prokaryotes. Many of them have the ability to produce hydrogen. It is also known as blue-green algae. It uses carbon dioxide as carbon source and produces ATP and oxygen through photosynthesis similar to eukaryotes. They have very simple nutritional requirements such as air (for CO_2 and N_2), water (for electrons), light (for energy), and mineral salts. Advantages of using cyanobacteria for hydrogen production are the necessity of simple reaction broth, vigorous growth rate, and lesser chances of contamination. In cyanobacteria, hydrogen evolution is separated from oxygen evolution either by compartmentalization (heterocysts) or by temporal separation. Heterocyst is a specialized cell for maintaining the anaerobic condition, for example *Nostoc, Anabaena*, and so on. Reserve carbon is transported inside the heterocyst, oxidized to release electrons that reduce protons inside the heterocyst. Nitrogenase is the key enzyme responsible for hydrogen generation within the heterocyst. However, nonheterocystus and nonnitrogen-fixing strains produce hydrogen mostly by bidirectional hydrogenase. The uptake hydrogenase is also present in cyanobacteria, which act against hydrogen production by oxidizing the molecular hydrogen. Some cyanobacteria use nitrogenase/uptake hydrogenase or bidirectional hydrogenase, and some use both. Cyanobacteria produce hydrogen by the process called biophotolysis. It could be direct, when electron comes directly from the splitting of water or indirect, when electrons derive from reserve carbon source. Hydrogen production has been studied in at least more than 14 cyanobacteria genera, under a wide range of culture conditions (Lopes Pinto et al., 2002). Dutta et al. (2005) reviewed hydrogen production from cyanobacteria highlighting the basic biology of cynobacterial hydrogen production, efficient strains, large-scale hydrogen production, and its future prospects.

2.4.1 Anabaena

Anabaena is nitrogen-fixing filamentous cyanobacteria. Under nitrogen-limiting conditions, vegetative cells differentiate into heterocysts at semiregular intervals along the filaments. Heterocysts are cells that are terminally specialized for nitrogen fixation. The interior of these cells is microoxic as a result of increased respiration, inactivation of O_2-producing PSII, and formation of a thickened envelope outside of the cell wall. During the starvation of nitrogen, *Anabaena* sp. is able to produce significant amount of hydrogen. Different species of H_2-producing Anabaena are well summarized in the review article of Dutta et al. (2005). In heterocysts, H_2 is produced by nitrogenase, but Spiller et al. described heterocyst-free ($NH4^+$-grown) cultures of *Anabaena variabilis* produced a hydrogenase which was reversibly inhibited by light and O_2 (Spiller et al., 1983). Two distinct types of hydrogenase have been found in *Anabaena* 7120. A reversible hydrogenase occurs both in heterocysts and vegetative cells (Houchins and Burris, 1981). Presence of uptake hydrogenases is also reported in different strains of *Anabaena* (Peterson and Wolk, 1978). A short-term decrease of the CO_2 concentration has also been observed in the air-suppressed H_2 evolution. Anoxygenic conditions over the dark periods had a negative effect on H_2 production. The peculiarity of hydrogen production and some physiological characteristics of *A. variabilis* PK84 during cultivation in the photobioreactor under a light–dark regime has been investigated by Borodin et al. (2000).

2.4.2 Nostoc

Nostoc is filamentous heterocystous cyanobacteria found in a variety of environmental niches, in soil, on moist rocks, at the bottom of lakes and springs (both fresh- and saltwater), and rarely in marine habitats. *Nostoc* sp. PCC 7422 has been found to have the highest nitrogenase activity. After disruption of uptake hydrogenase gene, mutant produced H_2 at the rate of 2.78 mol/kg (chlorophyll *a*), three times that of the wild type (Yoshino et al., 2007). Hydrogenase-deficient cyanobacteria *Nostoc punctiforme* NHM5 when incubated under high light for a long time, until the culture was depleted of CO_2, shows increase in hydrogen production (Lindberg et al., 2004).

2.4.3 Synechocystis

Synechocystis is a genus of cyanobacteria, which lives in freshwater and is capable of both phototrophic growth by oxygenic photosynthesis in sunlight and heterotrophic growth by glycolysis and oxidative phosphorylation during dark periods. It is able to effectively anticipate transitions of light and dark phases by using a circadian clock. In the *ndhB* mutant M55, which is defective in the type I NADPH-dehydrogenase complex (NDH-1), produces only low amounts of O_2 in the light, H_2 uptake was negligible during dark-to-light transitions, allowing several minutes of continuous H_2 production (Cournac et al., 2004). The nitrogen concentration and pH of culture media were optimized for increased fermentative H_2 production from the cyanobacterium, *Synechocystis* sp. PCC 6803. (Burrows et al., 2009).

2.5 Green Algae

Eukaryotes do not have much versatility in their metabolic pathways. They are mostly aerobes which have evolved to respire with oxygen. Hardly in their life cycle do they exhibit fermentative metabolism. The only potential eukaryotes that can produce hydrogen in anoxia are green algae. Not only can they carry out oxygenic photosynthesis like higher plants, but they can also survive in anoxia in their natural habitats. Hydrogen evolution in green algae is identified as a consequence of anaerobiosis (Happe and Naber, 1993) or nutrient deprivation (Melis et al., 2000). Higher plants in anoxic conditions can exhibit fermentation by depositing the excessive-reducing power; however, they are rarely exposed to these anoxic conditions. But most of the algal populations are aquatic and sometimes suffer from anoxia in stagnant water or during algal blooms. Gaffron and his co-researchers first observed that under aerobic condition unicellular green algae, *Scenedesmus obliquus*, is able to generate hydrogen in the presence of light (Gaffron and Rubin, 1942). It has been found that enzyme hydrogenase is involved for metabolizing hydrogen in algae.

2.5.1 Chlamydomonas

Chlamydomonas reinhardtii is used as model organism for studying hydrogen production from microalgae. The whole genome sequencing makes it ideal for genetic modification (Merchant et al., 2007). A sustained hydrogen production can be obtained under low sulfur conditions in *C. reinhardtii*. It reduces the net oxygen evolution by reducing the photosystem II activity and thereby overcoming the inhibition of the hydrogenases (Melis et al., 2000). The development of specially adapted hydrogen production strains led to higher yields and optimized biological process preconditions. Use of exact dosage of sulfur can eliminate the elaborate and energy intensive solid–liquid separation step and establish a process strategy to proceed further vs. large-scale production (Lehr et al., 2012).

Other genus of green algae, such as, *Chlorella, Scenedesmus*, and so on, is also investigated for hydrogen production. But the commercial success has to be achieved.

2.6 Concept of Consortia Development

In nature, symbiotic relationships between prokaryotes and eukaryotes may be made for co-evolution, whereby strong selective pressures for highly specific traits allow the pair to exploit a particular niche. This kind of associations with other organisms promote the protection from potentially inhibitory environmental factors. In a consortium, each species may bring unique metabolic pathways. This is very much useful to degrade the complex substrates, wastewater, sludge, and other pollutants for further hydrogen production. In a successful microbial consortium, bacterial cultures must be compatible with each other for their growth and produce necessary enzymes required for the degradation of the substrate. Taking this concept into consideration, researchers are generating defined microbial consortia to effectively utilize various complex organic substrates including wastes for H_2 production. In this context, peoples have isolated different mixed microflora from various sources, such as fermented soybean meal or sludges from anaerobic digesters of municipal sewage or organic waste and sludge from kitchen waste water. Sometimes these microflora contain unwanted methanogenic bacteria, which consume the produced hydrogen and convert it to methane. To solve this problem, forced aeration of the sludge or heat treatment has been done to inhibit the activity of the hydrogen consumers keeping the spores of anaerobes. Additionally in continuous fermentation, higher dilution rate is used to wash out the slow-growing methanogen and select for the acid-producing bacteria. The preliminary reports published on hydrogen production during waste water treatment showed inhibited methane production but low H_2 yields and lack of stability (Guwy et al., 1997; Sparling et al., 1997). Ueno et al. (1995) have found that the anaerobic microflora in sludge converted cellulose to hydrogen with high efficiency of 2.4 mol/mol hexose with the rate of 0.39 mmol/m³ s in batch experiments at 60°C. Furthermore, stable hydrogen production for 190 days from industrial wastewater from a sugar factory by the same microflora in a chemostat culture was reported (Ueno et al., 1996). Defined consortia comprising *Enterobacter cloacae* IIT BT-08, *Bacillus coagulans* IIT BT S1, and *Citrobacter freundii* IIT BT L139 in the ratio 2:1:1 were found suitable to utilize molasses and produced 16.66 mol H_2/kg COD reduced. Hydrogen yield was found to be more as compared to that obtained from the individual strains (Ghosh et al., 2010). Hiligsmann (2011) developed a method for the characterization of biochemical hydrogen potential of different pure strains and mixed cultures of hydrogen-producing bacteria (HPB) growing on glucose. The experimental results compared the hydrogen production yield 19 different pure strains and sludges; facultative and strict anaerobic; HPB strains along with anaerobic digester sludges thermally pre-treated or not. Yokoi et al. (1998, 2001) reported on a coculture in a continuous fermentation of *Clostridium butyricum* and *Enterobacter aerogenes* in which the higher H_2 yield of the strict anaerobe and the oxygen consumption by the facultative anaerobe were combined. The presence of *E aerogenes* was sufficient to restore anaerobic conditions in the fermentor upon short oxygen exposures. A continuous fermentation by immobilized mixed cells on porous glass beads and starch as the substrate showed a H_2 production rate of approximately 13.9 mmol/m^3 s and an H_2 yield of 2.6 on glucose at dilution rates of 2.78×10^{-4} s^{-1}. Use of different mixed culture from waste has been reported, such as Mizuno et al. (2000) improved the H_2 yield on glucose by mixed cultures isolated from fermented soybean meal. H_2 production rates of approximately 8 mmol/L · h were obtained. Noike and Mizuno et al. (2000) reported on hydrogen fermentations of organic waste, such as bean curd manufacturing waste, rice, and wheat bran by the same mixed culture in batch reactors. H_2 yield varied from 1.7 to 2.5 mmol/mol hexose.

2.7 Synthetic Microorganisms—Are They the Future?

Discovery of enormous microbial diversity and whole genome sequencing of a wide range of organisms enabled us to realize genetic variability, identify organisms with natural ability to acquire and transmit genes. Such organisms can be exploited through genome shuffling for transgenic expression and efficient generation of clean fuel and other diverse biotechnological applications. Synthetic biology has rapidly grown out of genetic and metabolic engineering into a new science. Genetic engineering modifies only the existing organisms by knock down or overexpressed few genes but the synthetic biology aims to design entirely new life forms with pre-selected functions. This strategy might involve designing an "optimal" organism using combinations of enzymes from other sources, or even completely new enzymes designed and created using protein engineering to have maximal catalytic activity. Strategies like these are most often being pursued to create microorganisms or algae optimized to produce biohydrogen. The J. Craig Venter Institute announced in May 2010 that it had successfully created the "first self-replicating synthetic bacterial cell" (Wade, 2010). According to M. Baker, scientists have not commercialized the genetic code. Few companies are using synthetic biological approaches to create novel organisms capable of producing ethanol and other biofuels from sunlight and CO_2 without depending on the biomass feedstock (Baker, 2011).

Although the creation of a potential synthetic cell is extremely ambitious, it is difficult to predict the associated risks and harms. Traditionally, the associated risks of a new genetically engineered organisms are often assessed by comparison with known, similar, and related organism. But synthetic organisms, most likely, will be one of its kind and will not have any relatives in nature. Using synthetic cell for generating biofuels thus might contain potential risks and harms to the environment. Moreover, these cells could share genes with other microorganism through horizontal gene transfer and/or evolve beyond their functionality. Until now, it is still unclear how to regulate this new science. Governments and civil societies are currently engaged with synthetic biology and on how to regulate under domestic and international law. But in future, we might need to depend on the taylor-made synthetic cell for our future fuel, and it will create a new era.

Glossary

ADP	Adenosine diphosphate
ATP	Adenosine triphosphate
COD	Chemical oxygen demand
CSTR	Continuous stirred-tank reactor
FeS protein	Iron sulfur protein
FHL	Formate hydrogenlyase
GlcNAc	N-acetyl-D-glucosamine
HPB	Hydrogen-producing bacteria
ICM	Intracytoplasmic membrane
LHI	Light-harvesting complex I
LHII	Light-harvesting complex II
NAD	Nicotinamide adenine dinucleotide
NADH	Nicotinamide adenine dinucleotide (reduced form)
NADPH	Nicotinamide adenine dinucleotide phosphate (reduced form)
NDH-1	NADPH-dehydrogenase complex (Type I)
PSII	Photosystem II
RC	Reaction center
ROS	Reactive oxygen species
RuBisCO	Ribulose-1,5-bisphosphate carboxylase oxygenase
SOD	Super oxide dismutase
TCA cycle	Tricarboxylic acid cycle

References

Ananyev, G., Carrieri, D., and Dismukes, G. C. 2008. Optimization of metabolic capacity and flux through environmental cues to maximize hydrogen production by the cyanobacterium *"Arthrospira (Spirulina) maxima"*. *Applied and Environmental Microbiology*, 74(19), 6102–6113.

Andel, J. G., van Zoutberg, G. R., Crabbendam, P. M., and Breure, A. M. 1985. Glucose fermentation by *Clostridium butyricum* grown under a self generated gas atmosphere in chemostat culture. *Applied Microbiology and Biotechnology*, *123*, 21–26.

Baker, M. 2011. The next step for the synthetic genome. *Nature*, *473*(7347), 403–408.

Bandyopadhyay, A., Stöckel, J., Min, H., Sherman, L. A., and Pakrasi, H. B. 2010. High rates of photobiological H_2 production by a cyanobacterium under aerobic conditions. *Nature Communications*, *1*, 139.

Berberoglu, H., Jay, J., and Laurent, P. 2008. Effect of nutrient media on photobiological hydrogen production by *Anabaena variabilis* ATCC 29413. *International Journal of Hydrogen Energy*, *33*, 1172– 1184.

Beresford, T. P., Fitzsimons, N. A., Brennan, N. L., and Cogan, T. M. 2001. Recent advances in cheese microbiology. *International Dairy Journal*, *11*(4), 259–274.

Bergquist, P. L., Gibbs M. D., Morris D. D., Te'o V. S., Saul D. J., and Moran, H. W. 1999. Molecular diversity of thermophilic cellulolytic and hemicellulolytic bacteria. *FEMS Microbiology Ecology*, *28*, 99–110.

Blankenship, R. E. 1992. Origin and early evolution of photosynthesis. *Photosynthesis Research*, *33*, 91–111.

Boichenko, V. A., Greenbaum, E., and Seibert, M. 2004. Hydrogen production by photosynthetic microorganisms. In *Photoconversion of Solar Energy: Molecular to Global Photosynthesis*, ed. Archer, M. D., and Barber, J., pp. 397–452. London: Imperial College Press.

Borodin, V. B., Tsygankov, A. A., Rao, K. K., and Hall, D. O. 2000. Hydrogen production by *Anabaena variabilis* PK84 under simulated outdoor conditions. *Biotechnology and Bioengineering*, *5*, *69*(5), 478–85.

Bott, M., Eikmanns, B., and Thauer, R. K. 1986. Coupling of carbon monoxide oxidation to CO_2 and H_2 with the phosphorylation of ADP in acetate-grown Methanosarcina barkeri. *European Journal of Biochemistry*, *159*(2), 393–398.

Brosseau, J. D., Kosaric, N., and Zajic, J. E. 1980. The effect of pH on hydrogen production with *Citrobacter intermedius*. *Biotechnology Letters*, *2*(3), 93–98.

Budiman, C., Koga, Y., Takano, K., and Kanaya, S. 2011. FK506-binding protein 22 from a psychrophilic bacterium, a cold shock-inducible peptidyl prolyl isomerase with the ability to assist in protein folding. *International Journal of Molecular Sciences*, *12*(8), 5261–5284.

Burrows, E. H., Chaplen, F. W., and Ely, R. L. 2008. Optimization of media nutrient composition for increased photofermentative hydrogen production by *Synechocystis* sp. PCC 6803. *International Journal of Hydrogen Energy*, *33*(21), 6092–6099.

Burrows, E. H., Wong, W. K., Fern, X., Chaplen, F. W., and Ely, R. L. 2009. Optimization of ph and nitrogen for enhanced hydrogen production by *Synechocystis* sp. pcc 6803 via statistical and machine learning methods. *Biotechnology Progress*, *25*(4), 1009–1017.

Calusinska, M., Happe, T., Joris, B., and Wilmotte, A. 2010. The surprising diversity of clostridial hydrogenases: A comparative genomic perspective. *Microbiology*, *156*(6), 1575–1588.

Chader, S., Hacene, H., and Agathos, S. N. 2009. Study of hydrogen production by three strains of *Chlorella* isolated from the soil in the Algerian Sahara. *International Journal of Hydrogen Energy*, *34*(11), 4941–4946.

Chen, J. S., Toth, J., and Kasap, M. 2001. Nitrogen-fixation genes and nitrogenase activity in *Clostridium acetobutylicum* and *Clostridium beijerinckii*. *Journal of Industrial Microbiology and Biotechnology*, *27*(5), 281–286.

Cheng,C. H., Hung, C. H., Lee, K. S. et al., 2008. Microbial community structure of a starch-feeding fermentative hydrogen production reactor operated under different incubation conditions. *International Journal of Hydrogen Energy, 33*(19), 5242–5249.

Claassen, P. A. M., de Vrije, T., and Budde, M. A. W. 2004. Biological hydrogen production from sweet sorghum by thermophilic bacteria. Paper presented at the 2nd World Conference on Biomass for Energy, Rome.

Cournac, L., Guedeney, G., Peltier, G., and Vignais, P. M. 2004. Sustained photoevolution of molecular hydrogen in a mutant of *Synechocystis* sp. strain PCC 6803 deficient in the type I NADPH-dehydrogenase complex. *Journal of Bacteriology, 186*(6), 1737–1746.

Das, D. and Veziroglu, T. N. 2001. Hydrogen production by biological processes: A survey of literature. *International Journal of Hydrogen Energy, 26*(1), 13–28.

de Vrije, T., Mars, A. E., Budde, M. A. W., et al. 2007. Glycolytic pathway and hydrogen yield studies of the extreme thermophile *Caldicellulosiruptor saccharolyticus. Applied Microbiology and Biotechnology, 74*(6), 1358–1367.

Dutta, D., De, D., Chaudhuri, S., and Bhattacharya, S. K. 2005. Hydrogen production by Cyanobacteria. *Microbial Cell Factories, 4*(1), 36.

Evvyernie, D., Morimoto, K., Karita, S., Kimura, T., Sakka, K., and Ohmiya, K. 2001. Conversion of chitinous wastes to hydrogen gas by *Clostridium paraputrificum* M-21. *Journal of Biosciences and Bioengineering, 91*, 339–343

Fedorov, A. S., Tsygankov, A. A., Rao, K. K., and Hall, D. O. 1998. Hydrogen photoproduction by *Rhodobacter sphaeroides* immobilised on polyurethane foam. *Biotechnology Letters, 20*(11), 1007–1009.

Fodor, B., Rákhely, G., Kovács, Á. T., and Kovács, K. L. 2001. Transposon mutagenesis in purple sulfur photosynthetic bacteria: Identification of *hypF*, encoding a protein capable of processing [NiFe] hydrogenases in α, β, and γ subdivisions of the proteobacteria. *Applied and Environmental Microbiology, 67*(6), 2476–2483.

Fox, G. E., Stackebrandt, E., Hespell, R. B. et al., 1980. The phylogeny of prokaryotes. *Science Wash, 208*, 457–463.

Friedrich, C. G., Bardischewsky, F., Rother, D., Quentmeier, A., and Fischer, J. 2005. Prokaryotic sulfur oxidation. *Current Opinion in Microbiology, 8*(3), 253–259.

Gaffron, H. 1935. Über den Stoffwechsel der Purpurbakterien. *Biochem Z, 275*, 301–319.

Gaffron, H. and Rubin, J. 1942. Fermentative and photochemical production of hydrogen in algae. *The Journal of General Physiology, 26*(2), 219–240.

Ghosh, S., Joy, S., and Das, D. 2010. Multiple parameters optimization for maximization of hydrogen production using defined microbial consortia. *Indian Journal of Biotechnology, 10*, 196–201.

Gibson, J., Stackebrandt, E., Zablen, L. B., Gupta, R., and Woese, C. R. 1979. A phylogenetic analysis of the purple photosynthetic bacteria. *Current Microbiology, 3*(1), 59–64.

Gogotov, I. N. 1984. Hydrogenase of purple bacteria: Properties and regulation of synthesis. *Archives of Microbiology, 140*(1), 86–90.

Gogotov, I. N., Zorin, N. A., and Serebriakova, L. T. 1991. Hydrogen production by model systems including hydrogenases from phototrophic bacteria. *International Journal of Hydrogen energy, 16*(6), 393–396.

Gray, C. T. and Gest, H. 1965. Biological Formation of Molecular Hydrogen A "hydrogen valve" facilitates regulation of anaerobic energy metabolism in many microorganisms. *Science, 148*(3667), 186–192.

Guwy, A. J., Hawkes, F. R., Hawkes, D. L., and Rozzi, A. G. 1997. Hydrogen production in a high rate fluidised bed anaerobic digester. _Water Research, 31_(6), 1291–1298.

Happe, T. and Naber, J. D. 1993. Isolation, characterization and N-terminal amino acid sequence of hydrogenase from the green alga _Chlamydomonas reinhardtii. European Journal of Biochemistry, 214_(2), 475–481.

Hawkes, F. R., Hussy, I., Kyazze, G., Dinsdale, R., and Hawkes, D. L. 2007. Continuous dark fermentative hydrogen production by mesophilic microflora: Principles and progress. _International Journal of Hydrogen Energy, 32_(2), 172–184.

Heda, G. D. and Madigan, M. T. 1986. Utilization of amino acids and lack of diazotrophy in the thermophilic anoxygenic phototroph _Chloroflexus aurantiacus. Journal of General Microbiology, 132_(9), 2469–2473.

Heyndrickx, M., Vansteenbeeck, A., De Vos, P., and De Ley, J. 1986. Hydrogen gas production from continuous fermentation of glucose in a minimal medium with _Clostridium butyricum_ LMG 1213t1. _Systematic and Applied Microbiology, 8_(3), 239–244.

Hiligsmann, S., Masset, J., Hamilton, C., Beckers, L., and Thonart, P. 2011. Comparative study of biological hydrogen production by pure strains and consortia of facultative and strict anaerobic bacteria. _Bioresource Technology, 102_(4), 3810–3818.

Holo, H. and Sirevåg, R. 1986. Autotrophic growth and CO_2 fixation of _Chloroflexus aurantiacus. Archives of Microbiology, 145_(2), 173–180.

Horvath, R., Orosz, T., Balint, B. et al., 2004. Application of gas separation to recover biohydrogen produced by _Thiocapsa roseopersicina. Desalination, 163_(1), 261–265.

Houchins, J. P. and Burris, R. H. 1981. Occurrence and localization of two distinct hydrogenases in the heterocystous cyanobacterium _Anabaena_ sp. strain 7120. _Journal of Bacteriology, 146_(1), 209–214.

Hügler, M., Huber, H., Stetter, K. O., and Fuchs, G. 2003. Autotrophic CO_2 fixation pathways in archaea (Crenarchaeota). _Archives of Microbiology, 179_, 160–173.

Hung, C. H., Chang, Y. T., and Chang, Y. J. 2011. Roles of microorganisms other than _Clostridium_ and _Enterobacter_ in anaerobic fermentative biohydrogen production systems—A review. _Bioresource Technology, 102_(18), 8437–8444.

Iannotti, E. L., Kafkewitz, D., Wolin, M. J., and Bryant, M. P. 1973. Glucose fermentation products of _Ruminococcus albus_ grown in continuous culture with _Vibrio succinogenes_: Changes caused by interspecies transfer of H_2. _Journal of Bacteriology, 114_(3), 1231–1240.

Imhoff, J. F. 2004. Taxonomy and physiology of phototrophic purple bacteria and green sulfur bacteria. In _Anoxygenic Photosynthetic Bacteria_, ed. R. E. Blankenship, M. T. Madigan, and C. E. Bauer, pp. 1–15. The USA: Kluwer Academic Publishers.

Jo, J. H., Lee, D. S., Park, D., and Park, J. M. 2008. Biological hydrogen production by immobilized cells of _Clostridium tyrobutyricum_ JM1 isolated from a food waste treatment process. _Bioresource Technology, 99_(14), 6666–6672.

Joyner, A. E., Winter, W. T., and Godbout, D. M. 1977. Studies on some characteristics of hydrogen production by cell-free extracts of rumen anaerobic bacteria. _Canadian Journal of Microbiology, 23_(3), 346–353.

Jung, G. Y., Kim, J. R., Jung, H. O., Park, J. Y., and Park, S. 1999. A new chemoheterotrophic bacterium catalyzing water-gas shift reaction. _Biotechnology Letters, 21_(10), 869–873.

Jung, G. Y., Kim, J. R., Park, J. Y., and Park, S. 2002. Hydrogen production by a new chemoheterotrophic bacterium _Citrobacter_ sp. Y19. _International Journal of Hydrogen energy, 27_(6), 601–610.

Kalia, V. C., Jain, S. R., Kumar, A., and Joshi, A. P. 1994. Fermentation of biowaste to H$_2$ by *Bacillus licheniformis. World Journal of Microbiology and Biotechnology*, 10(2), 224–227.

Kampf, C. and Pfennig, N. 1980. Capacity of Chromatiaceae for chemotrophic growth. Specific respiration rates of *Thiocystis violacea* and *Chromatium vinosum. Archives of Microbiology*, 127(2), 125–135.

Kataoka, N., Miya, A., and Kiriyama, K. 1997. Studies on hydrogen production by continuous culture system of hydrogen-producing anaerobic bacteria. *Water Science and Technology*, 36(6), 41–47.

Khanna, N., Kotay, S. M., Gilbert, J. J., and Das, D. 2011. Improvement of biohydrogen production by *Enterobacter cloacae* IIT-BT 08 under regulated pH. *Journal of Biotechnology*, 152(1), 9–15.

Kim, B. H., Bellows, P., Datta, R., and Zeikus, J. G. 1984. Control of carbon and electron flow in *Clostridium acetobutylicum* fermentations: Utilization of carbon monoxide to inhibit hydrogen production and to enhance butanol yields. *Applied and Environmental Microbiology*, 48(4), 764–770.

Kim, S., Seol, E., Mohan Raj, S., Park, S., Oh, Y. K., and Ryu, D. D. 2008. Various hydrogenases and formate-dependent hydrogen production in *Citrobacter amalonaticus* Y19. *International Journal of Hydrogen Energy*, 33(5), 1509–1515.

Kimble, L. K. and Madigan, M. T. 1992. Evidence for an alternative nitrogenase in *Heliobacterium gestii. FEMS Microbiology Letters*, 100(1), 255–260.

Koku, H., Eroğlu, I., Gündüz, U., Yücel, M., and Türker, L. 2002. Aspects of the metabolism of hydrogen production by *Rhodobacter sphaeroides. International Journal of Hydrogen Energy*, 27(11), 1315–1329.

Kovács, K. L., Kovacs, A. T., Maroti, G. et al., 2004. Improvement of biohydrogen production and intensification of biogas formation. *Reviews in Environmental Science and Bio/Technology*, 3(4), 321–330.

Kuhn, M., Steinbuchel, A., and Schlegel, H. G. 1984. Hydrogen evolution by strictly aerobic hydrogen bacteria under anaerobic conditions. *Journal of Bacteriology*, 159(2), 633–639.

Kumar, N. and Das, D. 2000. Enhancement of hydrogen production by *Enterobacter cloacae* IIT-BT 08. *Process Biochemistry*, 35(6), 589–593.

Kumar, A., Jain, S. R., Sharma, C. B., Joshi, A. P., and Kalia, V. C. 1995. Increased H$_2$ production by immobilized microorganisms. *World Journal of Microbiology and Biotechnology*, 11(2), 156–159.

Kumar, G. R. and Vatsala, T. M. 1989. Hydrogen production from glucose by *Citrobacter freundii. Indian journal of Experimental Biology*, 27(9), 824–825.

Kumazawa, S. and Mitsui, A. 1982. Hydrogen metabolism of photosynthetic bacteria and algae. In *CRC Handbook of Biosolar Resources*, ed. O. R. Zaborsky, pp. 299–316. Boca Raton: CRC press.

Lakaniemi, A. M., Hulatt, C. J., Thomas, D. N., Tuovinen, O. H., and Puhakka, J. A. 2011. Biogenic hydrogen and methane production from *Chlorella vulgaris* and *Dunaliella tertiolecta* biomass. *Biotechnology for Biofuels*, 4, 34.

Lambert, G. R. and Smith, G. D. 1977. Hydrogen formation by marine blue-green algae. *FEBS Letters*, 83(1), 159–162.

Laurinavichene, T. V., Fedorov, A. S., Ghirardi, M. L., Seibert, M., and Tsygankov, A. A. 2006. Demonstration of sustained hydrogen photoproduction by immobilized, sulfur-deprived *Chlamydomonas reinhardtii* cells. *International Journal of Hydrogen Energy*, 31(5), 659–667.

Lee, J. W. and Helmann, J. D. 2006. The PerR transcription factor senses H_2O_2 by metal-catalysed histidine oxidation. *Nature, 440*(7082), 363–367.

Lehr, F., Morweiser, M., Rosello Sastre, R., Kruse, O., and Posten, C. 2012. Process development for hydrogen production with *Chlamydomonas reinhardtii* based on growth and product formation kinetics. *Journal of Biotechnology, 168*(1), 89–96.

Lindberg, P., Lindblad, P., and Cournac, L. 2004. Gas exchange in the filamentous cyanobacterium *Nostoc punctiforme* strain ATCC 29133 and its hydrogenase-deficient mutant strain NHM5. *Applied and Environmental Microbiology, 70*(4), 2137–2145.

Lopes Pinto, F. A., Troshina, O., and Lindblad, P. 2002. A brief look at three decades of research on cyanobacterial hydrogen evolution. *International Journal of Hydrogen Energy, 27*(11), 1209–1215.

Lu, L., Ren, N., Zhao, X., Wang, H., Wu, D., and Xing, D. 2011. Hydrogen production, methanogen inhibition and microbial community structures in psychrophilic single-chamber microbial electrolysis cells. *Energy and Environmental Science, 4*(4), 1329–1336.

Madigan, M. 2004. Microbiology of nitrogen fixation by anoxygenic photosynthetic bacteria. In *Anoxygenic Photosynthetic Bacteria*, ed. R. E. Blankenship, M. T. Madigan and C. E. Bauer, pp. 915–928. USA: Kluwer Academic Publishers.

Madigan, M., Cox, S. S., and Stegeman, R. A. 1984. Nitrogen fixation and nitrogenase activities in members of the family *Rhodospirillaceae*. *Journal of Bacteriology, 157*(1), 73–78.

McCord, J. M., Keele, B. B., and Fridovich, I. 1971. An enzyme-based theory of obligate anaerobiosis: The physiological function of superoxide dismutase. *Proceedings of the National Academy of Sciences, 68*(5), 1024–1027.

Meinecke, B., Bertram, J., and Gottschalk, G. 1989. Purification and characterization of the pyruvate-ferredoxin oxidoreductase from *Clostridium acetobutylicum*. *Archives of Microbiology, 152*(3), 244–250.

Melis, A., Zhang, L., Forestier, M., Ghirardi, M. L., and Seibert, M. 2000. Sustained photobiological hydrogen gas production upon reversible inactivation of oxygen evolution in the green alga *Chlamydomonas reinhardtii*. *Plant Physiology, 122*(1), 127–136.

Merchant, S. S., Prochnik, S. E., Vallon, O. et al., 2007. The *Chlamydomonas* genome reveals the evolution of key animal and plant functions. *Science, 318*(5848), 245–250.

Mishra, J., Khurana, S., Kumar, N., Ghosh, A. K., and Das, D. 2004. Molecular cloning, characterization, and overexpression of a novel [Fe]-hydrogenase isolated from a high rate of hydrogen producing *Enterobacter cloacae* IIT-BT 08. *Biochemical and Biophysical Research Communications, 324*(2), 679–685.

Mitsui, A. 1975. The utilization of solar energy for hydrogen production by cell-free system of photosynthetic organisms. Paper presented at the Miami Energy Conference, Miami Beach, Fla. New York.

Mizuno, O., Dinsdale, R., Hawkes, F. R., Hawkes, D. L., and Noike, T. 2000. Enhancement of hydrogen production from glucose by nitrogen gas sparging. *Bioresource Technology, 73*(1), 59–65.

Mona, S., Kaushik, A., and Kaushik, C. P. 2011. Hydrogen production and metal-dye bioremoval by a *Nostoc linckia* strain isolated from textile mill oxidation pond. *Bioresource Technology, 102*(3), 3200–3205.

Nandi, R. and Sengupta, S. 1996. Involvement of anaerobic reductases in the spontaneous lysis of formate by immobilized cells of *Escherichia coli*. *Enzyme and Microbial Technology*, *19*(1), 20–25.

Nissilä, M. E., Tähti, H. P., Rintala, J. A., and Puhakka, J. A. 2011. Thermophilic hydrogen production from cellulose with rumen fluid enrichment cultures: Effects of different heat treatments. *International Journal of Hydrogen Energy*, *36*(2), 1482–1490.

Ohta, S., Miyamoto, K., and Miura, Y. 1987. Hydrogen evolution as a consumption mode of reducing equivalents in green algal fermentation. *Plant Physiology*, *83*(4), 1022–1026.

Pan, N. and Imlay, J. A. 2001. How does oxygen inhibit central metabolism in the obligate anaerobe *Bacteroides thetaiotaomicron*. *Molecular Microbiology*, *39*(6), 1562–1571.

Penfold, D. W., Forster, C. F., and Macaskie, L. E. 2003. Increased hydrogen production by *Escherichia coli* strain HD701 in comparison with the wild-type parent strain MC4100. *Enzyme and microbial technology*, *33*(2), 185–189.

Peterson, R. B. and Wolk, C. P. 1978. Localization of an uptake hydrogenase in *Anabaena*. *Plant physiology*, *61*(4), 688–691.

Pinchukova, E. E., Varfolomeev, S. D., and Kondrat'eva, E. N. 1979. Isolation, purification and study of thee stability of the soluble hydrogenase from *Alvaligenes eutrophus* Z-1. *Biokhimiya*, *44*, 605–615.

Rachman, M. A., Furutani, Y., Nakashimada, Y., Kakizono, T., and Nishio, N. 1997. Enhanced hydrogen production in altered mixed acid fermentation of glucose by *Enterobacter aerogenes*. *Journal of Fermentation and Bioengineering*, *83*(4), 358–363.

Rashid, N., Lee, K., and Mahmood, Q. 2011. Bio-hydrogen production by *Chlorella vulgaris* under diverse photoperiods. *Bioresource Technology*, *102*(2), 2101–2104.

Rashid, N., Lee, K., Han, J. I., and Gross, M. 2013. Hydrogen production by immobilized *Chlorella vulgaris*: Optimizing pH, carbon source and light. *Bioprocess and Biosystems Engineering*, *36* (7), 867–872

Ren, N. Q., Guo, W. Q., Wang, X. J. et al., 2008. Effects of different pretreatment methods on fermentation types and dominant bacteria for hydrogen production. *International Journal of Hydrogen Energy*, *33*(16), 4318–4324.

Robson, R. 2001. Biodiversity of hydrogenases. In *Hydrogen as a Fuel: Learning from Nature*, ed. R. Cammack, R. Robson and M. Frey. London: Taylor and Francis, 9–32.

Rosales-Colunga, L. M., Razo-Flores, E., and De León Rodríguez, A. 2012. Fermentation of lactose and its constituent sugars by *Escherichia coli* WDHL: Impact on hydrogen production. *Bioresource Technology*, *111*, 180–184.

Salminen, S., Von Wright, A., and Ouwehand, A. 2004. *Lactic Acid Bacteria: Microbiology and Functional Aspects*, No. Ed. 3, New York: Marcel Dekker, CRC press.

Sasikala, K., Ramana, C. V., Rao, R. P., and Kovacs, K. L. 1993. Anoxygenic phototrophic bacteria: Physiology and advances in hydrogen production technology. *Advanced Applied Microbiology*, *38*, 211–295.

Sawers, G. and Watson, G. 1998. A glycyl radical solution: Oxygen-dependent interconversion of pyruvate formate-lyase. *Molecular Microbiology*, *29*(4), 945–954.

Schröder, C., Selig, M., and Schönheit, P. 1994. Glucose fermentation to acetate, CO_2 and H_2 in the anaerobic hyperthermophilic eubacterium *Thermotoga maritima*: Involvement of the Embden-Meyerhof pathway. *Archives of Microbiology*, *161*(6), 460–470.

Sen, D. and Das, D. 2005. Multiple parameter optimization for the maximization of hydrogen production by *Enterobacter cloacae* DM11. *Journal of Scientific And Industrial Research, 64*(12), 984.

Seol, E., Kim, S., Raj, S. M., and Park, S. 2008. Comparison of hydrogen-production capability of four different *Enterobacteriaceae* strains under growing and non-growing conditions. *International Journal of Hydrogen Energy, 33*, 5169–5175.

Sirevåg, R. and Ormerod, J. G. 1977. Synthesis, storage and degradation of polyglucose in *Chlorobium thiosulfatophilum. Archives of Microbiology, 111*(3), 239–244.

Sparling, R., Risbey, D., and Poggi-Varaldo, H. M. 1997. Hydrogen production from inhibited anaerobic composters. *International Journal of Hydrogen Energy, 22*(6), 563–566.

Spiller, H., Bookjans, G., and Shanmugam, K. T. 1983. Regulation of hydrogenase activity in vegetative cells of *Anabaena variabilis. Journal of Bacteriology, 155*(1), 129–137.

Stickland, L. H. 1929. The bacterial decomposition of formic acid. *Biochemical Journal, 23*(6), 1187.

Suzuki, S. and Karube, I. 1981. Hydrogen production by immobilized whole cells of *Clostridium butyricum*. Paper presented at the International Hydrogen Energy Progress, Tokyo, Japan.

Taguchi, F., Chang, J. D., Mizukami, N., Saito-Taki, T., Hasegawa, K., and Morimoto, M. 1993. Isolation of a hydrogen-producing bacterium, *Clostridium beijerinckii* strain AM21B, from termites. *Canadian Journal of Microbiology, 39*(7), 726–730.

Taguchi, F., Hasegawa, K., Saito-Taki, T., and Hara, K. 1996. Simultaneous production of xylanase and hydrogen using xylan in batch culture of *Clostridium* sp. strain X53. *Journal of Fermentation and Bioengineering, 81*(2), 178–180.

Taguchi, F., Mizukami, N., Hasegawa, K., and Saito-Taki, T. 1994. Microbial conversion of arabinose and xylose to hydrogen by a newly isolated *Clostridium* sp. No. 2. *Canadian Journal of Microbiology, 40*(3), 228–233.

Taguchi, F., Takiguchi, S., and Morimoto, M. 1992. Efficient hydrogen production from starch by a bacterium isolated from termites. *Journal of Fermentation and Bioengineering, 73*(3), 244–245.

Tamagnini, P., Axelsson, R., Lindberg, P., Oxelfelt, F., Wünschiers, R., and Lindblad, P. 2002. Hydrogenases and hydrogen metabolism of cyanobacteria. *Microbiology and Molecular Biology Reviews, 66*(1), 1–20.

Tanisho, S. and Ishiwata, Y. 1994. Continuous hydrogen production from molasses by the bacterium *Enterobacter aerogenes. International Journal of Hydrogen Energy, 19*(10), 807–812.

Tanisho, S., Kuromoto, M., and Kadokura, N. 1998. Effect of CO_2 removal on hydrogen production by fermentation. *International Journal of Hydrogen Energy, 23*(7), 559–563.

Tanisho, S., Wakao, N., and Kosako, Y. 1983. Biological hydrogen production by *Enterobacter aerogenes. Journal of Chemical Engineering of Japan, 16*, 529–530.

Tsygankov, A. A., Laurinavichene, T. V., Gogotov, J. N., Asada, Y., and Miyake, J. 1996. Switching from light limitation to ammonium limitation in chemostat *of Rhodobacter capsulatus* grown in different types of photobioreactor. *Journal of Marine Biotechnology, 4*, 43–46.

Ueno, Y., Kawai, T., Sato, S., Otsuka, S., and Morimoto, M. 1995. Biological production of hydrogen from cellulose by natural anaerobic microflora. *Journal of Fermentation and Bioengineering, 79*(4), 395–397.

Ueno, Y., Otsuka, S., and Morimoto, M. 1996. Hydrogen production from industrial wastewater by anaerobic microflora in chemostat culture. *Journal of Fermentation and Bioengineering*, 82(2, 194–197.

van de Werken, H. J., Verhaart, M. R., van Fossen, A. L. et al., 2008. Hydrogenomics of the extremely thermophilic bacterium *Caldicellulosiruptor saccharolyticus*. *Applied and Environmental Microbiology*, 74(21), 6720–6729.

Wade, N. 2010. Researchers say they created a 'synthetic cell'. *The New York Times*, 20, 1–3.

Wang, A., Gao, L., Ren, N., Xu, J., and Liu, C. 2009. Bio-hydrogen production from cellulose by sequential co-culture of cellulosic hydrogen bacteria of *Enterococcus gallinarum* G1 and *Ethanoigenens harbinense* B49. *Biotechnology Letters*, 31(9), 1321–1326.

Warthmann, R., Cypionka, H., and Pfennig, N. 1992. Photoproduction of H_2 from acetate by syntrophic cocultures of green sulfur bacteria and sulfur-reducing bacteria. *Archives of Microbiology*, 157(4), 343–348.

Winter, J. and Zellner, G. 1990. Thermophilic anaerobic degradation of carbohydrates-metabolic properties of microorganisms from the different phases. *FEMS Microbiology Letters*, 75(2), 139–142.

Wu, L. F. and Mandrand, M. A. 1993. Microbial hydrogenases: Primary structure, classification, signatures and phylogeny. *FEMS Microbiology Letters*, 104(3), 243–269.

Yetis, M., Gündüz, U., Eroglu, I., Yücel, M., and Türker, L. 2000. Photoproduction of hydrogen from sugar refinery wastewater by *Rhodobacter sphaeroides* OU 001. *International Journal of Hydrogen Energy*, 25(11), 1035–1041.

Yokoi, H., Ohkawara, T., Hirose, J., Hayashi, S., and Takasaki, Y. 1995. Characteristics of hydrogen production by aciduric *Enterobacter aerogenes* strain HO-39. *Journal of Fermentation and Bioengineering*, 80(6), 571–574.

Yokoi, H., Saitsu, A., Uchida, H., Hirose, J., Hayashi, S., and Takasaki, Y. 2001. Microbial hydrogen production from sweet potato starch residue. *Journal of Bioscience and Bioengineering*, 91(1), 58–63.

Yokoi, H., Tokushige, T., Hirose, J., Hayashi, S., and Takasaki, Y. 1998. H_2 production from starch by a mixed culture of *Clostridium butyricum* and *Enterobacter aerogenes*. *Biotechnology Letters*, 20(2), 143–147.

Yoshino, F., Ikeda, H., Masukawa, H., and Sakurai, H. 2007. High photobiological hydrogen production activity of a *Nostoc* sp. PCC 7422 uptake hydrogenase-deficient mutant with high nitrogenase activity. *Marine Biotechnology*, 9(1), 101–112.

Yudkin, J. 1932. Hydrogenlyases: Some factors concerned in the production of the enzymes. *Biochemical Journal*, 26(6), 1859.

Zeidan, A. A., Rådström, P., and van Niel, E. W. 2010. Stable coexistence of two *Caldicellulosiruptor* species in a de novo constructed hydrogen-producing co-culture. *Microbial Cell Factories*, 9(1), 102–115.

3

Hydrogen Production Processes

3.1 Introduction

Most of the organisms can produce hydrogen naturally. In fact, for some of them, hydrogen production is a survival mechanism under stress conditions. Under anaerobic conditions, they can maintain the redox balance by releasing the electrons as molecular hydrogen. However, some organism such as methanogenic bacteria can consume hydrogen by oxidative metabolism to obtain energy from the H–H bond. They are involved in recycling the hydrogen in nature (Zeikus, 1977). The chemical equation for methanogenesis can be summarized as follows (Equation 3.1):

$$4H_2 + CO_2 \rightarrow CH_4 + 2H_2O \tag{3.1}$$

For every methane that is generated, one ATP is also generated (Conrad et al., 1985). They oxidize H_2 to a proton (H^+) and a hydride ion (H^-) via a hydrogenase enzyme. Besides, some hydrogen-producing microorganisms also possess bidirectional hydrogenase and uptake hydrogenase by which they can re-consume hydrogen and maintain the redox potential of the cell. To increase the yield of hydrogen, consumption of hydrogen is not desirable during hydrogen generation. Thus the reaction broth in a reactor should be devoid of methanogens and uptake hydrogenase activity of the hydrogen-producing organisms to increase the yield of hydrogen.

Over the years, different genera of organisms have adapted different kinds of mechanisms (based on the kind of the enzyme) to evolve hydrogen. The advantages and disadvantages of the various biohydrogen production processes are discussed in Table 3.1. Among the various methods of biohydrogen production known theoretically, biophotolysis of water appears to be an ideal concept in which solar energy is converted to molecular hydrogen by breakdown of water molecule by autotrophs. However, practically it appears less promising due to the low light conversion efficiency and the technical barriers of developing an ideal photobioreactor for maximum light absorption. However, heterotrophic photofermentative bacteria can reduce organic acids to produce mostly CO_2 and hydrogen in anoxic environment

TABLE 3.1

Overview of General Processes Involved in Biohydrogen Production Processes, the Biochemistry, Challenges, and Deployments

Process	Overall Process Reaction and Organism Involved	Challenges	Advantages
1. Dark fermentation	$C_6H_{12}O_6 + 2H_2O \rightarrow 2CH_3COOH + 2CO_2 + 4H_2$ Fermentative bacteria (*Clostridia* sp. and *Enterobacter* sp.)	• Low yields, which limits large-scale application. • With increase of yield, thermodynamic limitations occur. • Product contains mixture of CO_2 and H_2, which needs to be separated. This adds to the total cost of the process.	• As compared to other biological processes, it has the highest yields. • A variety of carbon sources including waste can be used as substrate. • It does not require light. This overcomes the restriction of day-and-night cycles required for photofermentative and biophotolytic processes. • No PSII systems are involved, hence no oxygen generation.
2. Photo-fermentation	$CH_3COOH + 2H_2O + light \rightarrow 4H_2 + 2CO_2$ Purple bacteria, Photofermentative bacteria (*Rhodobacter* sp.)	• Requires light, which adds to the cost of the process; If sunlight is used as source, then the process is limited by day-and-night cycles. • Total light conversion efficiency is extremely low, causing low yields.	• Wastewater from streams has been used to produce hydrogen using this process. • The bacteria are known to be able to utilize wide spectra of light.
3. Direct biophotolysis	$2H_2O + light \rightarrow 2H_2 + O_2$ Microalgae (*Chlorella* sp.)	• Process is limited by the requirement of light energy. Further, the total conversion of light is very low, which results in low yield of the process. • Due to the involvement of PSII, O_2 generation is a limitation. Different methods have to be adopted either to suppress PSII or to make it inactive. • Customized photobioreactors are needed which add to the cost of the process.	• The greatest advantage is the economy of the process since the organism can generate hydrogen using only water and natural light.
4. Indirect biophotolysis	$6H_2O + 6CO_2 + light \rightarrow C_6H_{12}O_6 + 6O_2$ $C_6H_{12}O_6 + 2H_2O \rightarrow 4H_2 + 2CH_3COOH + 2CO_2$ $2CH_3COOH + 4H_2O + light \rightarrow 8H_2 + 4CO_2$ Overall reaction: $12H_2O + light \rightarrow 12H_2 + 6O_2$ Microalgae, Cyanobacteria (*Synechocystis* sp.)	• Process is limited by the requirement of light energy. Further the total conversion of light is very low, which results in low yield of the process. • Presence of uptake hydrogenase lowers yield.	• The greatest advantage is the economy of the process since the organism can generate hydrogen using only water and natural light. • Has the ability to fix N_2 from atmosphere.

(photofermentation). However, they suffer from low light conversion efficiencies due to large antenna (chlorophyll) size, light shading, construction of efficient photobioreactor, and so on. Considering these scientific gaps, dark fermentative hydrogen production appears more promising as it can utilize a host of organic waste products as substrate for generation of clean energy. In addition, it does not require light and can be operated day and night. Absence of light requirement also reduces reactor design complexity. Overall, practically dark fermentation process appears to be more suitable for hydrogen production than photofermentative and photosynthetic processes. The biochemistry of the various processes will be dealt in detail in this chapter.

Recently, considerable number of research work has been conducted on two-stage dark and photofermentative process as also on combined dark, photoferemntative, and biophtolysis processes. Theoretically, hybrid processes can yield up to 12 mol of hydrogen per mol of glucose. However, practical culmination of these experiments presented technical barriers, which need to be overcome to make the process more attractive. More recently, a combination of microbial fuel cells (MFCs) with hydrogen production processes are underway to generate clean electricity along with green fuel. The possibilities seem endless. In this chapter, the technical barriers limiting large-scale use as well as the advancements made have been discussed in detail.

3.2 Photobiological Hydrogen Production

3.2.1 Basic Principles of Photobiological Hydrogen Production

Microalgae and cyanobacteria are photoautotrophic organisms because they can use light as the energy source and carbon dioxide as sole carbon source. Some bacteria are termed photoheterotrophic microorganisms because in spite of their ability use light as the energy source, they need organic carbon as the carbon source (Wijffels and Barten, 2003).

3.2.1.1 Photoautotrophic Production of Hydrogen

Microalgae and cyanobacteria utilize solar light and electrons derived from the splitting of water to metabolize CO_2 into energy-rich organic compounds $[C_n(H_2O)_n]$ as shown in the equation below:

$$CO_2 + H_2O + \text{"light energy"} \rightarrow [C_n(H_2O)_n] + O_2 \qquad (3.2)$$

Under anaerobic conditions, the organisms can produce hydrogen by direct photolysis of water

$$4H_2O + \text{"light energy"} \rightarrow 2O_2 + 4H_2 \ (\Delta G_0 = +1498 \text{ kJ}) \qquad (3.3)$$

3.2.1.2 Photoheterotrophic Production of Hydrogen

Microalgae and cyanobacteria require an external supply of carbon to metabolize in the presence of light. They produce hydrogen from exogenous available carbohydrate. The reaction is catalyzed primarily by nitrogenase (particularly in the absence of nitrogen) or hydrogenase.

$$C_2H_4O_2 + 2H_2 + \text{"light energy"} \rightarrow 2CO_2 + 4H_2 \ (\Delta G_0 = +75.2 \text{ kJ}) \qquad (3.4)$$

3.2.2 Fundamentals of Photosynthesis and Biophotolysis of Water

The action of light on a biological system that results in the dissociation of a substrate, usually water, into molecular hydrogen is called biophotolysis. Gaffron (1939) while working at the University of Chicago first discovered biological hydrogen production by photolysis of water. Gaffron and Rubin first reported that *Scenedesmus*, a green microalga, evolved molecular hydrogen under light conditions after being adapted in anaerobic and dark conditions (Gaffron and Rubin, 1942).

Hydrogen production in green algae and cyanobacteria is closely related to the process of photosynthesis. The reductants released during the breakdown of water molecules by solar energy may either be consumed in the Calvin Cycle or catalyzed by hydrogenase into molecular hydrogen. Photosynthesis is the fundamental biological process that converts the electromagnetic energy of sunlight into stored chemical energy that supports all life on the earth. In green algae as in higher plants, photosynthesis occurs in a specialized light-harvesting organelle, the chloroplast. The key components of the photosynthetic apparatus involved in light absorption and energy conversion are embedded in thylakoid membranes inside the chloroplast. Chloroplast contains photosystem (PS) I and II for capturing light energy to exhibit oxygenic photosynthesis. In photosystem II, solar energy is captured by the photosynthetic pigments present in light-harvesting complex (LHC) and transferred to the reaction center (RC) of photosystem II. Photosystem I (PSI) has a reaction center, P700 (absorbs the photons with a wavelength under 700 nm), and photosystem II (PSII) has another distinct reaction center, P680 (absorbs the photons with a wavelength under 680 nm). The LHC consists of several hundreds of molecules, primarily chlorophyll and other accessory pigments, which together are known as antenna chlorophylls. In the cell, light harvesting is primarily carried out by chlorophyll *a*. The antenna chlorophyll absorbs the photons and transfers and excites a special pair of chlorophyll *a* present in the reaction center. The chlorophyll molecules have a central magnesium atom. In the ground state, the magnesium ion (Mg^{2+}) has no unpaired electron (singlet state). However, after getting the solar energy, one electron from 2P orbital is transferred to the 3S orbital leading to the triplet state with two unpaired electrons. The excited state of chlorophyll is very unstable. The chlorophyll *a* molecule releases the energy by releasing an electron and generating a strong oxidant (photooxidation)

PS680* ($E = 0.82$ V) capable of splitting water into protons (H^+), electrons (e^-), and O_2 (Equation 3.5).

$$2H_2O \rightarrow 4H^+ + 4e^- + O_2 \qquad (3.5)$$

Consecutively, electron from the water is taken by the oxidized chlorophyll to return into ground state. Thus, using water as the source of electrons, light energy is predominantly stored as reduced CO_2. According to the Z-scheme of photosynthesis propounded by Hill and Bendall (1960), PSII can split water and reduce the plastoquinone (PQ) pool, the cytochrome (Cyt) b/f complex, and plastocyanin (PC), while PSI can reduce ferredoxin (Fd)/nicotinamide adenine dinucleotide phosphate (NADP) and oxidize PC, the Cyt b/f complex, and the PQ pool (Figure 3.1).

As a result, the electrons derived from water splitting are transferred to Fd/NADP, which provide the reducing power for reduction of CO_2 to carbohydrate in the stromal region of the chloroplast, by a series of enzymatic reactions collectively called the Calvin cycle. The electron transport in the membrane is coupled with the proton transport from the stroma into the lumen, generating a proton gradient across the thylakoid membrane. The proton gradient drives phosphorylation through the coupling factor CF_0-CF_1, to make essential adenosine triphosphate (ATP) for the reduction of CO_2. If the cells suffer shortage of energy, electrons flow back to plastoquinone center from ferredoxin. This cyclic flow of electron generates only ATP (Munekage et al., 2004).

$$2Fd_{red} + 2H^+ + NADP^+ \rightarrow 2Fd_{ox} + NADPH + H^+ \qquad (3.6)$$

Ferredoxin is located in the stroma, where its reduced form is specific to hydrogenase as well as ferredoxin: $NADP^+$ reductase and has bifunctional ability. Under aerobic conditions, reduced ferredoxin is involved in biomass

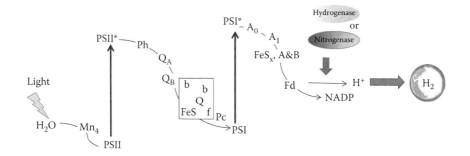

FIGURE 3.1
The Z-scheme of photosynthesis: The figure shows the energetic effect of photons exciting two photosystems to their respective excited states shown as PSI' and PSII'. (PS: photosystems, Ph: pheophytin, Q: quinone; Cyt: cytochrome; PC: plastocyanin, A_o: quinones, Fd: ferredoxin).

production by fixing carbon using reducing power ($NADPH_2$) via Calvin's cycle or pentose phosphate pathway (Horecker, 2002) to produce carbohydrates (CH_2O) and/or lipids (Hatch and Slack, 1970). Energy is driven from ATP produced by proton gradient across the thylakoid membrane via ATP synthase. In green algae, stored carbohydrate is starch granules present in the pyrenoid of the chloroplast while in cyanobacteria, the stored carbohydrate is glycogen. However, under anaerobic, dark conditions the photosynthetic organisms can survive by switching their metabolism to fermentation and utilize the reduced Fd to produce other reduced end metabolites including hydrogen (Gaffron and Rubin, 1942). However, high concentrations of reduced products like ethanol or formate are toxic, and organic acids (e.g., lactic or acetic acid) acidify the cell (Kennedy et al., 1992). Contrarily, under anaerobic conditions in the presence of light, low rate of photosynthesis occurs and the photosynthetically derived oxygen is consumed by the low rate of respiration. However, the partially active respiratory chain presumably does not suffice as an electron sink. Furthermore, under this physiological anaerobiosis, the electron-consuming Calvin cycle also fails to function as an electron sink. The hydrogenase plays an important role in physiological anaerobiosis (Happe et al., 2002). Photosynthetically generated electrons reduce protons and evolve molecular hydrogen. This is known as hydrogen production through biophotolysis as electrons are coming directly from water. There are two types of biophotolysis:

- *Direct biophotolysis*: This refers to the sustained hydrogen production under the light irradiation. The light is absorbed by the chlorophyll reaction centers PSI and PSII, which raise the energy of the electrons generated by the oxidation of the water molecules when they are transferred from PSII to ferredoxin via PSI. A portion of this light energy is directly stored in molecular hydrogen while the remaining is stored as carbohyhdrates and lipids.
- *Indirect biophotolysis*: In this process, the electrons or reducing equivalents for the generation of molecular hydrogen do not come directly from the splitting of water. Instead, they originate from the oxidation of intracellular energy reserve (mostly carbohydrates) that was produced by the Calvin cycle. Thus, in this process, stored energy is released indirectly through fermentation of the endogenous carbohydrates in dark conditions instead of from the splitting of water molecules as in the case of direct biophotolysis.

3.2.3 Biophotolysis

3.2.3.1 Direct Biophotolysis

A direct biophotolysis of H_2 production is a biological process, which utilizes solar energy and photosynthetic systems of the algae to convert water into

PHOTO-INHIBITION AND PHOTO-BLEACHING

The process of photosynthesis occurs in the chlorophyll, which has magnesium at its core. During photosynthesis, splitting of water molecule occurs in the presence of sunlight (photons). Two photons are required for releasing one electron from magnesium. However, as soon as the excitation energy exceeds the capacity of the reaction center (PSII), different undesirable products are formed such as triplet state of magnesium, singlet oxygen, and superoxide, which can be formed. The formation of these superoxides causes severe photodamages. One of the consequences of such a photodamage has been observed in the oxidation of pigments leading to *photobleaching* and cell death. Thus, photobleaching is defined as the photochemical bleaching of the flurophore. However, in cyanobacteria re-pigmentation after exposure of low irradiation has been observed (Nultsch and Agel, 1986). Thus, photobleaching is not exactly a phenomenon of photodamage rather it is considered to be a light adaptation process. There are some protective pathways to escape from the generation of harmful intermediates. This includes the direct quenching of chlorophyll, that is transfer of excess energy to the adjacent carotenoid molecules. Besides, carotenoids liberate excess energy by fluorescence and heat. Algal cells directly exposed to light energy for a longer time can overcome photodamages by another phenomenon called *photoinhibition* that is light-dependent inhibition of photosystem II (Aro et al., 1993). In high light intensities, chlorophyll can be damaged by the enhanced flow of electrons beyond that which it can process. This results in photoinhibition by decreasing the photosynthetic capacity. Algal cells that are exposed directly in light suffer from these bottlenecks, however, the cells floating under the surface may not be able to get sufficient solar energy. Overall, photosynthetic efficiency becomes lower as the surface cells are not able to manifest excess energy and deeper cells are not getting sufficient solar energy. Hydrogen production through biophotolysis in algae solely depends on efficiency of photosystem II so photobleaching and photoinhibition may hamper and/or even stop hydrogen production.

chemical energy. Encompassing a wide diversity in morphology and physiology, cyanobacteria and green alga are potential microbial species for hydrogen production via direct biophotolysis (Yu and Takahashi, 2007). Direct biophotolysis can occur in two ways as shown Sections 3.2.3.1.1 and 3.2.3.1.2:

3.2.3.1.1 One-Stage Direct Biophotolysis

In the course of direct biophotolysis, light energy absorbed by PSII and PSI helps to transport electrons linearly from water to ferredoxin. In case of

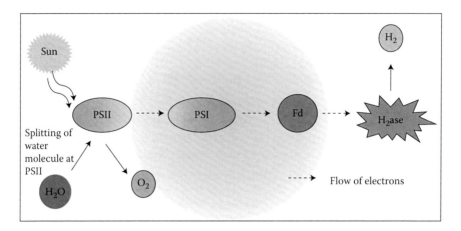

FIGURE 3.2
One-stage direct biophotolysis.

photosynthesis, reduced ferredoxin donates reducing equivalents to $NADP^+$. Reduced $NADPH_2$ enters the Calvin cycle, which continually produces and uses a five-carbon sugar called ribulose disphosphate (RuDP). However, under optimal conditions for hydrogen production, reduced ferredoxin may act as an electron donor to a hydrogenase enzyme, which reversibly catalyzes the reduction of protons (H^+) to molecular hydrogen (H_2), according to the following reaction (Figure 3.2) (Melis et al., 2000). Thus, if the reduced ferredoxin produced by photosynthesis during water splitting is used to directly reduce the H_2-producing hydrogenase or nitrogenase enzymes, without intermediate CO_2 fixation, then the process is termed direct biophotolysis (Figure 3.2).

Single stage, direct biophotolysis is a promising process in principle, however, practically, the process suffers from several limitations. Among other factors, the most critical is the strong inhibition of H_2 production by the simultaneously evolved O_2. One approach to overcoming this limitation is to remove the O_2 produced, for example, by respiration using endogenous or exogenous substrates as in two-stage direct biophotolysis. The fate of reduced ferredoxin may depend upon a number of considerations including oxidation reduction potentials, $NADP^+/NADPH_2$ ratio, ATP levels, etc.

3.2.3.1.2 Two-Stage Direct Biophotolysis

Direct biophotolysis can also occur under anaerobic conditions of low rate of photosynthesis (Figure 3.3).

The photosynthetically derived O_2 is consumed by low rate of respiration creating anaerobiosis. This occurs in green algae when its PSII activity is suppressed by artificial suppressors or by a sulfur deprivation technique developed by Melis and Coworkers (Melis, 2002). Such a direct biophotolysis process is essentially a two-stage process and rarely occurs naturally. In this

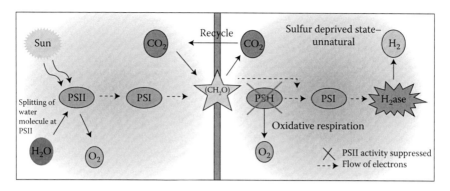

FIGURE 3.3
Two-stage direct biophotolysis.

process, at first, biomass is generated in open ponds. The reduced pyridoxine is dedicated toward the Calvin cycle and produces carbohydrates. In the second stage, the carbohydrate-rich algae is transferred into closed photobioreactors in a culture medium deprived of sulfur (S) as described by Melis et al. (2000). Depriving the medium of S partially suppresses the PSII activity. Sulfur is very crucial for biosynthesis of proteins present in photosystem II. Since PSII is concerned with splitting of water, oxygen evolution is reduced and an anaerobic condition is created in the reactor. Anaerobiosis activates hydrogenase, and electrons coming partly from water splitting and partly from catabolism of pyruvate are reduced to evolve hydrogen. In fact, hydrogen production by the sulfur-deprived method is a combination of direct and indirect biophotolysis (Figure 3.4).

In fact, if the cells are exposed to strong light, the PSII activity is completely suppressed but has little effect on PSI. This is known as "photoinhibition." Kyle et al. (1984) showed that photoinhibition was accompanied by a selective loss of a 32-kDa protein (later identified as the PSII reaction center protein D1) followed by activation of the reaction center through rapid in-built repair mechanism. In fact, only when the repair cycle is unable to match the rate of damage to PSII, photoinhibition is observed (Kyle et al., 1984). In the absence of sulfur (<0.45 mM), re-biosynthesis of the D1 protein after loss is inhibited, as the production of sulfur-containing amino acids cysteine or methionine becomes impossible. Thus, in the absence of sulfur, activity of photosystem II is hampered but has very little effect on PSI. Endogenous sulfur has been contributed to maintain the physiology of the cell. Partial inhibition of PSII can generate anaerobic condition for the cell within a photobioreactor, as there is less water oxidation activity to evolve O_2 and the residual O_2 is used by respiration (Wykoff et al., 1998). Anaerobiosis induces the expression of [FeFe]-hydrogenase in algal cells (Happe and Kaminski, 2002; Forestier et al., 2003), and sustained hydrogen production can be achieved (Melis et al., 2000, Ghirardi et al., 2000). In this condition, electrons are derived less from the water mostly from the oxidation of reserve carbohydrates.

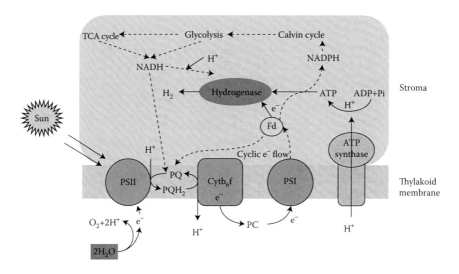

FIGURE 3.4
Direct and indirect biophotolysis carried out by green algae for H_2 production.

The process appears attractive. However, physical transfer of biomass from open ponds to close photobioreactors containing sulfur-deprived medium appears challenging. In this regard, Chen et al. (2003) reported a mutant with inefficient ability to transport sulfate into the chloroplast. This was obtained by affecting sulfate permease activity, and this mutant could evolve H_2 without depleting sulfate in the culture media. In addition, some photosystem II inhibitors have also been used to inhibit water oxidation activity. After a sufficient dark anaerobic incubation (for induction of hydrogenase) PSII inhibitor 3-(3,4-dichlorophenyl)-1,1-dimethylurea (DCMU) can be added in the same medium. This inhibitor binds at the Q_B site of PSII and blocks transport of electrons. This leads to high rate of H_2 production (Happe and Naber, 1993). Further, electron flow around photosystem I can be sustained and enhanced by adding external electron donor such as NAD(P)H into the cell culture media (Mus et al., 2005).

Direct biophotolysis is an inherently attractive process since solar energy is used to convert water into hydrogen and oxygen. It may be understood that hydrogen production in algae can be catalyzed by both nitrogenase and hydrogenase. However, direct biophotolysis can be mediated only through hydrogenase. In fact, in case of cyanobacteria, direct biophotolysis can occur only in single celled organisms which have no heterocysts. This is because direct biophotolysis directly employs the reducing equivalents from the photolysis of water (at PSII) to produce hydrogen. However, filamentous algae lack PSII in the heterocysts. Organisms which are involved in direct biophotolysis include cyanobacteria such as *Synechocystis* and green algae such as *Chlamydomonas reinhardtii* (Yu and Takahashi, 2007).

BIOPHOTOLYSIS IS THERMODYNAMICALLY FEASIBLE

Interestingly, unlike hydrogen production by dark fermentative processes, hydrogen production by biophotolysis is thermodynamically feasible. The photosynthetic system which consists of two photosystems operating in series can, by capturing two quanta of radiant energy, place an electron from the water–oxygen couple (0.8 V pH 7.0) to a negative value as low as −0.7 V, which is 0.3 V more negative than the hydrogen electrode (Pandu and Joseph, 2012).

However, though direct biophotolysis is an attractive process, it suffers from several technical limitations which have limited the rate and yield of hydrogen from this process. In addition, as compared to indirect biophotolysis (discussed in the next section), lower rates of the order of 0.02 mmol/m^3 s have been reported in the literature (Melis et al., 2000; Kosourov et al., 2002). Lower rate of hydrogen production in this process may be attributed to several technical challenges. The primary considerations involve the inhibition of H$_2$ production by O$_2$, photobioreactors, explosive H$_2$–O$_2$ mixtures. Each of these considerations has been elaborated in detail in Section 3.2.4 along with the plausible scientific breakthroughs and advancements.

3.2.3.2 Indirect Biophotolysis

In indirect biophotolysis, electrons are derived directly from the reserve carbon source that has been produced by fixing CO$_2$ via the Calvin cycle during photosynthesis (Figure 3.5). Thus the electrons are not directly coming from water as in case of direct biophotolysis. The stored carbohydrate is oxidized to produce H$_2$. The general reaction is as follows (Equations 3.7 and 3.8):

$$12H_2O + 6CO_2 \rightarrow C_6H_{12}O_6 + 6O_2 \qquad (3.7)$$

$$C_6H_{12}O_6 + 12H_2O \rightarrow 12H_2 + 6CO_2 \qquad (3.8)$$

Similar to direct biophotolysis, indirect biophotolysis may also occur as a one- or two-stage process. This essentially depends on the type of microorganism used. Filamentous microorganisms have heterocysts and can carry out single stage indirect biophotolysis, whereas green algae containing [FeFe] hydrogenase primarily carry out two-stage biophotolysis as shown in Sections 3.2.3.2.1 and 3.2.3.2.2.

3.2.3.2.1 One-Stage Indirect Biophotolysis

Like direct biophotolysis, indirect biophotolysis also occurs mainly in the presence of light. In one stage, indirect biophotolysis occurs due to

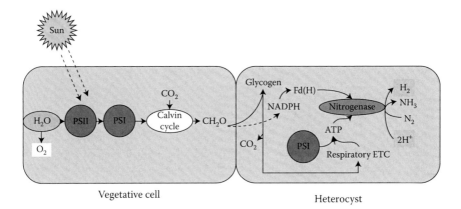

Indirect biophotolysis in filamentous cyanobacteria

FIGURE 3.5
One-stage indirect biophotolysis.

the spatial separation of hydrogen production from the oxygen-evolving photosynthesis. Spatial separation involves different physical locations of the photosynthetic apparatus and the hydrogen production apparatus. Mostly, in such organisms besides a vegetative cell, anaerobic heterocysts have evolved. Hydrogen by this process is produced by the filamentous cyanobacteria, where it occurs in specialized cells called the heterocysts. Heterocysts contain only PSI, the oxygen-evolving PSII is absent. Moreover, they contain nonoxygen permeable cell walls that further help in preventing oxygen diffusion from the adjacent vegetative cells. This helps in the creation of anaerobic conditions. The heterocysts contain the oxygen sensitive enzymes like nitrogenase. In the absence of nitrogen, nitrogenase is also involved in producing hydrogen in some cyanobacterial species such as *Anabena* and *Nostoc*. Nitrogenase is a metallo-enzyme and three different variants of the enzyme are known depending on the metals present. These include vanadium, molybdenom, or iron. All the three variants are known to produce hydrogen albeit with different stoichiometries as shown in Equations 3.9 through 3.11:

Iron enzyme:

$$N_2 + 21e^- + 21H^+ + 42ATP \rightarrow 2NH_3 + 7.5H_2 + 42ADP + 42\ P_i \qquad (3.9)$$

Vanadium enzyme:

$$N_2 + 12e^- + 12H^+ + 24ATP \rightarrow 2NH_3 + 3H_2 + 24ADP + 24\ P_i \qquad (3.10)$$

Molybdenom enzyme:

$$N_2 + 8e^- + 8H^+ + 16ATP \rightarrow 2NH_3 + H_2 + 16ADP + 16\ P_i \qquad (3.11)$$

Notable in the above reactions is the high energetic requirement of the over-all nitrogenase/hydrogenase catalysis (at least 2 ATP per electron). This substantial ATP requirement is met in the heterocysts of cyanobacteria via cyclic photophosphorylation, driven by light in the modified thylakoid membranes of these specialized cells. However, this appears as a wasteful process (Prince and Kheshgi, 2005). To overcome this loss of energy, the uptake hydrogenase has been found in almost all N_2-fixing strains examined thus far (Tamagnini et al., 2002). The uptake hydrogenase allows the microorganism to retrieve the electrons in the molecular bond of hydrogen and perhaps regain some of the lost energy (Tamagnini et al., 2002). Elimination of this uptake activity has been known to increase hydrogen production from most of the engineered cyanobacteria (Happe et al., 2002; Masukawa et al., 2002).

In cases in which a filamentous organism with heterocysts is present, the reducing equivalents produced by the splitting of water enter the Calvin cycle. The Calvin cycle generates carbohydrates which then diffuse into the heterocysts and are oxidized to produce electrons to reduce the plastoquinone pool (PQ). The two enzyme complexes, NAD(P) dehydrogenase (Ndh) and Ferredoxin-NADP-reductase (FNR), are involved in electron transfer from NAD(P)H and electron flow around photosystem I. In this situation, as the terminal electron acceptor (O_2) is absent, the excess electrons are deposited by another way that is the Fd-hydrogenase pathway (Mus et al., 2005) to produce H_2. Moreover, the presence and absence of O_2 act as a switch for the hydrogen production.

The existence of this anaerobic pathway in aerobic phototrophs probably focused the evolutionary linkage, and this has been exploited by researchers for production of hydrogen. Significant research has been done to improve H_2 production targeting this NAD(P)H plastoquinone (PQ) oxido-reductase pathway. Thus, hydrogen production is possible due to spatial separation of the hydrogen production process from oxygen-evolving photosynthetic process that destroys nitrogenase. Recently, the concept of replacing the energy intensive nitrogenase with hydrogenase in the heterocysts has been propounded by a few researchers as shown in Figure 3.6.

3.2.3.2.2 *Two-Stage Indirect Biophotolysis*

Besides, cyanobacteria that produce hydrogen by indirect biophotolysis can also be non-nitrogen fixing (*Synechocystis, Synechococcus, Gloebacter*) (Das et al., 2008). This type of cyanobacteria may possess two different kinds of [NiFe] hydrogenases with different properties and function. The first group is the so-called uptake [NiFe] hydrogenases, encoded by *hup* genes. These enzymes have primarily evolved to capture and recycle hydrogen produced by the nitrogenase. They are co-expressed and probably co-regulated in tandem with the nitrogenase. The second type of NiFe hydrogenase is a bidirectional hydrogenase. A multisubunits bidirectional [NiFe] hydrogenase has the capacity both to uptake and to generate hydrogen (Tamagnini et al., 2002). Non-nitrogen fixing cyanobacteria can produce hydrogen via the [NiFe] bidirectional (Tamagnini

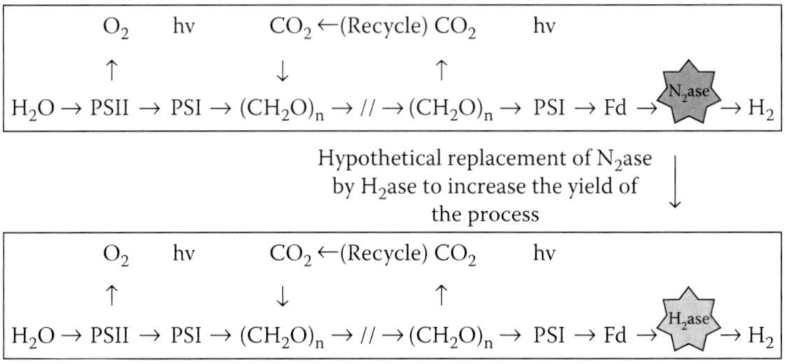

$$O_2 \quad\quad hv \quad\quad\quad CO_2 \leftarrow (Recycle)\ CO_2 \quad\quad\quad hv$$

$$\uparrow \quad\quad\quad\quad\quad\quad \downarrow \quad\quad\quad\quad\quad\quad \uparrow$$

$$H_2O \rightarrow PSII \rightarrow PSI \rightarrow (CH_2O)_n \rightarrow // \rightarrow (CH_2O)_n \rightarrow PSI \rightarrow Fd \rightarrow \boxed{N_2ase} \rightarrow H_2$$

Hypothetical replacement of N_2ase
by H_2ase to increase the yield of
the process

$$O_2 \quad\quad hv \quad\quad\quad CO_2 \leftarrow (Recycle)\ CO_2 \quad\quad\quad hv$$

$$\uparrow \quad\quad\quad\quad\quad\quad \downarrow \quad\quad\quad\quad\quad\quad \uparrow$$

$$H_2O \rightarrow PSII \rightarrow PSI \rightarrow (CH_2O)_n \rightarrow // \rightarrow (CH_2O)_n \rightarrow PSI \rightarrow Fd \rightarrow \boxed{H_2ase} \rightarrow H_2$$

FIGURE 3.6
Hypothetical replacement of nitrogenase by hydrogenase to increase the efficiency of one-stage indirect biophotolysis.

et al., 2002; Baebprasert et al., 2010). The advantage with this enzyme is that unlike nitrogenase, it does not require ATP. Moreover, they are exceptionally active enzymes with the turnover of 10^6/s. However, the free energy contained in the oxidation of ferredoxin is relatively small, and the enzyme is not able to produce much hydrogen against the existing backpressure.

In the two-stage process, hydrogen is produced by temporal separation (separation depending on time of expression) of hydrogen production from the oxygen-evolving photosynthesis. Naturally, this takes place by utilizing the diurnal (day-and-light cycle).

Even though biophotolysis is an attractive process due to minimum substrate requirements and generation of high energy hydrogen utilizing only the plentiful sunlight and water, the yield and rates of the processes are limiting. There are a number of technical challenges that must be addressed before the process can be considered for scale-up and subsequent commercialization.

3.2.4 General Considerations and Advancements Made in Biophotolysis

3.2.4.1 Explosive Hydrogen–Oxygen Mixture

Direct biophotolysis results in a mixture of hydrogen and oxygen in the reactor. The mixture is potentially harmful, dangerous, and explosive. Safe handling is required, besides, separation of this mixture would be expensive.

3.2.4.2 Oxygen Sensitivity of the Enzymes Involved in Hydrogen Production

Both algal and cyanobacterial hydrogenases are sensitive to oxygen, an essential product of photosynthesis. Hydrogen is produced only transiently when such cells are exposed to light. There are several strategies for extending the catalytic lifetime of such hydrogenases including: (i) computational simulation of pathways for oxygen gas diffusion into the catalytic site of hydrogenases and molecular engineering of these pathways, perhaps by narrowing the gas channels, to block O_2 from reaching the catalytic site; (ii) mutagenizing the hydrogenase gene and then screening for an oxygen-tolerant version of this enzyme; (iii) employing a metabolic switch such as sulfate deprivation to down regulate PSII-catalyzed oxygen evolution, inducing an anaerobic environment to sustain hydrogenase activity in the light; and (iv) searching for more oxygen-tolerant hydrogenases from nature and then transferring such genes into green algae and cyanobacteria. Except for the case in which oxygen production would be suppressed, it will be necessary to separate the hydrogen and oxygen being produced to avoid accumulating flammable or explosive mixtures.

Most of the above-stated solutions are still hypothetical and have not been proved experimentally. Previously, efforts to develop oxygen-tolerant hydrogenases have not met with much success. Though as early as 1995, McTavish et al. (1995) achieved some success in developing a site-directed oxygen insensitive hydrogenase of *Azotobacter vinelandii*. However, the hydrogenase lost its hydrogen evolution activity up to 78%.

3.2.4.3 Inefficiency of Biophotolysis Process Due to Large Antennae Size

The twin hearts of the photosynthetic process in algae, cyanobacteria, and higher plants are the photosystems PSI and PSII. The two photosystems are embedded in the thylakoid membrane and use the energy of the incoming photon to drive the electron across the membrane, generating both a strong reductant and oxidant. The actual photochemistry at the reaction center involves two or three chlorophylls but the membrane contains many more, typically about 300 per PS. These additional chlorophylls along with the other accessory pigments such as carotenoids are known as antennae pigments because they increase the likelihood that the light falling on the organism can be captured and transferred to the PS. The antennae are complex, made of several peptides. The large size of the antenna system is optimized for maximum photon absorption at low light intensities with the obvious corollary that far too many photons arrive in bright sunlight (Prince and Kheshgi, 2005). When grown on a mass scale, green algae and cyanobacteria are inefficient in their use of high-intensity light because their large light-absorbing antennae cause a saturation of electron transport at <10% full sunlight (light absorption is faster than the rate of electron transport at sunlight intensities above 10%). The excess absorbed light is therefore dissipated as heat or fluorescence. In this regard, Melis (2002) at the University of California has been able to generate PSI- or PSII-truncated mutants. They have also effectively generated a mutant with both the photosystems truncated. These mutants saturate at much higher light intensities as compared to their parent cell lines. However, in another similar study, Polle et al. (2003) showed that deletion of major antenna proteins in *Arabidopsis* failed to reduce the antenna size, as the plant utilized other antenna proteins to serve in its place, maintaining almost constant size of the antenna.

3.2.4.4 Quantum Efficiency

The solar conversion efficiency is the main bottleneck of biophotolysis. In the laboratory under low light intensities and relatively controlled conditions, the light conversion efficiency achieved is relatively higher (~10%) as compared to the same experiments conducted with solar light (<4%). However, solar energy conversion efficiencies depend on light intensity, irradiated area, reactor design, duration of hydrogen production, and amount of hydrogen accumulated.

Miyake and Kawamura (1987) reported solar-to-hydrogen energy conversion efficiencies reaching up to about 8%, while the theoretical maximum was reported as 10% (Akkerman et al., 2002; Kapdan and Kargi, 2006). Other studies showed that diurnal light–dark cycles help to increase the yield of hydrogen production (Koku et al., 2003; Eroglu et al., 2010), hence the solar-to-hydrogen energy conversion efficiency. The advantage of the diurnal light– dark cycles was originally reported by Meyer et al. (1978), as they achieved a stabler nitrogenase activity that translated into a greater H_2 production capacity.

Light and dark phases are the two phases of photosynthesis. In order to demonstrate the effect of illumination cycles, this natural phenomenon was mimicked as daily/hourly cycles. The reasons for studying the effects of these light cycles were to see the reaction of the cells to cycling illumination conditions, during hydrogen production. In previous studies, the researchers have first exposed the cells to light cycle and then have transferred them into hydrogen production conditions under continuous illumination (Melis et al., 2000; Tsygankov et al., 2002; Kima et al., 2006; Vijayaraghavan et al., 2009). In another study, circadian light–dark cycles were carried out all through the hydrogen production, under micro-aerobic conditions. The idea was to use the light phase as the storage phase and the dark phase as the catabolism phase, depleting O_2 leading to an increase of the hydrogenase activity to produce hydrogen in the dark (Miura et al., 1997). Moreover, since the top layer of cells in the reactor captures most of the incident photons, the remaining cells are shaded and do not contribute to hydrogen production resulting in low light conversion efficiencies.

3.2.4.5 Availability of More Reductant

In cells, the primary electron carriers for producing hydrogen, ferredoxin, and NAD(P)H also act as electron donors in other biochemical reactions, including those catalyzed by enzymes such as nitrate reductase and glutamate synthase. Culturing cells in media supplemented with ammonia instead of nitrate could suppress nitrate reductase activity. Further, depleting carbon dioxide would also leave more reduced ferredoxin available for producing hydrogen (Maness et al., 2009).

Another strategy for directing more reductant to hydrogenase is to prevent electron transfer around PSI to maintain a high ratio of NADPH to

ATP required for producing hydrogen. For example, a *C. reinhardtii* mutant with low cyclic photophosphorylation produces hydrogen more effectively than do ordinary cells, according to Ben Hankamer of the University of Queensland in Brisbane, Australia (Maness et al., 2009).

In yet another approach, researchers plan to link hydrogenase chemically to the reducing side of PSI, bypassing the competing pathways. In addition, deleting genes encoding components in competing electron-transfer pathways that are involved in respiration and nitrate assimilation in organisms such as *Synechocystis* sp. PCC 6803 are also postulated to improve the hydrogen production.

3.2.4.6 Natural Coupling of Photosynthetic Electron Transport to Proton Gradient

The photosynthetic electron flow is naturally coupled to the ATP synthase, which in turn participates in the CO_2 fixation along with the cellular reductant, obtained from cyclic electron transport by PSI. Under anaerobic conditions, during hydrogen production, the reducing equivalent thus generated is directed toward hydrogen production, and CO_2 fixation is naturally decreased. This affects the proton gradient and the rate of ATP synthesis. This affect naturally downregulates the rate of photosynthetic electron transport and thus further decreases the rate of hydrogen production by the hydrogenase. To resolve this issue, Lee and Greenbaum at the Oak Ridge National Laboratory in Oak Ridge suggest the introduction of inducible proton channel to disrupt that gradient during anaerobic conditions when nitrogen is produced (Lee and Greenbaum, 2003).

3.2.4.7 Photobioreactors

For hydrogen production purposes, the following types of bioreactors have been used: vertical column reactors, tubular ones, and flat panel photobioreactors. These reactors must comply with the following requirements (Prokop and Erickson, 1994; Dutta et al., 2005):

- Bioreactors must be enclosed systems such that hydrogen produced by the organisms can be collected without any loss
- Appropriate light arrangements either inside or outside the reactor
- Ease of sterilization
- Biorectors must have large surface to volume ratio to maximize the area of the incident light

However, these indoor photobioreactors have disadvantages too, namely high power consumption and high operation costs due to the need of artificial light sources (Chen et al., 2008a). Various types of photobioreactor with high illumination to volume ratios have been proposed, but most are limited by cost, mass transfer, contamination, scale-up, or a combination of these (Prokop and Erickson, 1994). Photobioreactors are discussed in detail in Chapter 8.

3.3 Photofermentation

Following the discovery of hydrogen production by photosynthetic organisms, photofermentation using photofermentative bacteria came into light. The concept was initially proposed by Benemann et al. (1973). Photofermentative bacteria essentially belong to two groups: Purple and Green. The purple bacteria can be further subdivided into purple sulfur (e.g., *Chromatium*) and purple nonsulfur bacteria (*Rhodobacter*), while the green bacteria are further subdivided into green sulfur (e.g., *Chlorobium*) and gliding bacteria (e.g., *Chloroflexus*). These photofermentative bacteria have evolved light-harvesting complexes akin to photosynthetic organism. Light energy is converted as chemical energy via photophosphorylation. However, the photosynthetic bacteria carry out photosynthesis with single photosystem (PSI) (Blankenship, 1992).

Purple bacteria contain PSII like reaction center and is thus incapable of reducing ferredoxin but can generate ATP via cyclic electron flow. The electrons desired for nitrogenase-mediated hydrogen evolution is derived from inorganic/organic substrates. Bacteriochlorophyll (P_{870}) in reaction center is excited and releases electron which reduces bacteriopheophytin (Bph) in the reaction center. Once reduced, the bacteriopheophytin reduces several intermediate quinone (Q) molecules to, finally, a quinone in "quinone pool." Electrons are now transported from the quinone through a series of iron–sulfur proteins (FeS) and cytochromes (Cyt) back to the reaction center (P_{870}). It is the cytochrome bC_1 complex that interacts with the quinone pool during photosynthetic electron flow as a proton motive force (PMF) used to derive ATP synthesis (Michel and Deisenhofer, 1988).

Purple sulfur bacteria are mostly photoautotrophs and obligate anaerobes. However, in aerobic, dark conditions, the organisms can undergo chemolithoautotrophic or chemo-organoheterotrophic growth. Members of Chromatiaceae use sulfide or organic substrate as electron donors (Kampf and Pfennig, 1980). In these bacteria, H_2 evolution is coupled with inorganic (sulfur-containing compounds) or organic substrate-driven reserve electron flow (Figure 3.7). A reversed electron flow operates in purple bacteria to reduce NAD^+ to NADH (McEwan, 1994). The reduced inorganic or organic substrates are oxidized by cytochromes and electrons from them eventually end up in quinone pool. However, the energy potential of quinone is insufficiently negative to reduce NAD^+ directly. Therefore, the electrons from the

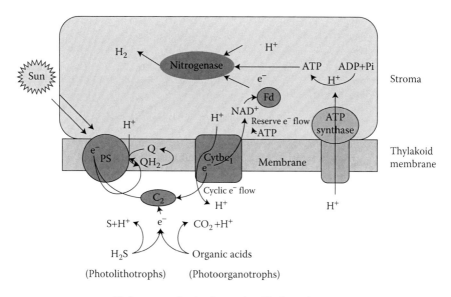

Hydrogen production in purple sulfur bacteria

FIGURE 3.7
Cyclic electron flow, hydrogen evolution, and ATP synthesis in purple bacteria.

quinone pool are forced backward to reduce NAD^+ to NADH. This energy requiring process is called reversed electron flow (McEwan, 1994) (Figure 3.7).

Electrons and ATP are used for CO_2 fixation as well as N_2 fixation. They contain Ribulose bisphosphate carboxylase (Rubisco) and phosphoribulokinase (Tabita, 1999) and fix CO_2 via Calvin cycle. Most of them do not have complete TCA cycle as they lack oxoglutarate dehydrogenase (Kondratieva, 1979). There is partial oxidation of substrate required for biosynthesis of some valuable products.

Hydrogen is evolved when electrons are transferred toward nitrogenase via ferredoxin by expense of ATP. The majority of purple sulfur bacteria are active diazotrophs. They are able to fix N_2 to ammonia (Madigan, 2004). Nitrogen fixation is linked with hydrogen production. Under nitrogen starvation, most of the purple sulfur bacteria are able to produce molecular hydrogen (Mitsui, 1975). No oxygen is evolved in this process. Hydrogen production through nitrogenase is very expensive process that requires 4 ATP for one mol hydrogen generation (Equation 3.12).

$$Fd \ (red) + 4ATP + 2H^+ \rightarrow Fd \ (ox) + 4ADP + 4Pi + H_2 \qquad (3.12)$$

Some species contain membrane-bound hydrogenase, which has hydrogen uptake capability instead of production of hydrogen. Hydrogen serves as photosynthetic electron donor for them. This capability was first detected in

Allochromatium vinosum strain D (Gaffron, 1939). Uptake hydrogenase resists excess loss of protons from the cell and maintains the redox potential of the cell (Equation 3.13).

$$H_2 \rightarrow H^+ + H^- \rightarrow 2H^+ + e^- \tag{3.13}$$

Hydrogen production is not well explored in purple sulfur bacteria. Most of the purple nonsulfur bacteria are well studied in this regard (Kumazawa and Mitsui, 1982). Purple nonsulfur bacteria exhibit aerobic respiration in photoautotrophic growth. Hydrogen or organic substrates are the electron donors for photosynthesis. Electrons and ATP (generated by photophosphorylation) are used for fixing CO_2 via Calvin cycle. TCA cycle and electron transport chain (ETC) operate completely, and oxygen is the terminal electron acceptor. However, in the presence of oxygen, there is no evolution of hydrogen as the oxygen is the electron acceptor instead of protons. Also the activity of enzyme nitrogenase is suppressed in the presence of oxygen.

Hydrogen evolution in purple nonsulfur bacteria has been observed in the absence of oxygen. They depend on external organic substrate for carbon and electron source. In photoheterotrophic growth, they use light energy only for ATP generation via cyclic electron flow. Reducing equivalents generated during oxidation of substrate reduce the ferredoxin, which transfers the reducing equivalents to nitrogenase to fix nitrogen. Protons are mainly used for formation of ammonia (Equation 3.14).

$$N_2 + 6H^+ + 6e^- \rightarrow 2NH_3 \tag{3.14}$$

Under nitrogen starvation, nitrogenase catalyzes the formation of molecular hydrogen from protons. The final amount of hydrogen produced is also influenced by the activity of uptake hydrogenase (Koku et al., 2003), which should be as low as possible. Highest yield of hydrogen was obtained at 303–310 K and with a high ratio of carbon to nitrogen in the medium (Eroglu et al., 1999). Photosynthetic nonsulfur purple bacteria are considered the best means of photobiological hydrogen production (Das and Veziroglu, 2001). The advantage of this process relies on application of organic wastes as the source of organic carbon (photofermentation). This adds extra advantage in the economy of hydrogen production. Some success has been found using industrial wastewater as substrate (Yetis et al., 2000). However, additional efforts have to be made for pretreatment of the industrial effluent before using within the reactor. This is deemed necessary as the wastewater may be toxic. Further, it could be opaque or have some color that could reduce light diffusion in the reactor. The overall reaction of the photofermentation is stated below (Equation 3.15):

$$CH_3COOH + 2H_2O \rightarrow 4H_2 + 2CO_2 \tag{3.15}$$

Green sulfur bacteria are obligate anaerobic photoautotrophic bacteria. In contrast to purple bacteria, they have the PSI type reaction center that can reduce ferredoxin. However, electron flow around the reaction center is insufficient for direct reduction of ferredoxin. Some sulfur compounds donate electrons to reaction center for the reduction of ferredoxin. Bacteriochlorophyll (P_{840}) absorbs light near 840 nm and resides at a significantly more negative reduction potential in comparison to purple bacteria (Figure 3.8). Unlike purple bacteria where the first stable electron acceptor molecule resides at about 0.0 reduction potential, the electron acceptors of green bacteria (FeS proteins) reside at about −0.6 reduction potential and have a much more electronegative reduction potential than NADH (Blankenship, 1985). In green bacteria, ferredoxin reduced by FeS protein directly serves as electron donor for the H_2 production.

Most of the green sulfur bacteria can fix nitrogen by the enzyme nitrogenase and in N_2 limited condition nitrogenase catalyzes the reduction of protons to hydrogen similar to purple bacteria. When H_2S donates electrons to green bacteria, sulfur globules remain outside of the cell. This is unlike purple bacteria where the globules of sulfur remain inside of the bacterial cell (Imhoff, 2004). Reaction center donates electrons only to nitrogenase for hydrogen production coupled with nitrogen fixation but not for CO_2 fixation. They lack Rubisco the key enzyme of Calvin cycle. Green sulfur bacteria fix CO_2 by reductive TCA cycle (Holo and Sirevåg, 1986; Hugler et al., 2005). Therefore, there is no competitive inhibition of hydrogen production by CO_2

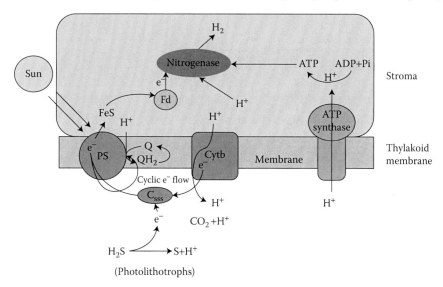

Hydrogen production in green sulfur bacteria

FIGURE 3.8
Cyclic electron flow, hydrogen evolution, and ATP formation in green bacteria.

fixation. CO_2-free environment within a reactor is not required but N_2-free environment is required to inhibit NH_3 formation. Storage product in green sulfur bacteria is mainly glycogen. In dark, they can catabolize storage product to produce different organic acids similar to dark fermentation. It has been suggested that storage product serves as energy source in dark and as reducing power in light (Sirevåg and Ormerod, 1977).

Very few studies have been conducted for the hydrogen production from green-gliding bacteria. They have photosystem II type of reaction center like purple bacteria but have light-harvesting structure chlorosome like green sulfur bacteria. Most of them probably do not have nitrogenase (Heda and Madigan, 1986). Hydrogenase is isolated from green-gliding bacteria *Chloroflexus aurantiacus* and is used as efficient biocatalyst for hydrogen production (Gogotov et al., 1991). This organism can survive in dark only in the presence of oxygen. Notably, the organism is orange colored in the dark but in presence of light it is green. *Chloroflexus* uses reduced sulfur compounds such as hydrogen sulfide, thiosulfate, or elemental sulfur and also hydrogen. *Chloroflexus aurantiacus* is thought to grow photoheterotrophically in nature, but it has the capability of fixing inorganic carbon through photoautotrophic growth. *Chloroflexus aurantiacus* has been demonstrated to use a novel autotrophic pathway known as the 3-hydroxypropionate pathway (Hügler et al., 2005). The complete electron transport chain for *Chloroflexus* spp. is not yet known.

Thus, photosynthetic bacteria undergo anoxygenic photosynthesis using light energy and reduced compounds. Some of these bacteria are also efficient in hydrogen production utilizing the reducing equivalents generated by the breakdown of organic acids such as lactic acid, ethanol, and so on. These acids are typically produced as competing end metabolites of dark fermentation. Therefore, to enhance the productivity of the process, hybrid processes have been suggested using dark fermentation in the first stage followed by photo fermentation. As compared to biophotolysis of water by algae, photofermentation offers several advantages. These include

- Light is not required for the splitting of water at PSII as in the case of biophotolysis of water in algae. Thus, the efficiency of light conversion to hydrogen is principally higher in photofermentation as compared to biophotolysis.
- Moreover, the absence of PSII automatically eliminates the problem of oxygen toxicity.
- Biophotolysis suffers from major drawback regarding the light spectrum utilization. However, bacteriocholorophyl has a wider range of spectrum utilization as compared to the algal and higher plants. However, low absorption by the bacteria in the 600–700 nm region and their strong absorption peak around 880 nm vs. a maximum absorption of 680 nm for algae reduce the energy content of the

photons absorbed. However, this also allows for the capture of solar photons in the 680–880 range, which cannot be harvested by algal photosynthesis.

- Photofermentative bacteria can utilize a wider range of substrates as compared to photosynthetic organisms. Keskin et al. (2011) in their review article have extensively discussed the wastes which are rich in organic acids that can be used by photofermenters. They have discussed a variety of feed-stocks, including the effluents of dark fermentations, leading to the development of various configurations of two-stage systems, or various industrial and agricultural waste streams rich in sugars or in organic acids.

3.3.1 General Considerations and Advancements Made in Photofermentation

Though photofermentation has several advantages over the biophotolysis process, it also has several bottlenecks, due to which even after extensive research over the last five decades the process is not close to commercialization. The bottlenecks include: (1) the use of the nitrogenase enzyme with its inherent high energy demand; (2) the low solar energy conversion efficiencies; and (3), as noted above, the requirement for elaborate anaerobic photobioreactors covering large areas. These bottlenecks are similar to those of biophotolysis. Initially, it was assumed that high solar-to-H_2 efficiencies should be achievable, as most of the energy in the H_2 would derive from the organic substrates. However, this assumption proved incorrect: H_2 production in outdoor photobioreactors by photosynthetic bacteria using organic acids as electron donor was similar to that of heterocystous cyanobacteria using water as substrate, with only about 0.2% of the sunlight energy converted to H_2 fuel (Benemann and Pedroni, 2008). Studies show that the very large light-harvesting chlorophyll structures in photosynthetic bacteria lead to an even stronger light saturation effect than in microalgae (Benemann and Pedroni, 2008).

Over the last two decades, a lot of intensive work has been carried out on photofermentation. These include improvement of the process design and also the organism by genetic manipulation. The recent advancements made in the field are discussed below.

3.3.1.1 Immobilization Approaches

A lot of work has been conducted with photofermentative bacteria using suspension cultures (Basak and Das, 2006; Nath and Das, 2008; Gilbert et al., 2011). Suspension cultures have both advantages and disadvantages. The major advantage of suspension type cultures is the efficient mass transfer due to stirring. However, in many cases, photofermentation with suspended cultures under continuous conditions has been characterized by low hydrogen production rates and low light conversion efficiencies. This is due to low HRT at which the continuous culture need to be operated to prevent cell washout. However, at such low HRT, hydrogen production decreases (Keskin et al., 2011). In view of this, different immobilization techniques have been applied to improve the yield and rate of hydrogen production from photofermentative bacteria. Immobilization techniques can generally be divided into three main categories: adsorption (biofilm formation), encapsulation, and entrapment. The most important criterion is to select a nontoxic, durable, and porous material with a high retention capacity. Every immobilization technique and each material have their own advantages and disadvantages (Hallenbeck, 1983). Several solid matrices have been successfully utilized for immobilization of photoheterotrophic bacteria, such as porous glass (Tsygankov et al., 1994), carrageenan (Francou and Vignais, 1984), agar gel (Zhu et al., 1999), and even clay surfaces (Chen and Chang, 2006). However, the bottleneck of these immobilized matrix is that hydrogen production potential of the organism is decreased. This is possibly due to the higher immobilization temperatures (40–45°C) at which the cell is immobilized. Later, immobilization on sodium alginate at room temperature was found to be more effective. Moreover, sodium alginate beads had smaller size which effectively increased the surface area and hence the biomass within the reactor (Basak and Das, 2006).

Among the various immobilization techniques, cell entrapment is a low-cost technique and easy to handle. However, the major disadvantage of this technique is light penetration. To overcome the bottleneck, light-penetrating transparent beads made of polyvinyl chloride were recently developed by Wang et al. (2010). The beads provide a transparent and effectively modified entrapment matrix. The matrix used for entrapment of *R. palustris* was composed of polyvinyl alcohol-124 (PVA) and carrageenan powder. Continuous operation lasted for 3 months with a hydrogen production rate of 0.73 mmol H_2/m^3 s and a yield of 3.59 mol H_2/mol glucose. This immobilization technique has many advantages, such as high mechanical strength, compared with other entrapment techniques.

In another technology, reverse micelles have been developed to entrap the bacteria for photofermentative hydrogen production (Pandey and Pandey, 2008). Reverse micelles are amphiphilic surfactant molecules comprising water and a nonpolar organic solvent. The orientation of the polar heads of the surfactant molecules is directed toward the interior of a water-containing sphere, whereas the aliphatic tails are oriented toward the nonpolar organic phase. The water structure within the reverse micelles may resemble that of water adjacent to biological membranes (Boicelli et al., 1982), and it has been suggested that the reverse micelle system reliably mimics the microenvironment that enzymes encounter in the intracellular milieu (Boicelli et al., 1982; Conner and Viggnais, 1988; Faeder and Ladanyi, 2000). The surfactants that have been used to develop the micellar system include single-chained anionic sodium dodecyl sulfate (SDS) and the cationic cetyltrimethyl ammonium chloride (CTAC) or bromide (CTAB), and the double-chained, anionic sodium bis-2-ethylhexyl-sulfosuccinate (Na [AOT]). The size of the reverse micelles is directly proportional to the amount of water added to the system. Reverse micelles of the size ranging from 4 to 18 nm diameter with 0.1 M have been reported using Na (AOT) surfactant (Luisi and Magid, 1986).

Reverse micelles have been popularly used to entrap the enzymes such as invertase and α-amylase. Immobilization of the enzymes within the micellar system is known both to increase the stability and the activity of the enzymes several folds (Gajjar et al., 1997). Several groups have used the reverse micelle system to entrap hydrogenase to increase the hydrogen production. Increased H_2 production has been reported by coupled system of *Halobacterium halobium* cells with chloroplast organelles (Singh et al., 1999a) and by *Rhodopseudomonas sphaeroides* in isolation when entrapped within different reverse micelles (Singh et al., 1999b). Genetic improvement of the organism has been discussed in detail in Chapter 6.

3.3.1.2 Scale-Up Considerations

Scale up of the process is also a tremendous challenge and requires a lot of consideration. However, recently, a few research groups have scaled up photofermentation from the lab reactors to pilot scale. Efforts have been made to test the productivity of the reactors by capturing the solar lights. However, design of such outdoor photobioreactors (PBRs) should meet several criteria to maintain anaerobic conditions as well as maximum light capture. Reactor design should allow an efficient mass transfer rate of produced hydrogen, and the material used should have no hydrogen diffusivity. Moreover, the material thickness and reactor size should be selected according to high visible light and near infrared transmissions (Asada and Miyake, 1999; Akkermann et al., 2002). A homogeneous substrate distribution within the reactor should be provided through continuous circulation, which would also keep the cells suspended, thereby, providing a better light distribution and stripping of the hydrogen produced. The reactor design should allow easy temperature control in outdoor conditions (Boran et al., 2012). In terms of the design geometry of the reactor, they can be classified as panel or tubular reactors. Tubular PBRs consist of long transparent tubes and when compared with the panel, tubular reactors have high surface to volume ratio (Akkermann et al., 2002). There are considerable data related to cyanobacteria cultivations (Torzillo et al., 1986; Torzillo et al., 1993; Watanabe et al., 1998; Zitelli et al., 1999) and reports on photofermentative hydrogen production in tubular PBRs (Delachapelle et al., 1991, Vincenzini et al., 1997; Modigell and Holle, 1998; Markov 1999; Carlozzi and Sacchi, 2001). Previously, a comparative operation of large-scale panel and tubular PBRs in outdoor conditions for photofermentative hydrogen production has been reported and their performance was evaluated in terms of ground area (GA) and illuminated reactor surface (IRS) (Gebicki et al., 2010). The mean hydrogen productivity for the panel and the tubular reactors was 0.2×10^{-5} mmol $H_2/(m^2_{IRS} \cdot s)$ and 0.17×10^{-5} mmol $H_2/(m^2_{IRS} \cdot s)$, respectively. A winter period operation of tubular PBR with *R. capsulatus* produced hydrogen at an average productivity of 0.86 mol $\times 10^{-4}$ $H_2/m^3 \cdot s$ on artificial medium with continuous circulation of the reactor (Boran et al., 2010). Similarly, 90 L pilot tubular PBR was operated with *R. capsulatus* by using thick juice dark fermenter effluent (DFE). The average molar productivity calculated according to daylight hour was 0.42×10^{-4} mol $H_2/(m^3 s)$ with regard to the total reactor volume, and the yield obtained was 0.5 mol H_2 per mole of acetic acid fed (Boran et al., 2010). More recently, Boran et al. (2012) successfully operated a pilot solar tubular photobioreactor for fed batch operation in outdoor conditions for photofermentative hydrogen production with *Rhodobacter capsulatus* (Hup$^-$) mutant. The feeding strategy was to keep acetic acid concentration in the photobioreactor in the range of 20 mM by adjusting feed acetate concentration. The maximum molar productivity obtained was 1.1×10^4 mol $H_2/(m^3 s)$ and the yield obtained was 0.35 mol H_2 per mole of acetic acid fed. Evolved gas contained 95–99% hydrogen, and the rest was carbon dioxide by volume. Photobioreactors have been discussed in detail in Chapter 8.

IS THE MONEY SPENT ON PHOTOBIOLOGICAL HYDROGEN PRODUCTION WORTHWHILE? ISSUES WITH PHOTOBIOLOGICAL AND PHOTOFERMENTATIVE HYDROGEN PRODUCTION

Hydrogen production using microbes was discovered in Gaffron in *C. reinhardtii*. Unknown, at that time he had hit on the process of Biophotolysis. Later in the 1970s, Benemann discovered photofermentation. 60 years down, the research on photosynthetic and photofermentative organisms as lucrative hydrogen producers has not ceased. Though abundant basic and applied research has not been able to enhance the productivity of the processes. However, still enormous efforts in terms of time and money continue to be spent. Is it justified? Although the process is lucrative, in terms of cost consideration, the yield and rate of the process still are a matter of great concern even at the lab scale. However, CO_2 sequestration appears a forte of photosynthetic algae and the CO_2-sequestered biomass can be used in new and innovative ways as cheap carbohydrate feedstock for dark fermentative bacteria in hybrid. Therefore, a few research groups focus on hybrid production by photosynthetic microorganisms than sole focus on biophotolysis.

Moreover, the consideration that microalgae photosynthesis could achieve 10% solar-to-H_2 efficiency (Kok, 1973; Bolton, 1996) is correct in theory, but practical processes will probably never reach this goal due to the many loss factors not considered in such theoretical estimates. Projections of 20–30% solar-to-H_2 efficiencies for direct biophotolysis (National Research Council, 2004) exceed known mechanisms and even challenge the very fundamentals of laws of thermodynamics (Hallenbeck and Ghosh, 2009).

In conclusion, the rates and efficiencies of hydrogen production by these, as well as all other systems directly involving photosynthesis to produce H_2 (photobiological hydrogen production), fall far short of even plausible economic feasibility. Since organic substrates are the ultimate source of hydrogen in photofermentations or indirect biophotolysis processes, it can be argued that it should be simpler and more efficient to extract the hydrogen from such substrates using a dark fermentation process (Hallenbeck and Ghosh, 2009).

3.4 Dark Fermentation

3.4.1 Anaerobic Fermentation

Dark fermentation is the fermentative conversion of organic substrate to biohydrogen. It is a complex process manifested by diverse group of bacteria by a series of biochemical reactions. Fermentative/hydrolytic microorganisms hydrolyze complex organic polymers to monomers, which are further converted to a mixture of lower molecular weight organic acids and alcohols by necessary H_2-producing acidogenic bacteria.

Fermentation is a metabolic process that occurs under anaerobic conditions to regenerate the cell's energy currency (ATP). Moreover, under anaerobic conditions, the TCA cycle is blocked. Therefore, fermentation disposes off the excess cellular reductant by the formation of reduced metabolic end products such as alcohol and acids. Similarly, hydrogen is also a reduced metabolic end product produced to maintain the cellular redox potential.

Carbohydrates, mainly glucose, are the preferred carbon sources for fermentation process that predominantly gives rise to acetic and butyric acids concomitantly with hydrogen gas. Complex organic polymers were converted into glucose by hydrolysis. Glucose produces pyruvate via the glycolytic pathway to regenerate ATP. Subsequently, pyruvate may be involved in two different biochemical reactions leading to the formation of hydrogen.

In obligate anaerobes such as *Clostridia* (McCord et al., 1971) and thermophilic bacteria (Zeikus, 1977), pyruvate is oxidized to acetyl coenzyme A (acetyl-CoA) by pyruvate-ferredoxin oxidoreductase (PFOR) (Uyeda and Rabinowitz, 1971). Acetyl-CoA subsequently converts to acetyl phosphate with concomitant generation of ATP and acetate. Oxidation of pyruvate to acetyl-CoA requires reduction of ferredoxin (Fd). Reduced Fd is oxidized by [FeFe] hydrogenase and catalyzes the formation of H_2 (Figure 3.9). The overall reaction is shown in Equations 3.16 and 3.18.

$$\text{Pyruvate} + \text{CoA} + 2\text{Fd(ox)} \rightarrow \text{Acetyl-CoA} + 2\text{ Fd(red)} + CO_2 \quad (3.16)$$

$$2H^+ + \text{Fd(red)} \rightarrow H_2 + \text{Fd(ox)} \quad (3.17)$$

Four moles of hydrogen per mole of glucose are formed when pyruvate is oxidized to acetate as the sole metabolic end product (Benemann, 1996). However, only 2 mol of hydrogen per mole of glucose are produced when pyruvate is oxidized to butyrate. Therefore, for the organisms that follow a mixed acid pathway, higher A/B ratio is critical for higher hydrogen production (Khanna et al., 2011). The overall biochemical reaction with acetic acid and butyric acid as the metabolic end products is shown in Equations 3.18 and 3.19, respectively.

$$C_6H_{12}O_6 + 2H_2O \rightarrow 2CH_3COOH + 2CO_2 + 4H_2 \quad (3.18)$$

$$C_6H_{12}O_6 \rightarrow CH_3CH_2CH_2COOH + 2CO_2 + 2H_2 \quad (3.19)$$

The second type of biochemical reaction involved in hydrogen production occurs in a few facultative anaerobic enteric bacteria such as *E. coli*. In this pathway, pyruvate is oxidized to acetyl CoA and formate. The reaction is catalyzed by the enzyme pyruvate formate lyase (PFL) (Knappe and Sawers, 1990) (Equation 3.20, Figure 3.9):

$$\text{Pyruvate} + \text{CoA} \rightarrow \text{Acetyl-CoA} + \text{formate} \quad (3.20)$$

Subsequently, formate is cleaved by formate hydrogen lyase (FHL) to produce carbon dioxide and hydrogen (Equation 3.21, Figure 3.9). Stephenson and Stickland first described this pathway in the 1930s (Stephenson and Stickland, 1932).

$$\text{HCOOH} \rightarrow CO_2 + H_2 \quad (3.21)$$

In the lactic acid fermentation, pyruvate is directly oxidized to lactate. No hydrogen is produced when ethanol, lactic, or propionic acid are the sole metabolic end products. Moreover, some facultative anaerobes such as enteric bacteria may carry out anaerobic respiration instead of fermentation using nitrate, fumarate, so on. as terminal electron acceptors. Anaerobic respiration may hamper hydrogen production. Therefore, for carrying out studies on hydrogen production, the media should be devoid of these electron acceptors.

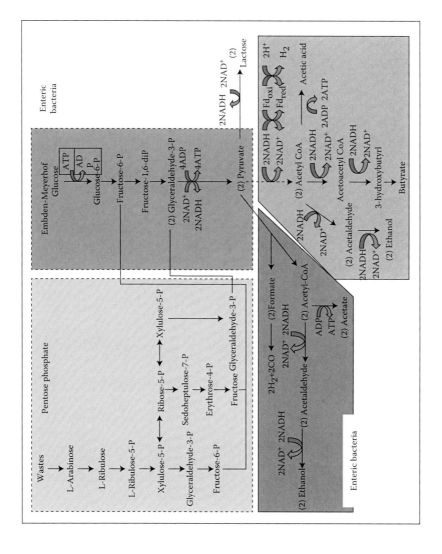

FIGURE 3.9
Overall biochemical pathways for dark fermentative metabolism.

Dark fermentation has gained popularity primarily because it can utilize a wide variety of substrate for hydrogen production. Moreover, it can utilize different wastewaters and thus serves the dual purpose of waste bioremediation and clean energy generation. Industrial wastewater as fermentative substrate for H_2 production addresses most of the criteria required for substrate selection viz. availability, cost, and biodegradability (Cai et al., 2009). Chemical wastewater (Cakir et al., 2010), cattle wastewater (Chen et al., 2008b), and starch hydrolysate wastewater (Chen et al., 2008c) have been reported to produce biohydrogen apart from wastewater treatment from dark fermentation process using selectively enriched mixed culture under acidophilic conditions. Various wastewaters viz., paper mill wastewater (Cheng et al., 2011), food processing wastewater (Cheong and Hansen, 2006), rice winery wastewater (Nath and Das, 2004), distillery- and molasses-based wastewater (Das et al., 2008), wheat straw wastes (Das and Vezrioglu, 2001), and palm oil mill wastewater were also studied as fermentable substrates for H_2 production along with wastewater treatment (Pandu and Joseph, 2012). Utilizing mixed culture is extremely important and well suited to the nonsterile, ever-changing, complex environment of wastewater treatment. However some of these wastes cannot be utilized directly and must undergo pretreatment processes such as plant biomass which are rich in lignocellulose. Physicochemical pretreatment of lignocellulosic material, such as the application of acid, alkaline, or oxidative conditions at ambient or elevated temperatures, yields a mixture of pentoses and hexoses. Efficient microbial fermentation of hexoses and pentoses is, therefore, the key step for hydrogen production from plant biomass. However, combined fermentation of mixtures of hexoses and pentoses is often prevented due to catabolic repression; in the presence of glucose, pentoses might be converted to a lesser extent thereby decreasing overall fermentation yields (Strobel, 1993; Aberu et al., 2010). Moreover, efficient hydrogen production from sugars is dependent on the different possible fermentation pathways (Abreu et al., 2012).

Organic acids produced during dark fermentation can be used as a substrate for photofermentation. Photofermentative bacteria can oxidize organic acids to produce CO_2 and H_2. Therefore, higher hydrogen production can be obtained by following a two-stage process of dark fermentation followed by photofermentation (Das et al., 2008). Theoretically, 12 mol of hydrogen per mole of glucose can be produced from this hybrid process (Nath and Das, 2008).

$$2CH_3COOH + 4H_2O \rightarrow 8H_2 + 4CO_2 \qquad (3.22)$$

3.4.2 General Considerations to Commercialization of the Technology

Major challenges need to be overcome for the smooth transition from the fossil fuel based economy to the H_2 energy-based economy and may be outlined as follows.

3.4.2.1 Low Yield and Rate of Production

The yield of H_2 from any of the processes defined above is far too low for commercial application. The pathways of H_2 production have not been identified completely and the reaction remains energetically unfavorable. In fact, till date, complete knowledge of the chief enzyme, hydrogenase involved in hydrogen production, is lacking. The crystal structures of only two [FeFe] enzymes with respect to dark fermentation have been determined. These include hydrogenase I enzyme from *Clostridium pasteurianum* (*CpI*) and crystal structure of structure of the [FeFe]-hydrogenase from *Desulfovibrio desulfuricans* (*Dd*HydAB). This has limited the understanding of the catalysis process and the deficiency in the development of engineered hydrogenases. Moreover, the thermodynamic limitation of the process makes it technologically and economically challenging as compared to the process of bioethanol or methane. H_2 partial pressure severely disrupts the efficiency of the processes and adds to the problems of lower yields. However, integrated processes may help overcome the problem of low yield.

3.4.2.2 Processing of Some Biomass Feed Stock Is Too Costly

There is a need to develop low-cost methods for growing, harvesting, transporting, and pretreating energy crops and/or biomass waste products. The added advantage to the dark fermentation process is the use of waste for generation of clean energy. However, most of the wastes need to be pretreated before it can be used effectively. Pretreatment is a crucial process step for the biochemical conversion of lignocellulosic biomass into biohydrogen. It is required to alter the structure of cellulosic biomass to make cellulose more accessible to the enzymes that convert the carbohydrate polymers into fermentable sugars (Mosier et al., 2005). Pretreatment has been recognized as one of the most expensive processing steps in cellulosic biomass-to-fermentable sugar conversion, and several recent review articles provide a general overview of the field (Hendriks and Zeeman, 2008; Taherzadeh and Karimi, 2008; Alvira et al., 2009; Carvalheiro et al., 2008). A number of physical, chemical, and biological pretreatment processes are available (Harmsen et al., 2010) for pretreatment; however, currently they result in sugar losses and add to the operation costs. Therefore, there is an immediate need to address the issue to develop rapid and cost-effective processes.

3.4.2.3 Incomplete Substrate Degradation

Dark fermentation is fraught with incomplete substrate conversion challenges. This limits the total cumulative production. Introduction of novel pathways and hybrid processes may improve the process yields.

3.4.2.4 Lack of Robust Industrial Strain

There is no clear contender for a robust, industrially capable microorganism that can be metabolically engineered to produce more than 4 mol H_2/mol of glucose. Currently, there is no verified evidence that any naturally occurring microbe can or will produce more than 4 mol of hydrogen per mole of glucose. Achieving higher yields is a critical make-or-break for direct hydrogen production via fermentation. Therefore, a crucial research objective is to create such an organism through metabolic engineering. The ideal outcome would be a microorganism that would be able to produce high hydrogen yields from an inexpensive feedstock. A number of high-priority technical breakthroughs and R&D activities are needed, including genetic tools to overcome the metabolic barrier by manipulating electron flux in hydrogen-producing organisms (e.g., single host organism for transgenic expression of hydrogen pathways; genomic database; tailored genome shuffling/combinatorial tools, etc.).

3.4.2.5 Engineering Issues

Several engineering issues need to be addressed which include

- Long term steady continuous H_2 production rate
- Scale-up of the process
- Separation/purification of H_2

Current reactors do not perform optimally, especially under conditions required for industrial hydrogen production (i.e., robust, reliable performance, and high-sustained hydrogen yields). Research is needed to improve reactor designs and process parameters, including membrane technology to lower hydrogen concentrations within the reactor and improved techniques for mixing, pH, and temperature control and cell harvesting.

3.4.2.6 Sensitivity of Hydrogenase to Oxygen

Oxygen sensitivity is a critical issue in hydrogen process development. Over the years, several scientists have attempted to develop an oxygen insensitive hydrogenase. However, the results thus far have not been encouraging. In particular [FeFe] hydrogenases are irreversibly destroyed by oxygen, whereas oxygen does not affect the structural integrity of [NiFe] hydrogenases but reversibly inactivates their catalytic function. It was shown by various spectroscopic techniques that an oxygen species is bound between nickel and iron (Brecht et al., 2003). This bridging ligand occupies the position that is required for binding of a formal hydride under turnover conditions. The bridging oxygen ligand is removed reductively, hence giving hydrogen access to the catalytic site.

Recently, Bingham et al. (2012) presented a paper in AIChE Annual Meeting Topical Conference on Hydrogen Production and Storage. In the paper, he described the discovery of a mutant strain of *Clostridium pasteurianum* (CpI) with increased hydrogen sensitivity of [FeFe] hydrogenase I. A cell-free protein synthesis-based screening platform was used to identify an initial mutant from a randomly mutated CpI library. They identified a mutant strain, which showed significantly enhanced tolerance toward oxygen. Further studies revealed three mutations responsible for the oxygen sensitivity. They showed further improvement in tolerance by PCR mutagenesis at influential sites. They found that after oxygen exposure under conditions where the enzyme is in the resting state, the mutant hydrogenase retained significant enzyme activity as compared to the wild-type strain. However, surprisingly, when the enzyme is actively catalyzing hydrogen production during oxygen exposure, the mutant hydrogenase showed no improved oxygen tolerance. This and similar experiments show the extreme complexity of the hydrogenase enzyme toward catalysis and oxygen sensitivity.

However, several theoretical models and explanations have been proposed by different groups of how theoretically it is possible to overcome the oxygen instability. Goldet et al. (2009) have proposed that two steps may be rate limiting in oxygen inhibition: (1) the diffusion of oxygen through the protein to the active site pocket, and (2) the binding of oxygen to the $[2Fe]_H$ subcluster (Roseboom et al., 2006). Further, it has been proposed that the accessory [FeS] clusters may play a crucial role in the oxygen inactivation mechanism (Lemon and Peters, 1999). Tard et al. (2005) described that the oxygen molecule initially binds to the distal Fe of the $[2Fe]_H$ subcluster, then forms a reactive oxygen species that destroys the $[4Fe4S]_H$ subcluster. The distal Fe, the Fe farthest from the $[4Fe4S]_H$ cluster, is believed to be the hydrogen-binding site and also the site of reversible carbon monoxide-binding and inhibition (Peters et al., 1998; Pandey et al., 2008).

A few other groups have studied oxygen inactivation using protein film electrochemistry. In this technique, the enzyme, hydrogenase, is adsorbed to an electrode, and its activity is directly measured by electron transfer through the electrode under oxidizing or reducing potentials during oxygen exposure (Nicolet et al., 2001; Silakov et al., 2007; Lubitz et al., 2007). Protein film electrochemistry experiments have demonstrated that reversible carbon monoxide inhibition protects against oxygen inactivation and has provided further evidence favoring O_2 binding at the distal Fe atom (Pandey et al., 2008). It has been proposed that oxygen diffusion occurs at the hydrogenase active site causing the inactivation of the enzyme. In view of this, another method proposed to overcome the oxygen sensitivity suggested to limit the diffusion of oxygen molecules to the active site (Stripp et al., 2009a,b). Based on this hypothesis, a screen developed by Boyer et al. (2008) would retain more activity after oxygen exposure than the wild-type protein. In this screen, mutated hydrogenases were first expressed and activated in cell-free protein synthesis reactions. The hydrogenase activities were then measured before and after oxygen exposure. It was believed that if the protein structure could be modified to better exclude oxygen, the active site would be protected whether or not the enzyme was actively catalyzing the hydrogen conversion reaction at the time of oxygen exposure (Bingham et al., 2012).

3.4.2.7 Mixed Consortia Have Methanogens: Suppression of Methanogen Activity

Mixed consortia are reported to be better hydrogen producers as compared to pure strains especially if some anaerobic-mixed cultures cannot produce H_2 as it is rapidly consumed by the methane-producing bacteria. Successful biological H_2 production requires inhibition of H_2-consuming microorganisms, such as methanogens, and pretreatment of parent culture is one of the strategies used for selecting the requisite microflora. The physiological differences between H_2-producing bacteria and H_2-consuming bacteria (methanogenic bacteria) form the fundamental basis behind the development of various methods used for the preparation of H_2-producing seeds. The problem can be solved using the traditional dilution methods. Methanogens from mixed culture could also be eliminated/repressed by maintaining short HRT (2–10 h) as H_2-producing bacteria grow faster than methanogens (Fang and Liu, 2002; Lin and Jo, 2003; Fan et al., 2006).

3.4.2.8 Low Gaseous Energy Recovery

Considering the lower-heating values (LHV) of hydrogen and glucose as 238 kJ/ mol and 2976 kJ/mol, respectively, following energy analysis was performed according to Nath et al. (2006) (Equation 3.23).

Energy recovery as hydrogen from substrate may be calculated as follows:

$$\left[\frac{LHV H_2}{LHV \, glucose} \times process \, yield \right] \times 100 \tag{3.23}$$

Considering, acetic acid as the only end metabolite, which produces 4 mol H_2/mol glucose, energy recovered from the substrate is 32%. This is low in comparison to ethanol and methanol as biofuels where the product recovery is 80–90% near the theoretical maximum (Claassen et al., 2000). A competitive goal for H_2 fermentations would be to produce a yield of about 10 mol H_2/ mol glucose, close to conventional ethanol yields from corn starch. A recent techno-economic analysis concluded that if such a yield were achievable, H_2 production costs would be similar to those of ethanol fermentations, or only slightly higher due to higher fuel handling (H_2 purification and storage) costs (Eggeman, 2004). The fundamental questions in H_2 fermentations is, therefore, how to achieve this goal (Benemann and Pedronni, 2008).

Moreover, Benemann and Pedroni (2008) are of the opinion that to maximize hydrogen production from the bacteria, growth of the microbes must be arrested. This is contrary to methane and ethanol fermentations, where the product is growth associated. To increase H_2 production yields, it would be necessary to decouple biomass growth from product formation, as used in several other industrial processes. However, this is still a nascent area and needs to be worked on. This is essential because both the biomass production and hydrogen generation compete for the same reducing equivalents. Thus, how to channel cellular metabolism into low redox potential ferredoxin reduction, required for H_2 production, is one of the key issues that needs to be addressed to increase the yield and rate of the process.

3.4.2.9 Biomass and End Metabolite Formation Compete with Hydrogen Production

Besides biomass, other end metabolites, formed during anaerobic fermentation, also compete with hydrogen production. It must be understood that all the end metabolites are not related to hydrogen production. Principally, the production of acetate is associated with 4 mol H_2/mol glucose. Butyrate produces 2 mol of hydrogen. However, end products such as propionate, lactate, and solvents like ethanol are not associated with hydrogen production. In fact, formation of these reduced products drives the reducing equivalent and the protons away from hydrogen production and decreases the final yield of the desired product. Therefore, recently, researchers have focused their attention to redirect the metabolic pathway by blocking or deleting other competing end metabolites.

3.4.2.10 Thermodynamic Limitations

Besides dark fermentation, the yield of 4 mol of hydrogen per mol of glucose is limiting. The overall theoretical yield achievable is low as compared to the total yield that can be achieved in terms of ethanol as fuel or methane. Thermodynamic limitations restrict the overall yield. More fundamentally, a near-stoichiometric yield is only achievable under near-equilibrium conditions, which implies very slow rates, and/or at very low partial pressures of H_2.

3.4.2.11 Integration of Processes

A lack of understanding on the improvement of economics of the process by integration of H_2 production with other processes has been discussed in detail in the next section.

3.4.3 Progress Made in the Field of Dark Fermentation

Technological approaches to circumvent these problems might be to combine enzymes or pathways with desirable properties from different organisms to generate a robust hybrid pathway. Other approaches will involve understanding and controlling the cellular regulation of the gene products required for hydrogen production.

3.4.3.1 Overcoming Techno-Engineering Barriers

It is possible that fermentative hydrogen production systems may be made more economical by combining them with other processes that add to the yield of the process. Examples might include a dark fermentation reactor followed by a photobiological reactor; a two- or three-stage system that utilizes all of the biomass feedstock (cellulose, hemicellulose, and lignin); or a fermentation system that includes an another suitable microbe for conversion of fermentation by-products (e.g., organic acids) to hydrogen. Such hybrid systems would no doubt enhance the total yield of the process.

3.4.3.2 Molecular Advancements

Several molecular advancements have been made in the field. The molecular advancements were targeted both at the organism level and the enzyme level. At the organism level, attempts have been made to enhance hydrogen production by heterologous overexpression of [FeFe] hydrogenase from more potent organisms, homologous overexpression of the enzymes, and elimination of competing pathways. At the enzymatic level, attempts have been made to overcome the oxygen sensitivity of the enzyme. All these factors have been discussed in detail in Chapter 5.

3.4.3.3 Modeling and Optimization of the Process

Modeling and optimization have been carried out in attempts to improve biohydrogen production. The optimal yields and rates of any bioprocess depend up several factors, primarily, pH, temperature, substrate concentration, and nutrient availability, among others. Although many studies have reported the effects of varying individual parameters one at a time, modeling and analysis could be used to determine the optimal values of the important relevant parameters. Today, a number of modeling methods have been developed to find the most suitable values. A widely used model that is used in hydrogen production processes is the Gompertz model. A Gompertz curve or Gompertz function, named after Benjamin Gompertz, is a sigmoid function. It is a type of mathematical model for a time series, where growth is slowest at the start and end of a time period. The right-hand or future value asymptote of the function is approached much more gradually by the curve than the left-hand or lower-valued asymptote, in contrast to the simple logistic function in which both asymptotes are approached by the curve symmetrically. It was designed to study growth, however modified form of the Gompertz equation is now widely used to study experimental and simulated results obtained from hydrogen set-ups (Nath et al., 2006; Khanna et al., 2011). Although the metabolic pathways and fluxes are relatively well understood for a pure culture fermenting a defined substrate, this type of analysis can nevertheless have some value for modeling fermentations of complex substrates by microbial consortia that involve multiple metabolic types with unknown interactions.

3.4.3.4 Pilot Scale Demonstration of the Technology

Several research groups have attempted to scale up the dark fermentation technology to the bench scale and the pilot scale. Ren et al. (2006) demonstrated the successful operation of a 1.48 m³ capacity pilot scale reactor using molasses as substrate. They showed the operation of the reactor for 200 days. They reported an optimal organic-loading rate (OLR) of 3.6–99 × 10⁻⁵ kg of COD/m³ s during which hydrogen yields increased with increase in OLR, however, higher OLR decreased the yields drastically. The biogas mainly comprises CO_2 and H_2 and contain 40–52% (v/v) hydrogen. A maximum H_2 production rate of 645 × 10⁻⁵ m³ of H_2/m³ s with a specific H_2 production rate of 0.87 × 10⁻⁵ m³ of H_2/kg (of MLVSS) s was obtained. One pilot plant of capacity 0.8 m³ is in operation at the Indian Institute of Technology Kharagpur using immobilized whole cell (*E. cloacae* IIT-BT 08) bioreactor (Figure 3.10).

FIGURE 3.10
(**See color insert.**) Pilot plant of 0.8 m³ capacity installed at IIT, Kharagpur, India.

3.5 Hybrid Processes

The above brief review of the known biological methods for hydrogen production demonstrates that each process has both significant advantages and shortcomings. Moreover, the yield of any individual process cannot alone meet the standards of commercialization. However, integration of two or more steps may hold a better promise. Some of the practical and hypothetical hybrid processes are described below.

A combination of two biological processes implies that the by-products of one stage will be substrates for the organisms performing the next stage with the corresponding increase in the efficiency of energy source (organic compounds or light) utilization (Tekucheva and Tsygankov, 2012). Possible integration ways for hydrogen (and electric current) production include

1. Integration of dark fermentative process with photofermentation
2. Integration of biophotolysis with dark fermentative process and photofermentation
3. Integration of biohydrogen with microbial fuel cells

3.5.1 Integration of Dark Fermentative Process with Photofermentation

This has been one of the traditional approaches. Initially dark fermentation of the carbohydrates takes place that generates extracellular-reduced end metabolites such as acetic acid, butyric acid, and so on. along with hydrogen. Reducing equivalents disposed off by the cell as low molecular weight organic acids cannot be further utilized by the cell. However, photosynthetic bacteria have the capacity to overcome the free energy barrier in the utilization of small organic acids by utilizing sunlight energy to bring organic acids into their metabolism, while fixing nitrogen and producing hydrogen under anaerobic conditions. Integration of the two processes reduces the overall energy demand of the photosynthetic bacteria and increases the amount of hydrogen production per substrate source (Das and Veziroglu, 2001) (Figure 3.11). The main advantage of such a scheme is an increase of theoretically possible hydrogen yield to 12 mol/mol glucose. Moreover, it is possible to utilize organic wastewaters and solid waste as substrates. Currently, this scheme is the most evident and widely studied one. The overall reactions in an integrated dark- and photofermentation system are shown in Equations 3.24 and 3.25:

Dark fermentation stage:

$$C_6H_{12}O_6 + 2H_2O \rightarrow 2CH_3COOH + 2CO_2 + 4H_2 \qquad (3.24)$$

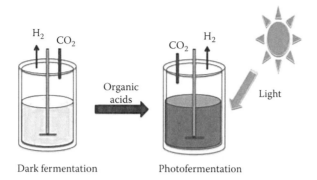

FIGURE 3.11
(**See color insert.**) Integrated dark fermentation and photofermentation to increase the overall yield of the process.

Photofermentation stage:

$$2CH_3COOH + 4H_2O \rightarrow 4CO_2 + 8H_2 \quad (3.25)$$

The following factors should be taken into account when selecting a substrate for such a two-stage system (Tekucheva and Tsygankov, 2012):

1. The organic substrates contained in the waste should be fermented during the dark stage to produce by-products which can be readily utilized by purple bacteria.

2. The nitrogenase activity of purple bacterial should not be inhibited or decreased by an excess concentration of nitrogenous compounds contained in the dark bioreactor effluent after the first stage.

3. The presence and concentration of mineral components in the waste should not limit or inhibit the hybrid processes.

When selecting bacteria for the two-stage process, it is necessary to take into account that four combinations of bacterial metabolic pathways can theoretically provide the yield of 12 mol H_2/mol hexose.

3.5.1.1 Lactic Acid Fermentation Integrated with Photofermentation

In this process, one molecule of glucose ferments to give 2 mol of lactic acid. However, fermentation of glucose to lactic acid produces no hydrogen as shown in Equation 3.26. Lactic acid can be utilized as a substrate by *Rhodobacter* sp. Theoretically, 1 mol of lactic acid can produce the theoretical maximum of 12 mol of hydrogen as shown in the equations below:

$$\text{Dark fermentation: 1 glucose} \rightarrow 2 \text{ lactic acid} + OH_2 \quad (3.26)$$

$$\text{Photofermentation: 2 lactic acid} \rightarrow 12H_2 \quad (3.27)$$

3.5.1.2 Acetic Acid Fermentation Integrated with Photofermentation

Glucose ferments to produce 4 mol of hydrogen per mol of glucose. This is the theoretical maximum moles of hydrogen that can be produced from glucose. Acetic acid can be utilized in the second stage by reverse electron flow by *Rhodobacter* sp. to produce 8 mol of hydrogen. This is an attractive process and can theoretically produce the maximum hydrogen yield of 12 mol of hydrogen per mole of glucose.

$$\text{Dark fermentation: 1 glucose} \rightarrow 2 \text{ acetic acid} + 4H_2 \qquad (3.28)$$

$$\text{Photofermentation: 2 acetic acid} \rightarrow 8H_2 \qquad (3.29)$$

3.5.1.3 Mixed Acid Fermentation

Most of the bacteria including obligate microorganism like *Clostridia* or enteric like *Enterobacter* undergo mixed acid fermentation producing primarily acetic acid with butyric acid and ethanol. Spent media containing a mixture of volatile fatty acids can also be used as substrate for the second stage photofermentation as shown below:

$$\text{Dark fermentation: 1 glucose} \rightarrow 1 \text{ acetic acid} + 1 \text{ butyric acid}$$
$$+ 1 \text{ ethanol} + 2H_2 \qquad (3.30)$$

$$\text{Photofermentation: 2 acetic acid} \rightarrow 8H_2 \qquad (3.31)$$

Recently, Chen et al. (2008a) developed a two-stage process combining dark and photofermentation to increase the overall hydrogen yield from sucrose and also to reduce the chemical oxygen demand (COD) in the effluent. Dark-H_2 fermentation was carried out using *Clostridium pasteurianum* CH4. It produced a maximum H_2 production yield of 3.80 mol H_2/ mol sucrose. They used the spent media from the dark fermentation stage primarily comprising of acetate and butyrate for a subsequent second stage photofermentation. Using soluble products from dark fermentation as substrate, *Rhodopseudomonas palustris* WP3-5 could produce H_2 phototrophically, elevating the total hydrogen yield to 10.02 mol H_2/mol sucrose (dark/photofermentation). Meanwhile, a 72.0% COD removal was also achieved. When the photobioreactor was illuminated with side-light optical fibers and was supplemented with 2.0% (w/v) of clay carriers, the overall H_2 yield of the two-stage process was further enhanced to 14.2 mol H_2/mol sucrose with a nearly 90% COD removal. This demonstrates the feasibility of using the two-stage process combining dark and photofermentation for simultaneous hydrogen production and COD removal.

Similarly, Nath et al. (2005) carried out studies on two stage combined dark and photofermentation processes to enhance the total yield of the process. In dark fermentation, hydrogen was produced by *Enterobacter cloacae* strain DM11 using glucose as substrate. This was followed by a photofermentation process. Here, the spent medium from the dark process (containing unconverted metabolites, mainly acetic acid, etc.) underwent photofermentation by *Rhodobacter sphaeroides* strain O.U.001 in a column photobioreactor. This combination could achieve higher yields of hydrogen by complete utilization of the chemical energy stored in the substrate. Dark fermentation was studied in terms of several process parameters, such as initial substrate concentration, initial pH of the medium, and temperature, to establish favorable conditions for maximum hydrogen production. Also, the effects of the threshold concentration of acetic acid, light intensity, and the presence of additional nitrogen sources in the spent effluent on the amount of hydrogen produced during photofermentation were investigated. The light conversion efficiency of hydrogen was found to be inversely proportional to the incident light intensity. In a batch system, the yield of hydrogen in the dark fermentation was about 1.86 mol H_2/ mol glucose; and the yield in the photofermentation was about 1.5–1.72 mol H_2/mol acetic acid. The overall yield of hydrogen in the combined process, considering glucose as the preliminary substrate, was found to be higher than that in a single process.

More recently, Laurinavichene et al. (2010) demonstrated H_2 production in an integrated process utilizing potato homogenate for dark fermentative H_2 production, followed by H_2 photoproduction using purple nonsulfur bacteria. They reported maximal production yields of 11.5 m^3 gas/m^3 culture in the dark fermentation stage and H_2 photoproduction yield of 40 m^3 gas/ m^3 of the fermentation effluent, with a total H_2 yield of 5.6 mol/mol glucose equivalent for the two-stage integrated process.

3.5.2 Integration of Biophotolysis with Dark Fermentative Process and Photofermentation

This is a more hypothetical approach and has only recently been given attention (Figure 3.12). The microalgae are excellent sources of carbohydrates. They are rich in stored starch. This approach considers the utilization of these algae as substrate for dark fermentation and further the utilization of the end metabolites in a regular two-stage photofermentation process. This promises to give yields close to the theoretical maximum.

3.5.3 Integration of Biophotolysis with Photofermentation

Integration of biophotolysis and photofermentation works on a similar principle of integration of dark fermentation with photofermentation. It entails a two-stage process in which in the first stage, biophotolytic production of hydrogen occurs by algae. Further, the spent media rich in organic acids is filtered and further used as the media for photofermentation. Photofermentative bacteria can utilize the electrons present in the organic solvents by reverse electron flow. The technology has already been demonstrated at the pilot scale.

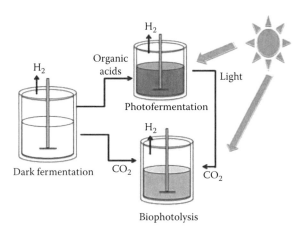

FIGURE 3.12
Hypothetical integrated dark fermentation, photofermentation, and biophotolysis processes to increase the overall process yield.

More than a decade ago, Miura et al. (1997) demonstrated the technology at the pilot scale. They used tubular reactors for indirect biophotolysis using *Chlamydomonas MGA* 161. The accumulated starch was fermented *in situ* to produce organic acids in the natural day and night cycle. The conversion yield of organic compounds from starch of *Chlamydomonas* MGA 161 was 80–100% of the theoretical yield. In a second step, photofermentation was carried out by *Rhodovulum sulfidophilum* W-1S from organic compounds produced by *Chlamydomonas* MGA 161. The molar yield of hydrogen photoproduction of *R. sulfidophilum* W-1S was approximately 40%. The low yield of hydrogen in the second stage was attributed to the competition between the accumulation of intracellular Poly (3-hydroxybutyric acid), PHB, and the hydrogen production under light anaerobic conditions. To overcome the bottleneck, the PHB accumulation was repressed and the hydrogen photoproduction was promoted by enhancing the nitrogenase activity of *R. sulfidophilum* W-1S.

3.5.4 Biohydrogen Production Integrated with Anaerobic Methane Production

This approach aims at integrating hydrogen production with methane production. This is feasible because methane can be produced by acetoclastic bacteria. The acetoclastic bacteria feed on organic acids, while the spent media obtained after biohydrogen production is rich in organic acids. Thus, it is possible to integrate the two technologies in a two-step process: biohydrogen production followed by biomethane production. In fact, this technology is well established and has already been demonstrated at the pilot scale (Cavinato et al., 2012).

HYTHANE ON THE RUN

Recently, San Francisco International Airport, in conjunction with Hythane Company (a subsidiary of Eden Energy Ltd. listed as EDE on the Australian Securities Exchange) and BAF Technologies, received a huge grant to acquire airline crew shuttles to run on Hythane(R) fuel (a blend of 20% hydrogen and 80% natural gas). Concerned about the increasing pollution levels, even a small-scale implementation of hythane fuel would help drastically reduce the pollutants emanating from the existing gasoline- and diesel-powered airport shuttles. It is assumed that the hythane-powered fuel cell will specifically decrease nitrous oxides by 56%, nonmethane hydrocarbons by 30%, and carbon dioxide by 40% over the existing versions. The Hythane(R) model is also expected to outperform comparable natural gas shuttles by emitting 30% less hydrocarbons and 20% less carbon dioxide.

The blend of natural gas and hydrogen is popularly known as hythane. It is a mixture of 20% hydrogen and 80% natural gas. Although less desirable than a pure hydrogen stream, hydrogen/methane mixtures are of some utility since they burn considerably cleaner in an internal combustion engine than methane alone. They release upto 20% lower concentrations of carbon dioxide. In addition, on combustion, it offers significant benefits in the emissions performance and efficiency of engines. These improvements in combustion not only reduce the tailpipe emissions of pollutants but also allow for increased engine speed operation, enabling the downsizing of engines.

3.6 Microbial Electrolysis Cell

In yet another approach, research is going on to integrate the microbial fuel cells to hydrogen production. This is more popularly known as microbial electrohydrogenesis cells (MECs), in which electricity applied to a microbial fuel cell provides the necessary energy to convert organic acids (present in the spent media), to hydrogen (Ditzig et al., 2007; Cheng and Logan, 2007; Rozendal et al., 2008). The versatility of the process is in the fact that the microbial community present in the anodic chamber can completely ferment glucose. Thermodynamic restrictions can be overcome in this process to achieve near stoichiometric yields. MECs primarily employ mixed consortia. The mixed community carries out dark fermentation *in situ*, and the resulting fermentation products are used by the electrogenic members of the community. In MECs, the electrogenic bacteria catabolize their substrate and the electrode (anode) acts as the terminal electron acceptor. Interestingly, to drive hydrogen production at the cathode, a supplementary voltage (>200 mV) is also provided in addition to the voltage generated from the oxidation of the substrate. Thus, in principle, a second stage MEC after an initial fermentative hydrogen stage could completely convert a substrate to hydrogen, achieving 12 mol H_2/mol glucose with only a small investment of electricity (Hallenbeck and Ghosh, 2009). From its innovation, several developments and improvements have been carried out to improve their performance. Currently reported volumetric rates are much higher than those originally obtained. However, as compared to dark fermentations, the rate of hydrogen production in MECs still lags behind. Moreover, additional voltage needs to be supplied to the MECs to generate voltage close to the dark hydrogen fermentation process. However, with the ongoing research at a high pace, these technical barriers may soon be overcome. However, in consideration with all the other technologies, MEC is one of the most promising technologies and has the potential to bring about large-scale commercialization of the process. Thus, substantial challenges to the practical implementation of this promising technology remain, including the replacement of expensive platinum for the cathode, the increase of current densities and the reduction of the electrical input required (Hallenbeck and Ghosh, 2009).

3.7 Thermodynamic Limitations

If biohydrogen production is to become a commercial reality, the thermodynamic limitation of the processes must be overcome. Due to this thermodynamic limitations the theoretical yield from glucose is restricted to 4 mol per mole glucose. However, for commercialization, around 10 mol H_2 per mole glucose are required (Benemann, 2008). The theoretical limit of to 4 mol H_2 per mole glucose is today known as the "Thauer limit" after it was first proposed by Thauer and coworkers in 1977. The group also showed that in practice only 2–3 mol of hydrogen could be recovered from glucose. After nearly 50 years of research in this field, we find that "Thauer limit" still exists due to which large-scale commercialization of the technology has been prevented. Thus, only 20–25% of the energy can be recovered as gaseous energy, the rest is processed by the cell in the release of other reduced end products like alcohol, butyric acid, acetic acid, propionic acid, and so on. The release of some of these competing metabolites drastically reduces the hydrogen yield. However, bioethanol and biomethane, which have already received significant recognition as biofuels, show nearly 90% energy conversion from glucose in a process that is already commercialized.

The "Thauer limit" is imposed due to thermodynamic limitations of the process. The highest Gibbs-free energy is obtained during the conversion of glucose to 2 mol of acetate and 4 mol of molecular hydrogen. The Gibbs-free energy for the reaction is −215 kJ under STP. This free energy is utilized in generating ATP for the cell. However, besides acetate, other pathways like butyrate can also generate ATP. However, which pathway the organism prefers depends on the availability of free energy.

During molecular hydrogen formation, NADH or Fd are also employed as electron carriers to hydrogenase. Fermentation of glucose generates, NADH and Fd as the principle-reducing equivalents. The midpoint redox potentials of NAD^+/NADH and oxidized ferredoxin/reduced ferredoxin are −320 and −398 mV, respectively. To maintain the continuity of the process, the reducing equivalents must be recycled. Recycling of these electron carriers can be accomplished by a variety of reactions, like the production of ethanol by reduction of pyruvate by NADH, but the likelihood of these reactions occurring is determined by the standard Gibbs-free energy change ($\Delta G0'$). Researchers have shown that, under conditions where glucose is not a limiting substrate, bacteria prefer to undergo mixed acid fermentation. This preference is primarily due to enhanced growth with mixed acid fermentation as compared to slow but efficient growth (in terms of ATP) with production of only acetate.

Under standard conditions the reduction of protons with NADH, resulting in molecular hydrogen, has a redox potential of the couple H^+/H_2 being −414 mV. This potential is higher than that of NAD^+/NADH couple. Therefore, formation of molecular hydrogen by oxidation of NADH is thermodynamically constrained. Comparatively, oxidation of ferredoxin to produce molecular hydrogen is more favorable.

In case of hydrogen production by thermophiles, the Thauer limit of 4 H_2 and 2 acetates/glucose is typically approached. This is possible owing to the higher temperature of operation where H_2 production is more favorable by a large entropic factor (also accounting for its higher redox potential, allowing use of NADPH as reductant, as noted above).

For commercialization of the process, nearly 10 mol of glucose are required. Stoichiometrically, a mole of glucose can generate 12 mol of hydrogen. However, the Gibbs-free energy of the process is extremely low: only ~26 kJ/mol glucose, about ~2 kJ/mol H_2, at 25°C, 1 bar of H_2 and physiological pH and bicarbonate concentrations. Thus, this provides insufficient driving force for the reaction, which thus does not take place. However, in an interesting *in vitro* experiment conducted by Woodward and coworkers (2000), they converted glucose-6-phosphate was converted to 11.6 mol of H_2 by action of the enzymes of the pentose phosphate pathway, which generates NADPH, and the hydrogenase of the hyperthermophile *Pyrococcus furiosus*, which is reduced by NADPH (Woodward et al., 2000). However, as presented by Benemann (2008), this yield is not possible from glucose. In fact, the authors have used an energy-rich substrate, Glucose-6-phosphate and possibly operated the system at low partial pressures. Thus, for such a reaction to be feasible, at a high rate and, most important, at a high partial pressure of H_2 (as would be present in a practical fermenter), a much greater energy investment would be required than that contained in the glucose-6-phosphate substrate (equivalent to only ~1 kJ/H$_2$).

Glossary

ATP	Adenosine triphosphate
Bph	Bacteriopheophytin
COD	Chemical oxygen demand
CpI hydrogenase	*Clostridium pasteuranium* hydrogenase
Cyt	Cytochrome
ETC	Electron transport chain
Fd	Ferredoxin
FeS	Iron–sulfur clusters
FNR	Ferredoxin NADP reductase
Hup$^-$	Hydrogenase uptake mutant
LHV	Lower-heating value
MEC	Microbial electrohydrogenesis cell
NADP$^+$	Nicotinamide adenine dinucleotide phosphate
NADPH	Nicotinamide adenine dinucleotide phosphate dehydrogenase
Ndh	NAD(P) dehydrogenase
OLR	Organic-loading rate
PFOR	Pyruvate ferredoxin oxidoreductase
PMF	Proton motive force
PQ	Plastoquinone
PSI	Photosystem I
PSII	Photosystem II
TCA	Tricarboxylic acid cycle

References

Abreu, A., Karakashev, D., Angelidaki, I., Sousa, D. Z., and Alves, M. M. 2012. Biohydrogen production from arabinose and glucose using extreme thermophilic anaerobic mixed cultures. *Biotechnology for Biofuels*, 5, 6.

Abreu, A. A., Alves, J. I., Pereira, M. A., Karakashev, D., Alves, M. M., and Angelidaki, I. 2010. Engineered heat treated methanogenic granules: A promising biotechnological approach for extreme thermophilic biohydrogen production. *Bioresource Technology*, 101, 9577–86.

Akkerman, I., Janssen, M., Rocha, J., and Wijfels, R. H. 2002. Photobiological hydrogen production: Photochemical efficiency and bioreactor design. *International Journal of Hydrogen Energy*, 27, 1195–1208.

Alvira, P., Tomas-Pejo, E., Ballesteros, M., and Negro, M. J. 2009. Pretreatment technologies for an efficient bioethanol production process based on enzymatic hydrolysis: A review. *Bioresource Technology*, 101, 4851–4861.

Aro, E., Virgin, I. and Ersson, B. 1993. Photoinhibition of photosystem II. Inactivation, protein damage and turnover. *Biochemical and Biophysical Acta-Bioenergetics*, 11, 113–134.

Asada, Y. and Miyake, J. 999. Review: Photobiological hydrogen production. *Journal of Bioscience Bioengeneering*, 88, 1–6.

Baebprasert, W., Lindblad, P. and Incharoensakdi, A. 2010. Response of H_2 production and Hox-hydrogenase activity to external factors in the unicellular cyanobacterium *Synechocystis* sp. strain PCC 6803. *International Journal of Hydrogen Energy*, 35, 6611–6616.

Basak, N. and Das, D. 2006. The prospect of purple non-sulfur (PNS) photosynthetic bacteria for hydrogen production: The present state of the art. *World Journal of Microbiology and Biotechnology*, 23, 31–42.

Benemann, J. R. 1996. Hydrogen biotechnology: Progress and prospects. *Nature Biotechnology*, 14, 1101–1103.

Benemann, J. R. 2008. Biological production of H_2: Mechanisms and processes. *Trends in biotechnology*, 3, 337–360.

Benemann, J. R., Berenson, J. A., Kaplan, N. O., and Kamen, M. D. 1973. Hydrogen evolution by chloroplast-ferredoxin-hydrogenase system. *Proceedings of Natural Academy Science*, 70, 2317–2320.

Benemann, J. R. and Pedroni, P. M. 2008. Biological production of H_2: Mechanisms and processes. *Energy Carriers*, 3, 337–360.

Bingham, A. S., Smith, P. R. and Swartz, J. R. 2012. Evolution of an [FeFe] hydrogenase with decreased oxygen sensitivity. *International Journal of Hydrogen Energy*, 37, 2965–2976.

Blankenship, R. E. 1985. Electron transport in green photosynthetic bacteria. *Photosynthesis Research*, 6, 317–333.

Blankenship, R. E. 1992. Origin and early evolution of photosynthesis. *Photosynthesis Research*, 33, 91–111.

Boicelli, C. A., Conti, F., Giomini, M., and Guiliani, A. M. 1982. The influence of phosphate buffers on the 31P longitudinal relaxation time in inverted micelles. *Spectrochemistry Acta Part A Molecular Biomolecular Spectroscopy*, 38, 299–300.

Bolton J. R. 1996. Solar photoproduction of hydrogen, Technical report, IEA Agreement on the production and utilization of hydrogen, Springfield (VA), NTIS.

Boran, E., Özgür, E., Yücel, M., Gündüz, U., and Eroglu, I. 2010. Biological hydrogen production of tubular photo bioreactor by *Rhodobacter capsulatus* on a real dark fermenter effluent of thick juice. *Journal of Cleaner Production*, 18, 29–35.

Boran, E., Özgür, E., Yücel, M., Gündüz, U., and Eroglu, I. 2012. Biohydrogen production by *Rhodobacter capsulatus* Hup mutant in pilot solar tubular photobioreactor. *International Journal of Hydrogen Energy*, 37, 16437–16445.

Boyer, M., Stapleton, J. A., Kuchenreuther, J. M., Wang, C. and Swartz, J. R. 2008. Cell-free synthesis and maturation of hydrogenases. *Biotechnology and Bioengeneering*, 99, 59–67.

Brecht, M., van Gastel, M., Buhrke, T., Friedrich, B. and Lubitz, W. 2003. *Journal of American Chemical Society*, *125*, 13075–13083.

Cai, J. L., Wang, G. C., Li, Y. C., Zhu, D. L. and Pan, G. H. 2009. Enrichment and hydrogen production by marine anaerobic hydrogen producing Microflora. *Chinese Science Bulletin*, *54*, 2656–2661.

Cakir, A., Ozmihci, S. and Kargi F. 2010. Comparison of biohydrogen production from hydrolyzed wheat starch by mesophilic and thermophilic dark fermentation. *International Journal of Hydrogen Energy*, *35*, 13214–13218.

Carlozzi, P. and Sacchi, A. 2001. Biomass production and studies on *Rhodopseudomonas palustris* grown in an outdoor, temperature controlled, underwater tubular photobioreactor. *Journal of Biotechnology*, *88*, 239–249.

Carvalheiro, F., Duarte, L. C. and Gírio, F. M. 2008. Hemicellulose biorefineries: A review on biomass pretreatments. *Journal of Scientific and Industrial Research*, *67*, 849–864.

Cavinato, C., Giuliano, A., Bolzonella, D., Pavan, P., and Cecchi, F. 2012. Bio-hythane production from food waste by dark fermentation coupled with anaerobic digestion process: A long-term pilot scale experience. *International Journal of Hydrogen Energy*, *37*, 11549–11555.

Chen, C. Y. and Chang, J. S. 2006. Enhancing phototrophic hydrogen production by solid-carrier assisted fermentation and internal optical-fiber illumination. *Process Biochemistry*, *41*, 2041–2049.

Chen, S. D., Lee, K. S., Lo, Y. C. et al., 2008c. Batch and continuous biohydrogen production from starch hydrolysate by *Clostridium* species. *International Journal of Hydrogen Energy*, *33*, 1803–1812.

Chen, C. Y., Saratale, G. D., Lee, C. M., Chen., P. C. and Chang J. S. 2008b. Phototrophic hydrogen production. *International Journal of Hydrogen Energy*, *33*, 6886–6895.

Chen, C. Y., Yang, M. H., Yeh, K. L., Liu, C. H., and Chang, J. S. 2008a. Biohydrogen production using sequential two-stage dark and photo fermentation processes, *International Journal of Hydrogen Energy*, *33*, 4755–4762.

Chen, H., Yokthongwattana, K., Newton, A. and Melis, A. 2003. SulP, a nuclear gene encoding a putative chloroplast-targeted sulfate permease in *Chlamydomonas reinhardtii*. *Planta*, *218*, 98–106.

Cheng, J., Su, H., Zhou, J., Song, W., and Cen, K. 2011. Hydrogen production by mixed bacteria through dark and photo-fermentation. *International Journal of Hydrogen Energy*, *36*, 450–457.

Cheng, S. and B. E. Logan. 2007. Sustainable and efficient biohydrogen production via electrohydrogenesis. *Proceedings of National Academy of Sciences, USA*, *104*, 18871–18873

Cheong, D. Y. and Hansen, C. L. 2006. Bacterial stress enrichment enhances anaerobic hydrogen production in cattle manure sludge. *Applied Microbiology Biotechnology*, *72*, 635–643.

Classen, P. A. M., van Groenestijn J. W., Janssen, A. J. H., van Niel, E. J. W. and R. H. Wijffels. 2000. Feasibility of biological hydrogen production from biomass for utilization in fuel cells. *Proceedings of the 1st World Conference and Exhibition on Biomass for Energy and Industry*, Sevilla, Spain.

Conner, C. J. O. and Viggnais, P. M. 1988. Ordered water structure and enhanced reactivity. *Biocatalysis*, *1*, 249–256.

Conrad, R., Phelps, N. R. and Zeikus, J. G. 1985. Gas metabolism evidence in support of the juxtaposition of hydrogen producing and methanogenic bacteria in sewage sludge and lake sediments. *Applied Environment Microbiology*, *50*, 595–601.

Das, D., Khanna, N. and Veziroglu, T. N. 2008. Recent developments in biological hydrogen production processes. *Chemical Industry Chemical Engineering Quarterly*, *14*, 57–67.

Das, D. and Veziroglu, T. N. 2001. Hydrogen production by biological process: A survey of literature, *International Journal of Hydrogen Energy*, *26*, 13–28.

Delachapelle, S., Renaud, M. and Vignais, P. M. 1991. Hydrogen production in bioreactor by a photosynthetic bacterium *Rhodobacter capsulatus*, Photobioreactor and optima conditions of hydrogen production. *Revue des Sciences de lEau*, *4*, 83–99.

Ditzig, J., Liua, H. and Logan, B. E. 2007. Production of hydrogen from domestic wastewater using a bioelectrochemically assisted microbial reactor. *International Journal of Hydrogen Energy*, *32*, 2296–2304.

Dutta, D., De, D., Chaudhuri, S. and Bhattacharya, S. 2005. Hydrogen production by Cyanobacteria. *Microbial Cell Factories*, *4*, 36.

Eggeman, T. 2004. Boundary analysis for H_2 production by fermentation, Final report, National Renewable Energy Laboratory, Neoterics International.

Eroglu, I., Aslan, K., Gündüz, U., Yücel, M., and Türker, L. 1999. Substrate consumption rates for hydrogen production by *Rhodobacter sphaeroides* in a column photobioreactor. *Journal of Biotechnology*, *70*, 103–113.

Eroglu, E., Gunduz, U., Yucel, M., and Eroglu, I. 2010. Photosynthetic bacterial growth and productivity under continuous illumination or diurnal cycles with olive mill wastewater as feedstock. *International Journal of Hydrogen Energy*, *35*, 5293–5300.

Faeder, J. and Ladanyi, B. M. 2000. Molecular dynamics simulations of the interior of aqueous reverse micelles. *Journal of Physical Chemistry*, *104*, 1033–1046.

Fan, Y. T., Zhang, Y. H., Zhang, S. F., Hou, H. W., and Ren, B. Z. 2006. Efficient conversion of wheat straw wastes into biohydrogen gas by cow dung compost. *Bioresource Technology*, *97*, 500–505.

Fang, H. H. P. and Liu, H. 2002. Effect of pH on hydrogen production from glucose by a mixed culture. *Bioresource Technology*, *82*, 87–93.

Forestier, M., King, P., Zhang, L. et al., 2003. Expression of two [Fe]-hydrogenases in *Chlamydomonas reinhardtii* under anaerobic conditions. *European Journal of Biochemistry*, *270*, 2750–2758.

Francou, N. and Vignais, P. M. 1984. Hydrogen production by *Rhodopseudomonas capsulata* cells entrapped in carrageenan beads. *Biotechnology Letters*, *6*, 639–644.

Gaffron, H. 1939. Reduction of carbon dioxide with molecular hydrogen in green algae. *Nature*, *143*, 204.

Gaffron, H. and Rubin, J. 1942. Fermentative and photochemical production of hydrogen in algae. *Journal of General Physiology*, *26*, 219–240.

Gajjar, L., Singh, A., Dubey M. and Srivastava, R. C. 1997. Enzymic activity of the whole cells entrapped into reversed micelles-studies on alpha amylase and invertase in the entrapped yeast cells. *Applied Biochemistry and Biotechnology*, *66*, 159–172 .

Gebicki, J., Modigell, M., Schumacher, M., van der Burg, J., and Roebroeck, E. 2010. Comparison of two reactor concepts for anoxygenic H_2 production by *Rhodobacter capsulatus*. *Journal of Cleaner Production*, *18*, 36–42.

Ghirardi, M. L., Zhang, L., Lee, J. W. et al, 2000. Microalgae: A green source of renewable H_2. *Trends in Biotechnology*, *18*, 506–511.

Gilbert, J. J., Ray, S. and Das, D. 2011. Hydrogen production using *Rhodobacter sphaeroides* (O.U.001) in a flat panel rocking photobioreactor. *International Journal of Hydrogen Energy*, *36*, 3434–3441.

Gogotov, I. N., Zorin, N. A. and Serebriakova, L. T. 1991. Hydrogen production by model systems including hydrogenase from phototrophic bacteria. *International Journal of Hydrogen Energy*, 16, 393–396.

Goldet, G., Brandmayr, C., Stripp, S. T. et al., 2009. Electrochemical kinetic investigations of the reactions of [FeFe]-hydrogenases with carbon monoxide and oxygen: Comparing the importance of gas tunnels and active-site electronic/redox effects. *Journal of the American Chemical Society*, 131(41), 14979–89.

Hallenbeck, P. C. 1983. Immobilized microorganisms for hydrogen and ammonia production. *Enzyme Microbial Technology*, 5, 171–180.

Hallenbeck, P. C. and Ghosh, D. 2009. Advances in fermentative biohydrogen production: The way forward? *Trends in Biotechnology*, 27, 287–297.

Happe, T., Hemschemeier, A., Winkler, M., and Kaminski, A. 2002. Hydrogenases in green algae: Do they save the algae's life and solve our energy problems? *Trends in Plant Science*, 7, 246–250.

Happe, T. and Kaminski, A. 2002. Differential regulation of the Fe-hydrogenase during anaerobic adaptation in the green alga *Chlamydomonas reinhardtii. European Journal of Biochemistry*, 269, 1022–1032.

Happe, T. and Naber J. D. 1993. Isolation, characterization and N-terminal amino acid sequence of hydrogenase from the green alga *Chlamydomonas reinhardtii. European Journal of Biochemistry*, 214, 475–481.

Harmsen, P., Huijgen, W., Bermudez, L., and Bakker, R. 2010. Literature review of physical and chemical pretreatment processes for lignocellulosic biomass. *Biosynergy Report*, 1184, 1–54.

Hatch, M. D. and Slack, C. R. 1970. Photosynthetic CO_2-fixation pathways. *Annual Review in Plant Physiology*, 21, 141–162.

Heda, G. D. and Madigan M. T. 1986. Utilization of amino acids and lack of diazotrophy in the thermophilic anoxygenic phototroph *Chloroflexus aurantiacus. Journal of General Microbiology*, 132, 2469–2473.

Hendriks, W. M. and Zeeman, G. 2008. Pretreatments to enhance the digestibility of lignocellulosic biomass. *Bioresource Technology*, 100, 10–18.

Hill, R. and Bendall, F. A. Y. 1960. Function of the two cytochrome components in chloroplasts: A working hypothesis. *Nature*, 186(4719), 136.

Holo, H. and Sirevåg, R. 1986. Autotrophic growth and CO_2 fixation *of Chloroflexus aurantiacus. Archives in Microbiology*, 145, 173–180.

Horecker, B. L. 2002. The pentose phosphate pathway. *Journal of Biological Chemistry*, 277, 47965–47971.

Hugler, M., Wirsen, C. O., Fuchs, G., Taylor, C. D. and Sievert, S. M. 2005. Evidence for Autotrophic CO_2 fixation via the reductive tricarboxylic acid cycle by members of the ε subdivision of proteobacteria. *Journal of Bacteriology, 187*, 3020–3027.

Imhoff, J. 2004. Taxonomy and physiology of phototrophic purple bacteria and green sulfur bacteria. In *Anoxygenic Photosynthetic Bacteria*, eds. R. E. Blankenship, M. T. Madigan, and C. E. Bauer, pp. 1–15. The USA: Kluwer Academic Publishers.

Kampf, C. and Pfennig, N. 1980. Capacity of chromatiaceae for chemotrophic growth. Specific respiration rates of *Thiocystis violacea* and *Chromatium vinosum. Archives in Microbiology*, 127, 125–135.

Kapdan, I. K. and Kargi, F. 2006. Bio-hydrogen production from waste materials. *Enzyme Microbiology Technology*, 38, 569–582.

Kennedy, R. A., Rumpho, M. E. and Fox, T. C. 1992. Anaerobic metabolism in plants. *Plant Physiology, 100,* 1–6.

Keskin, T., Abo-Hashesh, M. and Hallenbeck, P. C. 2011. Photofermentative hydrogen production from wastes. *Bioresource Technology, 102,* 8557–68.

Khanna, N., Kotay, S. M., Gilbert, J. J., and Das, D. 2011. Improvement of biohydrogen production by *Enterobacter cloacae* IIT-BT 08 under regulated pH. *Journal of Biotechnology, 152,* 15–30.

Kima, J. P., Kang, C. D., Park, T. Y., Kim, M. S., and Sim, S. J. 2006. Enhanced hydrogen production by controlling light intensity in sulfur-deprived *Chlamydomonas reinhardtii* culture. *International Journal of Hydrogen Energy, 31,* 1585–1590.

Knappe, J. and Sawers, G. 1990. A radical-chemical route to acetyl-CoA: The anaerobically induced pyruvate formate-lyase system of *Escherichia coli. FEMS Microbiology Letters, 75,* 383–398.

Kok, B. 1973. Photosynthesis. *Proceedings of the Workshop on Biosolar Hydrogen Conversion,* Bethesda, MD, pp. 22–30.

Koku, H., Gunduz, U., Yucel, M., and Turke, L. 2003. Kinetics of biological hydrogen production by the photosynthetic bacterium *Rhodobacter sphaeroides* O.U.001. *International Journal of Hydrogen Energy, 28,* 381–388.

Kondratieva, E. N. 1979. Interrelation between modes of carbon assimilation and energy production in phototrophic purple and green bacteria. In *Microbial Biochemistry,* ed. J. R. Quayle, pp. 117–175. Baltimore: University of Park Press.

Kosourov, S., Tsygankov, A., Seibert, M., and Ghirardi, M. L. 2002. Sustained hydrogen photoproduction by *Chlamydomonas reinhardtii:* Effects of culture parameters. *Biotechnology and Bioengineering, 78,* 731–740.

Kumazawa, S. and Mitsui, A. 1982. Hydrogen metabolism of photosynthetic bacteria and algae. In *CRC Handbook of Biosolar Resources,* ed. O. R. Zaborsky, pp. 299–316. Boca Raton: CRC press.

Kyle, D. J., Ohad, I. and Arntzen, C. J. 1984. Membrane protein damage and repair: Selective loss of a quinone-protein function in chloroplast membranes. *Natural Academy of Sciences, 81,* 4070–4074.

Laurinavichene, T. V., Belokopytov, B. F., Laurinavichius, K. S., Tekucheva, D. N., Seibert, M., and Tsyganko, A. A. 2010. Towards the integration of dark and photo-fermentative waste treatment: Potato as substrate for sequential dark fermentation and light-driven H_2 production. *International Journal of Hydrogen Energy, 35,* 8536–8543.

Lee, J. and Greenbaum, E. 2003. A new oxygen sensitivity and its potential application in photosynthetic H_2 production. *Applied Biochemistry and Biotechnology, 106,* 303–313.

Lemon, B. J. and Peters, P. W. 1999. Binding of exogenously added carbon monoxide at the active site of the iron-only hydrogenase (CpI) from *Clostridium pasteurianum. Biochemistry, 38,* 12969–12973.

Lin, C. Y. and Jo, C. H. 2003. Hydrogen production from sucrose using an anaerobic sequencing batch reactor process. *Journal of Chemical Technology and Biotechnology, 78,* 678–684.

Lubitz, W., Reijerse, E. and van Gastel, M. 2007. [NiFe] and [FeFe] hydrogenases studied by advanced magnetic resonance techniques. *Chemical Reviews, 107,* 4331–4365.

Luisi, P. L. and Magid, L. J. 1986. Solubilization of enzymes and nucleic acids in hydrocarbon micellar solutions. *CRC Critical Reviews in Biochemistry, 20,* 409–474.

Madigan, M. 2004. Microbiology of nitrogen fixation by anoxygenic photosynthetic bacteria. In *Anoxygenic Photosynthetic Bacteria*, eds. R. E. Blankenship, M. T. Madigan and C. E. Bauer, pp. 915–928. The USA: Kluwer Academic Publishers.

Maness, P. C., Yu, J., Eckert, C., and Ghirardi, M. L. 2009. Photobiological hydrogen production—prospects and challenges. *Microbe*, 4, 275–280.

Markov, S. A. 1999. *Bioreactors for Hydrogen Production Biohydrogen*. The USA, Springer.

Masukawa, M., Mochimaru, M. and Sakurai, H. 2002. Disruption of the uptake hydrogenase gene, but not of the bidirectional hydrogenase gene, leads to enhanced photobiological hydrogen production by the nitrogen-fixing cyanobacterium *Anabaena* sp. PCC 7120. *Applied Microbiology and Biotechnology*, 58, 618–624.

McCord, J., Keele, M. and Fridovich, I. 1971. An enzyme-based theory of obligate anaerobiosis: The physiological function of superoxide dismutase. *Natural Academy of Sciences*, 68, 1024–1027.

McEwan, A. G. 1994. Photosynthetic electron transport and anaerobic metabolism in purple non-sulfur phototrophic bacteria. *Antonie van Leeuwenhoek*, 66,151–164.

McTavish, H., Sayavedra-Soto, L. A. and Arp, D. J. 1995. Substitution *of Azotobacter vinelandii* hydrogenase small-subunit cysteines by serines can create insensitivity to inhibition by O_2 and preferentially damages H_2 oxidation over H_2 evolution. *Journal of Bacteriology*, 177, 3960–3964.

Melis, A. 2002. Green alga hydrogen production: Progress, challenges and prospects. *International Journal of Hydrogen Energy*, 27, 1217–1228.

Melis, A., Zhang, L., Forestier, M., Ghirardi, M. L., and Seibert, M. 2000. Sustained photobiological hydrogen gas production upon reversible inactivation of oxygen evolution in the green alga *Chlamydomonas reinhardtii*. *Plant Physiology*, 122, 127–136.

Meyer, J., Kelley, B. C. and Vignais, P. M. 1978. Effect of light on nitrogenase function and synthesis in *Rhodopseudomonas capsulata*. *Journal of Bacteriology*, 136, 201–208.

Michel, H. and Deisenhofer, H. 1988. Relevance of the photosynthetic reaction center from purple bacteria to the structure of photosystem II. *Biochemistry*, 27, 1–7.

Mitsui, A. 1975. The utilization of solar energy for hydrogen production by cell-free system of photosynthetic organisms. *Paper presented at the Miami Energy Conference*, Miami Beach, Fla. New York.

Miura, Y., Akano, T., Fukatsu, K. et al., 1997. Stably sustained hydrogen production by biophotolysis in natural day/night cycle. *Energy Conservation Management*, 38, 533–537.

Miyake, J. and Kawamura, S. 1987. Efficiency of light energy conversion to hydrogen by photosynthetic bacteria *Rhodobacter sphaeroides*. *International Journal of Hydrogen Energy*, 12, 147–149.

Modigell, M. and Holle, N. 1998. Reactor development for a biosolar hydrogen production process. *Renewable Energy*, 14, 421–426.

Mosier, N., Wyman, C., Dale, B. et al., 2005. Features of promising technologies for pretreatment of lignocellulosic biomass. *Bioresource Technology*, 96, 673–686.

Munekage, Y., Hashimoto, M., Miyake, C. et al., 2004. Cyclic electron flow around photosystem I is essential for photosynthesis. *Nature*, 429, 579–582.

Mus, F., Cournac, L., Cardettini, V., Caruana, A., and Peltier, G. 2005. Inhibitor studies on non-photochemical plastoquinone reduction and H_2 photoproduction in *Chlamydomonas reinhardtii*. *Biochemistry and Biophysics Acta Bioenergetics*, 170, 322–332.

Nath, K. and Das, D. 2004. Improvement of fermentative hydrogen production: Various approaches. *Applied Microbiology and Biotechnology*, 65, 520–529.

Nath, K. and Das, D. 2008. Effect of light intensity and initial pH during hydrogen production by an integrated dark and photofermentation process. *International Journal of Hydrogen Energy*, *34*, 7497–7501.

Nath, K., Kumar, A. and Das, D. 2005. Hydrogen production by *Rhodobacter sphaeroides* strain O.U.001 using spent media of *Enterobacter cloacae* strain DM11. *Applied Microbiology and Biotechnology*, *68*, 533–541.

Nath, K., Kumar, A. and Das, D. 2006. Effect of some environmental parameters on fermentative hydrogen production by *Enterobacter cloacae* DM11. *Canadian Journal of Microbiology Research*, *532*, 525–532.

National Research Council—Committee on alternatives and strategies for future hydrogen production and use. 2004. *The Hydrogen Economy: Opportunities, Costs, Barriers and R&D Needs*. Washington, DC: National Academy Press.

Nicolet, Y., De Lacey, A. L., Vernede, X., Fernandez, V. M., Hatchikian, E. C., and Fontecilla-Camps, J. C. 2001. Crystallographic and FTIR spectroscopic evidence of changes in Fe coordination upon reduction of the active site of the Fe-only hydrogenase from *Desulfovibrio desulfuricans*. *Journal of American Chemical Society*, *123*, 1596–1601.

Nultsch, W. and Agel, W. 1986. Fluence rate and wavelength dependence of photobleaching in the cyanobacterium *Anabaena variabilis*. *Archives in Microbiology*, *144*, 268–271.

Pandey, A. S., Harris, T. V., Giles, L. J., Peters, J. W., and Szilagyi, R. K. 2008. Dithiomethylether as a ligand in the hydrogenase H-cluster. *Journal of American Chemical Society*, *130*, 4533–4540.

Pandey, A. and Pandey, A. 2008. Reverse micelles as suitable microreactor for increased biohydrogen production. *International Journal of Hydrogen Energy*, *33*, 273–278.

Pandu, K. and Joseph, S. 2012. Comparisons and limitations of biohydrogen production processes: A review. *International Journal of Advances in Engineering Technology*, *2*, 342–356.

Peters, J. W., Lanzilotta, W. N., Lemon, B. J., and Seefeldt, L. C. 1998. X-ray crystal structure of the Fe-only hydrogenase (CpI) from *Clostridium pasteurianum* to 1.8 angstrom resolution. *Science*, *282*, 1853–1858.

Polle, J. E. W., Kanakagiri, S. D. and Melis, A. 2003. Tla1, a DNA insertional transformant of the green alga *Chlamydomonas reinhardtii* with a truncated light-harvesting chlorophyll antenna size. *Planta*, *217*, 49–59.

Prince, R. C. and Kheshgi, H. S. 2005. The Photobiological production of hydrogen: Potential efficiency and effectiveness as a renewable fuel. *Critical Reviews in Microbiology*, *31*, 19–31.

Prokop, A. and Erickson, L. E.. 1994. Photobioreactors. In *Bioreactor System Design*, eds. J. A. Asenjo and J. C. Merchuk, pp. 441–477. New York: Marcel Dekker, Inc.

Ren, N., Li, J., Li, B., Wang, Y., and Liu, S. 2006. Biohydrogen production from molasses by anaerobic fermentation with a pilot-scale bioreactor system. *International Journal of Hydrogen Energy*, *31*, 2147–2157.

Roseboom, W., Lacey, A. L., Fernandez, V. M., Hatchikian, E. C., and Albracht, S. P. J. 2006. The active site of the [FeFe]-hydrogenase from *Desulfovibrio desulfuricans*. II. Redox properties, light sensitivity, and CO-ligand exchange as observed by infrared spectroscopy. *Journal of Biological and Inorganic Chemistry*, *11*, 102–118.

Rozendal, R. A., Jeremiasse, A. D., Hamelers, H. V. M., and Buisman, C. J. N. 2008. Hydrogen production with a microbial biocathode. *Environmental Science and Technology*, *42*, 629–634.

Silakov, A., Reijerse, E. J., Albracht, S. P. J., Hatchikian, E. C., and Lubitz, W. 2007. The electronic structure of the H-cluster in the [FeFe]-hydrogenase from *Desulfovibrio desulfuricans*: A Q-band Fe-57-ENDOR and HYSCORE study. *Journal of American Chemical Society, 129*, 11447–11458.

Singh, A., Pandey, K. D. and Dubey, R. S. 1999a. Enhanced hydrogen production by coupled system of *H. halobium* and chloroplast after entrapment within reverse micelles. *International Journal of Hydrogen Energy, 24*, 693–698.

Singh, A., Pandey, K. D. and Dubey, R. S. 1999b. Reverse micelles: A novel tool for H_2 production. *World Journal of Microbiology and Biotechnology, 15*, 277–282.

Sirevåg, R. and J. G. Ormerod. 1977. Synthesis, storage and degradation of polyglucose in *Chlorobium thiosulfatophilum*. *Archives in Microbiology, 111*, 239–244.

Stephenson, M. and Stickland, L. H. 1932. Hydrogenlyases. *Journal of Biochemistry, 26*, 712–724.

Stripp, S. T., Goldet, G., Brandmayr, C. et al., 2009b. How oxygen attacks [FeFe] hydrogenases from photosynthetic organisms. *Proceedings of National Academy Sciences, USA, 106*, 1–6.

Stripp, S., Sanganas, O., Happe, T. and Haumann, M. 2009a. The structure of the active site H-cluster of [FeFe] hydrogenase from the green alga *Chlamydomonas reinhardtii* studied by X-ray absorption spectroscopy. *Biochemistry, 48*, 5042–5049.

Strobel, H. J. 1993. Evidence for catabolite inhibition in regulation of pentose utilization and transport in the ruminal bacterium *selenomonas-ruminantium*. *Applied Environmental Microbiology, 59*, 40–46.

Tabita, F. R. 1999. Microbial ribulose 1,5-bisphosphate carboxylase/oxygenase: A different perspective. *Photosynthesis Research, 60*, 1–28.

Taherzadeh, M. J. and Karimi, K. 2008. Pretreatment of lignocellulosic wastes to improve ethanol and biogas production: A review. *International Journal of Molecular Sciences, 9*, 1621–1651.

Tamagnini, P., Axelsson, R., Lindberg, P., Oxelfelt, F., Wunschiers, R., and Lindblad, P. 2002. Hydrogenases and hydrogen metabolism of cyanobacteria. *Microbiol Molecular Biology Reviews, 66*, 1–20.

Tard, C. Liu, X., Ibrahim, S. K. et al., 2005. Synthesis of the H-cluster framework of iron-only hydrogenase. *Nature, 433*, 610–613.

Tekucheva, D. N. and Tsygankov, A. A. 2012. Combined biological hydrogen-producing systems: A review. *Applied Biochemistry and Microbiology, 48*, 319–337.

Torzillo, G., Carlozzi, P. and Chini Zitelli, R. 1993. A vertical alveolar panel (VAP) for outdoor mass cultivation of microalgea and cyanobacteria. *Bioresource Technology, 38*, 153–159.

Torzillo, G., Pushparaj, B., Bocci, W., Balloni, R., and Materassi, G. 1986. Production of spirulina biomass in closed photobioreactor. *Biomass, 11*, 61–74.

Tsygankov, A. A., Hall, D. O., Liu, J., and Rao, K. K. 1994. An automated helical photobioreactor incorporating cyanobacteria for continuous hydrogen production. In *Biohydrogen*, ed. O. R. Zaborsky, pp. 431–440. London: Plenum Press.

Tsygankov, A., Kosourov, S., Seibert, M. and Ghirardi, M. L. 2002. Hydrogen photoproduction under continuous illumination by sulfur-deprived, synchronous *Chlamydomonas reinhardtii* cultures. *International Journal of Hydrogen Energy, 27*, 1239–1244.

Uyeda, K. and Rabinowitz, J. C. 1971. Pyruvate-ferredoxin oxidoreductase. *Journal of Biology and Chemistry, 246*, 3111–3119

Vijayaraghavan, K., Karthik, R. and Kamala, S. P. 2009. Hydrogen production by *Chlamydomonas reinhardtii* under light driven sulfur deprived condition. *International Journal of Hydrogen Energy, 34*, 7964–7970.

Vincenzini, M., Marchini, A., Ena, R., and De Philippis. 1997. H_2 and polyhydroxybutyrate, two alternative chemicals from purple non-sulfur bacteria. *Biotechnology Letters, 19*, 759–762.

Watanabe, Y., Morita, M. and Saiki, S. M. 1998. Photosynthetic CO_2 fixation performance by a helical tubular photobioreactor incorporating *Chlorella* sp. under outdoor culture conditions. In *Studies in Surface Science and Catalysis*, eds. T. Inui, M. Anpo, K. Izui, S. Yanagida, and T. Yamaguchi, pp. 483–486. Amsterdam: Elsevier.

Wang, Y. Z., Liao, Q., Zhu, X., Tian, X., and Zhang, C. 2010. Characteristics of hydrogen production and substrate consumption of *Rhodopseudomonas palustris* CQK 01 in an immobilized-cell photobioreactor. *Bioresource Technology, 101*, 4034–4041.

Wijffels, R. H. and Barten, H. 2003. Bio-methane and bio-hydrogen. In *Energy*, eds. R. H. W. and H. B. J. H. Reith. The Netherlands: Dutch Biological Hydrogen Foundation, c/o Energy research Centre, pp. 165.

Woodward, J., Orr, M., Cordray, K., and Greenbaum, E. 2000. Enzymatic production of biohydrogen. *Nature, 405*, 1014–1015.

Wykoff, D. D., Davies, J. P., Melis, A., and Grossman, A. R. 1998. The regulation of photosynthetic electron transport during nutrient deprivation in *Chlamydomonas reinhardtii*. *Plant Physiology, 117*, 129–139.

Yetis, M., Gunduz, U., Eroglu, I., Yucel, M., and Turker, L . 2000. Photoproduction of hydrogen from sugar refinery wastewater by *Rhodobacter sphaeroides* O.U.001. *International Journal of Hydrogen Energy, 25*, 1035–1041.

Yu, J. and P. Takahashi. 2007. Biophotolysis-based hydrogen production by cyanobacteria and green microalgae. *Trends in Applied Microbiology, 1*, 79–89.

Zeikus, J. G. 1977. The biology of methanogenic bacteria. *Bacteriology Reviews, 41*, 514–541.

Zhu, H., Suzuki, T., Tsygankov, A. A., Asada, Y., and Miyake, J. 1999. Hydrogen production from tofu wastewater by *Rhodobacter sphaeroides* immobilized in agar gels. *International Journal of Hydrogen Energy, 24*, 305–310.

Zitelli, G., Lavista, F., Bastianini, A., Rodolfi, L., Vincenzini, M., and Tredici, M. R. 1999. Production of eicosapentaenoic acid by *Nannochloropsis* sp. cultures in outdoor tubular photobioreactors. *Journal of Biotechnology, 70*, 299–312.

4

Biohydrogen Feedstock

4.1 Introduction

Hydrogen can be produced from a wide spectrum of feedstocks (Figure 4.1). Almost 80% of investigation on hydrogen production by dark fermentation was from pure sugars: mono, di, or polysaccharides. However, the cost-effective biohydrogen should be produced from renewable feedstocks (Show et al., 2012). The potential feedstocks could be biomass of certain woody plants, aquatic plants, and algae. Furthermore, they could be the agricultural waste by-products, waste from food processing, livestock effluents, and other industrial waste. In this section, different types of feedstocks are discussed in terms of their availability, applicability, and operational challenges, as well as the motivation for their use in fermentative hydrogen production. Different types of feedstocks with their concentration use for the fermentation along with the process parameters, yield, and rate of evolution of H_2 have been presented in Table 4.1.

4.2 Simple Sugars as Feedstock

Simple sugars such as the mono and disaccharides are mostly used as model substrate for the research to standardize the fermentation process. Sugars contain molecules of carbon, hydrogen, and oxygen in their backbone.

The widely found monosaccharides in nature are pentose and hexose sugars, commonly found in plant and animal-based feedstocks. The most common 5-carbon sugars, xylose, and arabinose are also present in fruits. Ribose is the part of the backbone of RNA and available in all plants. Among the 6-carbon monosaccharides, glucose is the simplest one found in all plants, galactose is found in milk and some plants, fructose is found in fruit and honey, and mannose is in some fruits. Disaccharide sugars include sucrose (a combination of glucose and fructose), which is obtained from sugar cane or sugar beets, maltose (two glucose molecules) present in germinating grain and in a small

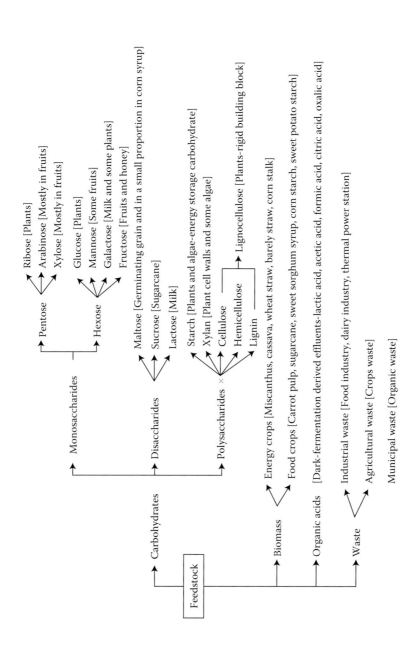

FIGURE 4.1
Feedstock used for biohydrogen production and their probable sources.

TABLE 4.1

Different Types of Feedstocks for Hydrogen Production are Tabulated with Respective Process Parameters, Yield, and Rate of Evolution of H_2

Substrate Type	Substrate Conc. (g/L)	Organism	Reactor Type	Temp.	pH	Hydrogen Yield (mol H_2/mol Glucose, Hexose or Equivalent)	H_2 Production Rate (L/L/d)	Reference
Simple sugar								
Glucose	31	*Caldicellulosiruptor saccharolyticus*	Batch	72	7	2.8	8.8	Mars et al. (2010)
	20	*Caldicellulosiruptor saccharolyticus*	Batch	70	7	3.4	6.5	de Vrije et al. (2010)
	10	*Caldicellulosiruptor saccharolyticus*	Batch	72	7	3.4	6.4	Mars et al. (2010)
	27	*Thermotoga neapolitana* DSM 4359	Batch	75	7	3	8.5	Ngo et al. (2012)
	10	*Thermotoga neapolitana* DSM 4359	Batch	75	7	2.9	8.43	Ngo et al. (2012)
	2.5	*Thermotoga neapolitana* DSM 4359	Batch	85, 77, 70, 65	7.5	3.75, 3.85, 3.18, 3.09, 2.04	0.54, 0.56, 0.32, 0.19, 0.033	Munro et al. (2009)
	10	*Clostridium* DMHC-10	Batch	37	5	3.35	2.14	Kamalaskar et al. (2010)
	10	*Enterobacter Cloacae* IIT-BT08	Batch	36	6	2.2		Kumar and Das (2000)
	10	*Enterobacter Cloacae* DM11	Batch	37	6.5	3.31		Nath et al. (2006)

continued

TABLE 4.1 (continued)

Different Types of Feedstocks for Hydrogen Production Are Tabulated with Respective Process Parameters, Yield and Rate of Evolution of H_2

Substrate Type	Substrate Conc. (g/L)	Organism	Reactor Type	Temp.	pH	Hydrogen Yield (mol H_2/mol Glucose, Hexose or Equivalent)	H_2 Production Rate (L/L/d)	Reference
	3	*Clostridium beijerinckii* L9	Batch	35	7.2	2.81	1.9	Lin et al. (2007)
	10	*Clostridium beijerinckii* Fanp 3	Batch	35	6.47–6.98	2.52	9.36	Pan et al. (2008)
	10	*Clostridium beijerinckii* AM21B	Batch	36	6.5	2	15.84	Taguchi et al. (1992)
	3	*Clostridium butyricum* ATCC19398	Batch	36	7.2	2.29	1.52	Lin et al. (2007)
	10	*Clostridium butyricum* EB6	Batch	37	6	0.6	4.128	Chong et al. (2009)
	15.7	*Clostridium butyricum* EB6	Batch	37	5.6	2.2		Chong et al. (2009)
	3	*C. butyricum* and *E. coli*	Batch	37	6.5	1.65	0.52	Seppala et al. (2011)
	10	*Thermoanaerobacterium thermosaccharolyticum* PSU-2	Batch	60	6.5	2.53		O-Thong et al. (2008)
	10	*Ethanoligenens harbinese* YUAN-3	Batch	35	5	1.91	1.66	Xing et al. (2008)
	10	*Ethanoligenens harbinese* YUAN-3	CSTR	35	5	1.93	19.6	Xing et al. (2008)
	9, 12, and 6	*Ethanoligenens harbinese* B49	Batch	37	7	1.83, 1.71, and 1.36		Liu et al. (2009)

14.5	*Ethanoligenens harbinese* B49	Batch	35	6	2.2		Guo et al. (2009)
10	*Ethanoligenens harbinese* B49	Batch	35	6	2.26	9.91	Xu et al. (2008)
10, 20, and 20	*Pantoea agglomerans*	Batch	37	7.2	3.8, 4.2, and 3.3	1.82, 0.78, and 0.61	Zhu et al. (2008)
20	*Clostridium tyrobutyricum* JM1	Batch	37	6.3	3.24		Jo et al. (2008)
5	*Clostridium tyrobutyricum* JM1	CSTR	37	6.7	1.81	7.21	Jo et al. (2008)
50	*Clostridium tyrobutyricum* ATCC 25755	Fed batch	37	5.7	2.33	16.7	Mitchell et al. (2009)
3	*Clostridium acetobutylicum* M121	Batch	37	7.2	1.8	1.42	Lin et al. (2007)
10	*Clostridium acetobutylicum* ATCC 824	Cont. Trickling bed reactor	30	6.2	0.9	5.34	Zhang et al. (2006)
5	*Escherichia coli* S3 and S6	Batch	30	6.8	0.84 and 0.49	0.39 and 0.39	Junyapoon et al. (2011)
15	*Escherichia coli* WDHL	Batch	37	6	0.3	0.45	Rosales-Colunga et al. (2012)
10	*Escherichia coli* DJT135	Batch	35	6.5	1.51		Ghosh and Hallenbeck (2009)
10	*Enterobacter aerogenes* HO-39	Batch	38	6.5	1.0	7.2	Yokoi et al. (1995)
0.2	*Enterobacter aerogenes* ATCC29007	Batch	38	7.0		4.8	Shigeharu and Noriaki (1983)

continued

TABLE 4.1 (continued)

Different Types of Feedstocks for Hydrogen Production Are Tabulated with Respective Process Parameters, Yield and Rate of Evolution of H_2

Substrate Type	Substrate Conc. (g/L)	Organism	Reactor Type	Temp.	pH	Hydrogen Yield (mol H_2/mol Glucose, Hexose or Equivalent)	H_2 Production Rate (L/L/d)	Reference
	5	Enterobacter aerogenes W23 and Candida maltosa HY-35	Batch	35	6.5	2.19	6.27	Lu et al. (2007)
Sucrose	10	Caldicellulosiruptor saccharolyticus	Batch	75	7	2.96	4.5	Van Niel et al. (2002)
	5	Thermotoga neapolitana DSM 4359	Fed batch-CSABR	75	7.5	2.5	1.3	Ngo et al. (2011)
	10	Enterobacter Cloacae IIT-BT08	Batch	36	6	3.014	15.84	Kumar and Das (2000)
	17.8	Clostridium butyricum CGS5	Batch	37	5.5	1.39	3.9	Chen et al. (2005)
	22.3	Clostridium butyricum TISTR 1032	Batch	37	6.5	1.34	3.11	Plangklang et al. (2012)
	10	Clostridium butyricum TM-9A	Batch	37	8	1.49		Junghare et al. (2012)
	10	Thermoanaerobacter mathranii A3N	Batch	70	8	2.69	2.4	Jayasinghearachchi et al. (2012)
	10	Escherichia coli DJT135	Batch	35	6.5	0.35		Ghosh and Hallenbeck (2009)

Feedstock	Conc.	Organism	Mode	Temp	pH			Reference
Xylose	5	*Thermotoga neapolitana* DSM 4359	CSABR	75	7	3.36	2.66	Ngo et al. (2012)
	5	*Thermotoga neapolitana* DSM 4359	Batch	75	7.5	1.31	0.4	Ngo et al. (2012)
	5	*Thermotoga neapolitana* DSM 435	Fed batch-CSABR	75	7.5	2.66	2.3	Ngo et al. (2011)
	10	*Clostridium butyricum* TM-9A	Batch	37	8	0.59		Junghare et al. (2012)
	10	*Thermoanaerobacter mathranii* A3N	Batch	70	8	2.5	1.8	Jayasinghearachchi et al. (2012)
	10	*Thermoanaerobacterium thermosaccharolyticum* W16	Batch	60	6.5	2.62	5.71	Ren et al. (2008)
	10	*Thermoanaerobacterium thermosaccharolyticum* W16	Batch	60	7		7.4	Ren et al. (2010)
	12.24	*Thermoanaerobacterium thermosaccharolyticum* W16	Batch	60	6.8	2.84		Cao et al. (2010)
	10	*Escherichia coli* DJT135	Batch	35	6.5	0.68		Ghosh and Hallenbeck (2009)
	5	*Enterobacter aerogenes* IAM 1183	Batch	37	6.3	2.64	5.97	Ren et al. (2009)
Arabinose	10	*Clostridium butyricum* TM-9A	Batch	37	8	0.06		Junghare et al. (2012)
	10	*Escherichia coli* DJT135	Batch	35	6.5	1.2		Ghosh and Hallenbeck (2009)
	10	*Enterobacter aerogenes* IAM 1183	Batch	37	6.3	1.56	2.88	Ren et al. (2009)

continued

TABLE 4.1 (continued)

Different Types of Feedstocks for Hydrogen Production Are Tabulated with Respective Process Parameters, Yield and Rate of Evolution of H_2

Substrate Type	Substrate Conc. (g/L)	Organism	Reactor Type	Temp.	pH	Hydrogen Yield (mol H_2/mol Glucose, Hexose or Equivalent)	H_2 Production Rate (L/L/d)	Reference
Galactose	10	*Clostridium butyricum* TM-9A	Batch	37	8	0.86		Junghare et al. (2012)
	10	*Escherichia coli* DJT135	Batch	35	6.5	0.69		Ghosh and Hallenbeck (2009)
	10	*Enterobacter aerogenes* IAM 1183	Batch	37	6.3	2.82	6.96	Ren et al. (2009)
Fructose	10	*Clostridium butyricum* TM-9A	Batch	37	8	0.84		Junghare et al. (2012)
	10	*Escherichia coli* DJT135	Batch	35	6.5	1.27		Ghosh and Hallenbeck (2009)
Mannose	10	*Clostridium butyricum* TM-9A	Batch	37	8	0.67		Junghare et al. (2012)
	25	*Enterobacter aerogenes* IAM 1183	Batch	37	6.3	0.96	12.4	Ren et al. (2009)
Ribose	10	*Clostridium butyricum* TM-9A	Batch	37	8	0.84		Junghare et al. (2012)
Complex substrate								
Starch	10	*Clostridium beijerinckii* AM21B	Batch	36	6.5	1.8	9.84	Taguchi et al. (1992)

	Feedstock	Microorganism	Mode					Reference
10		Clostridium beijerinckii RZF-1108	Batch	36 and 41	6.8	2.0 and 2.6	8.64 and 9.48	Taguchi et al. (1994)
		Enterobacter aerogenes and C. butyricum	Batch and Repeated batch	36	6.5 and 5.5			Yokoi et al. (1998)
10	Cellulose (micro-crystalline)	Ethanoligenens harbinese B49 and C. acetobutylicum X9	Batch	37	5	1.32	11.08	Wang et al. (2008)
4, 3, and 2	Cellulose	Clostridium thermocellum ATCC 27405	CSTR	60	7	1.29, 1.53 and 1.65	0.67, 0.5 and 0.4	Magnusson et al. (2009)
Biomass								
10	Pretreated wheat straw	Caldicellulosiruptor saccharolyticus	Batch	70	7.2	3.8		Ivanova et al. (2009)
20	Pretreated barely straw	Caldicellulosiruptor saccharolyticus	Batch	70	7		0.6	Panagiotopoulos et al. (2012)
10	Carrot pulp hydrolysate	Caldicellulosiruptor saccharolyticus	Batch	70	7	2.8	8.4	de Vrije et al. (2010)
10 and 20		Thermotoga neapolitana DSM 4359	Batch	75	7	2.7 and 2.4	8.6 and 6.2	de Vrije et al. (2010)
10, 14, and 28	Miscanthus hydrolysate	Caldicellulosiruptor saccharolyticus	Batch	72	7	3.4, 3.3, and 2.4	8.64, 7.13, and 4.25	de Vrije et al. (2009)
10, 14, and 28		Thermotoga neapolitana DSM 4359	Batch	80	7	2.9, 3.2, and 2	9.0, 8.4, and 3.7	de Vrije et al. (2009)
20 g COD/L	Sugarcane bagasse hemicellulose hydrolysate	Clostridium butyricum EB6	Batch	37	5.5	1.73 Mol H_2/mol total sugar	1.611	Pattra et al. (2008)

continued

TABLE 4.1 (continued)

Different Types of Feedstocks for Hydrogen Production Are Tabulated with Respective Process Parameters, Yield and Rate of Evolution of H_2

Substrate Type	Substrate Conc. (g/L)	Organism	Reactor Type	Temp.	pH	Hydrogen Yield (mol H_2/mol Glucose, Hexose or Equivalent)	H_2 Production Rate (L/L/d)	Reference
Corn hydrolyzed		*Thermoanaerobacterium thermosaccharolyticum* W16	Batch	60	7	2.24 Mol H_2/mol total sugar		Cao et al. (2009)
		Enterobacter aerogenes NCIMB 10102	Continuous packed column	40	5.5	2.55	6.3	Palazzi et al. (2000)
	14.2	*Clostridium butyricum* CGS5	Batch	37	7.5	0.84	1.5	Lo et al. (2010)
Pretreated straw hydrolysate	9.2	*Clostridium butyricum* CGS5	Batch	37	7.5	0.91	0.64	Lo et al. (2010)
Chlorella vulgaris ESP6 (microalgal hydrolysate)	9.0	*Clostridium butyricum* CGS5	Batch	37	5.5		5.8	Liu et al. (2012)
Sugarcane juice	22.3	*Clostridium butyricum* TISTR 1032	Batch	37	6.5	1.33	3	Plangklang et al. (2012)
Organic acid (Corn stover fermentation effluent)		*Rhodobacter sphaeroides*		30	7			Zhu et al. (2010)

Waste		Microorganism		Temp	pH			Reference
Corn stalk waste	10	*Clostridium thermocellum and Clostridium thermosaccharolyticm*	Batch and CSTR	55	7.2		0.34 and 0.44	Li and Lin (2012)
Molasses	100	*Clostridium butyricum W5*	Batch	37	7	1.63		Wang et al. (2008)
	100	*Clostridium butyricum W5*	Batch	39	6.5	1.85	11.9	Wang and Jin (2009)
Cassava wastewater	5 g COD/L	*Clostridium acetobutylicum ATCC 824*	Batch	36	7	2.41	1.32	Cappelletti et al. (2011)
Palm oil mill effluent		*Clostridium butyricum EB6*	Batch	37	5.5	0.22	24.8	Chong et al. (2009)

proportion in corn syrup, and lactose (a combination of glucose and galactose) mostly found in milk and to a lesser extent in milk-derived dairy products.

They are readily biodegradable, easy to study the bacterial kinetics, assess adequate nutrients, and identify the optimized operational parameters for the process (Pan et al., 2008; Xu et al., 2008; Kamalaskar et al., 2010; Niu et al., 2010; Zhao et al., 2011). It needs short time for the fermentation process for production of hydrogen (Lee et al., 2003; Ferchichi et al., 2005). The control experiments assessed the fermentation preferences of the bacteria among the different types of sugars. In majority of studies, researchers utilized pure carbohydrates as substrate, including monosaccharides (glucose, xylose, fructose, arabinose, mannose, and ribose) and disaccharides (sucrose, cellobiose, maltose, and lactose). Glucose, sucrose, and lactose are the most commonly used as pure substrates due to their ease biodegradability, relatively simple structures, and are present in real industrial effluents. However, pure carbohydrate sources are expensive raw materials for real-scale hydrogen production (Hu and Chen, 2007; Hafez et al., 2010). There is very limited study that has been done on pure monosaccharides such as glucose (Horiuchi et al., 2001) and other simple sugars such as sucrose, xylose, lactose, maltose, and so on. Hydrogen conversion capability from different substrates is related to the characteristic of microbial species. For example, Ferchichi et al. (2005) found that one of the most important acetone–butanol–ethanol (ABE)-generating industrial microorganism *Clostridium saccharoperbutylacetonicum* ATCC 27021 when grown on disaccharides produces 2.81 mol H_2/mol-sugar, which is more than twice as hydrogen produced by the monosaccharides (1.29 mol H_2/mol-hexose). So it is not always true that bacteria prefer monosaccharides than disaccharides. The hydrogen production also varies from substrate to substrate (2.2 mol H_2/mol-glucose, 6 mol H_2/mol-sucrose, and 5.4 mol H_2/mol-cellobiose) as reported by Kumar and Das (2000) working with immobilized *Enterobacter cloacae* IIT-BT08. Panagiotopoulos et al. (2010) conducted four experiments to evaluate the hydrogen production by *Caldicellulosiruptor saccharolyticus* from a mixture of glucose, sucrose, and xylose. They observed that, in fermentations of 10 g/L glucose and 10 g/L fructose, *C. saccharolyticus* could virtually completely consume all substrates with almost identical rates of consumptions. In contrast, *Thermotoga neapolitana* consumption trend was different for glucose than fructose, suggesting a preference for glucose. However, in higher substrate concentrations, lactate production increased dramatically at the expense of hydrogen production (de Vrije et al., 2009). To determine the fermentative behavior of the *Thermoanaerobacterium thermosaccharolyticum* W16, a set of control experiments supplemented with glucose, xylose, and a mixture of glucose, xylose, and arabinose at a fixed total sugar quantity of 10 g/L were undertaken. The concentrations of glucose, xylose, and arabinose in the mixture were at the same levels as found in the corn stover hydrolysate. It was observed that in contrast to the above-described experiments the bacterium showed preference for glucose over the other types of sugars. The bacterium grew well on the hydrolysate and reached a similar optical density and maximum hydrogen production rate as on simulated medium, although hydrogen yield was slightly higher on hydrolysate (Ren et al., 2010).

4.3 Complex Substrates as Feedstock

In real life mostly the complex sugars (polysaccharides) are available, to provide relevant insight into the feedstock utilization. It is indispensable to employ these substrates in the dark fermentation (Lo et al., 2010; Cheng and Liu, 2011). The processes of H_2 production can only be viable when based on renewable feedstocks, and low-cost sources include polysaccharides (starch, cellulose, and xylan).

Starch and cellulose are very large polymer molecules composed of many hundreds or thousands of glucose molecules (polysaccharides). Starch is an energy storage carbohydrate used by plants as an energy source. Although cellulose is also composed of glucose sugar molecules, cellulose is a rigid building block molecule used to make cell walls for the leaves, stems, stalks, and woody portions of plants. Xylan is a type of highly complex polysaccharide found in plant cell walls and in some algae. Xylans are polysaccharides made from units of xylose (a pentose sugar). Xylans are almost as ubiquitous as cellulose in plant cell walls and contain predominantly β-D-xylose units linked as in cellulose. Other than the more easily degradable feedstocks such as starch and cellulose, the main components of future feedstocks will, most probably to a large extent, be derived from lignocellulosic raw materials. Lignocellulose is a biopolymer consisting of tightly bound lignin, cellulose, and hemicellulose. Whereas cellulose and hemicellulose can be feedstocks for hydrogen fermentation, lignin is not degraded under anaerobic conditions. Moreover, lignin strongly hampers the utilization of cellulose and hemicellulose because (a) the bonding in lignocellulose resists mobilization and (b) chemically degraded lignin is often inhibitory to microbial growth. In Table 4.1, in this section, the H_2 yield and rate of production from polysaccharides are tabulated. From where it can be observed that the co-culture of different bacteria can be more effective to use the starchy material to produce H_2. As for example, *Enterobacter aerogenes* and *Clostridium butyricum* are more efficient than mono-cultures for yielding H_2 from starch (Yokoi et al., 1998).

4.4 Biomass Feedstock

Biomass could greatly enhance production of hydrogen. Since biomass is renewable and consumes atmospheric CO_2 during growth, it can have a small net CO_2 impact compared to fossil fuels. However, the yield of hydrogen is low from biomass since the hydrogen content in biomass is low to begin with (approximately 6% versus 25% for methane), and the energy content is low due to the 40% oxygen content of biomass (Milne et al., 2002). Biomass can be converted to H_2 by different conversion technologies to break the complex sugar to simple one through thermochemical (gasification, high pressure aqueous, and pyrolysis) and biological routes. Biological route of H_2 production includes anaerobic digestion and fermentation.

In most recent studies, sugar-containing crops (e.g., sugar cane, sweet sorghum and sugar beet), starch-based crops (e.g., corn and wheat), lignocellulosics (e.g., plant materials, fodder grass, and *Miscanthus*), and food industry by-products are mostly used for fermentation process (Yokoi et al., 2002; de Vrije et al., 2009; de Vrije et al., 2010; Lo et al., 2010; Panagiotopoulos et al., 2010; Ren et al., 2010). However, use of food crops for energy production is publically not acceptable as it will increase the crisis of food. In view of that, energy crops (e.g., wheat straw, barley straw, corn stalk, *Miscanthus*, and cassava) are employed as feedstock for energy generation (Elsharnouby et al., 2013). They are commonly referred to as second-generation cellulosic biomass. However, some energy crops compete with food crops for land. Microalgae would be a better option for energy production as they utilize lesser space for biomass production. Additionally, utilizing the waste of industries includes food industry (e.g., carrot pulp, potato steam peels, sugarcane waste, sweet sorghum syrup, corn starch, and sweet potato starch) and agricultural waste products (e.g., delignified wood fibers and corn stalk waste) (Elsharnouby et al., 2013).

4.5 Organic Acids

In fermentation process huge amount of organic acids have been produced and could be used as the substrate for photofermentative bacteria. It includes the mixer of lactic acid, acetic acid, formic acid, citric acid, oxalic acid, and so on. Combination of dark- and photofermentation has shown to be the most efficient method to produce hydrogen through fermentation (Das and Veziroglu, 2001). One of the main bottlenecks is that fermentation effluent contains excess nitrogenous compounds and ammonia, which inhibit nitrogenase activity by wild-type photosynthetic bacteria (Redwood and Macaskie, 2006). This problem can be solved by downregulation of nitrogenase in response to nitrogen excess (Zinchenko et al., 1997) or selective electroseparation of organic acids from an active fermentation (Redwood et al., 2012). The energetic cost of electroseparation of organic acids was found to be acceptable in a combined fermentation.

4.6 Waste as Feedstock

Biohydrogen generation from agricultural wastes is slowly gearing up and there is a lot of scope to work on these wastes and get patents in this area. We can also concentrate more on generating biohydrogen from various industrial wastewaters, which not only helps in getting renewable energy source but also helps in the reduction of the toxic effects of the wastewater on the receiving land or water bodies.

The food manufacturing industry produces high-strength organic wastes or wastewater. These are potential and important feedstock for biohydrogen production as they contain concentrated carbohydrates, mainly simple sugar, starch, and cellulose. Various studies have been conducted to investigate the feasibility of biohydrogen production using waste as feedstock including agricultural waste (Noike and Mizuno, 2000; Hussy et al., 2005), food and beverage industries waste, as for example, bean curd, noodle manufacturing, sugar factory, rice winery (Ueno et al., 1996; Mizuno et al., 2000; Yu et al., 2002), molasses (Tanisho and Ishiwata, 1995), cheese whey (Khanna et al., 2011), dehydrated brewery mixture (Fan and Chen, 2004), starch manufacturing (Yokoi et al., 2002), and organic wastewater (Show et al., 2010). Complete solid waste, such as waste from kitchen, and municipal waste (OFMSW) have also been tested as feedstock for fermentative hydrogen production (Ntaikou et al., 2010). These wastes usually have quite high contents of carbohydrates, proteins, and fats, and thus hydrogen production is comparatively lower than those obtained from carbohydrate-based waste. But it is found to be cost-effective waste than others (Lay and Fan, 2003).

4.7 Assessment of Cost Components for Several Feedstocks for Dark Hydrogen Fermentation

Use of carbohydrate as a source for producing hydrogen is often expensive, which can only be viable when based on renewable and low-cost sources (Das and Veziroglu, 2001). Production of hydrogen by biotransformation of wastes and wastewater is economical as well as environment friendly (Ntaikou et al., 2010). Thermodynamically, conversion of carbohydrates to hydrogen and organic acids is preferred as it yields the maximum amount of hydrogen per mole of substrate. A wide range of carbohydrates ranging from simple monosaccharide to complex polymers such as starch, cellulose, or xylan may be used as raw materials. The dark hydrogen production itself has a great potential as recognized by many workers in this field. Noike and Mizuno (2000) and Yu et al. (2002) used several forms of organic waste streams ranging from solid wastes like rice straw to wastewater from a sugar factory and rice winery to generate hydrogen through dark fermentation. Besides carbohydrate and other organic substrates, CO in syngas has been successfully used as feedstock for biological H_2 production. Syngas or fuel gas is a mixture of CO and H_2, which can be produced easily and cheaply by thermochemical gasification of coal or wood.

4.8 Conclusion

According to the reviewed literature pure monosaccharides 59%, pure disaccharides 10%, pure polysaccharides 11%, and sustainable feedstocks 20% have been used as feedstock for H_2 production (Elsharnouby et al., 2013). However, the cost-effective production is only possible when using the waste materials. Presently, wheat straw, barely straw, Algal biomass, and different biomass hydorlysate and waste are being used as sustainable feedstock. Hydrogen production from waste is found to be economically more feasible than the other feedstocks.

Glossary

COD	Chemical oxygen demand
CSTR	Continuous stirred-tank reactor
OFMSW	Organic fraction of municipal solid waste
RNA	Ribonucleic acid

References

Cao, G., Ren, N., Wang, A. et al., 2009. Acid hydrolysis of corn stover for biohydrogen production using *Thermoanaerobacterium thermosaccharolyticum* W16. *International Journal of Hydrogen Energy*, 34(17), 7182–7188.

Cao, G. L., Ren, N. Q., Wang, A. J. et al., 2010. Statistical optimization of culture condition for enhanced hydrogen production by *Thermoanaerobacterium thermosaccharolyticum* W16. *Bioresource Technology*, 101(6), 2053–2058.

Cappelletti, B. M., Reginatto, V., Amante, E. R., and Antônio, R. V. 2011. Fermentative production of hydrogen from cassava processing wastewater by *Clostridium acetobutylicum*. *Renewable Energy*, 36(12), 3367–3372.

Chen, W. M., Tseng, Z. J., Lee, K. S., and Chang, J. S. 2005. Fermentative hydrogen production with *Clostridium butyricum* CGS5 isolated from anaerobic sewage sludge. *International Journal of Hydrogen Energy*, 30(10), 1063–1070.

Cheng, X. Y. and Liu, C. Z. 2011. Hydrogen production via thermophilic fermentation of cornstalk by *Clostridium thermocellum*. *Energy and Fuels*, 25(4), 1714–1720.

Chong, M. L., Rahim, R. A., Shirai, Y., and Hassan, M. A. 2009. Biohydrogen production by *Clostridium butyricum* EB6 from palm oil mill effluent. *International Journal of Hydrogen Energy*, 34(2), 764–771.

Das, D. and Veziroglu, T. N. 2001. Hydrogen production by biological processes: A survey of literature. *International Journal of Hydrogen Energy*, 26(1), 13–28.

de Vrije, T., Budde, M. A., Lips, S. J., Bakker, R. R., Mars, A. E., and Claassen, P. A. 2010. Hydrogen production from carrot pulp by the extreme thermophiles *Caldicellulosiruptor saccharolyticus* and *Thermotoga neapolitana*. *International Journal of Hydrogen Energy*, 35(24), 13206–13213.

de Vrije, T., Bakker, R. R., Budde, M. A., Lai, M. H., Mars, A. E., and Claassen, P. A. 2009. Efficient hydrogen production from the lignocellulosic energy crop *Miscanthus* by the extreme thermophilic bacteria *Caldicellulosiruptor saccharolyticus* and *Thermotoga neapolitana*. *Biotechnology for Biofuels*, 2(1), 12.

Elsharnouby, O., Hafez, H., Nakhla, G., and El Naggar, M. H. 2013. A critical literature review on biohydrogen production by pure cultures. *International Journal of Hydrogen Energy*, 38(12), 4945–4966

Fan, K. S. and Chen Y. Y. 2004. H₂ production through anaerobic mixed culture: Effect of batch S-0/X-0 and shock loading in CSTR. *Chemosphere*, 57(9), 1059–1068.

Ferchichi, M., Crabbe, E., Gil, G.H., Hintz, W., and Almadidya, A. 2005. Influence of initial pH on hydrogen production from cheese whey. *Journal of Biotechnology*, 120, 402–409.

Ghosh, D. and Hallenbeck, P. C. 2009. Fermentative hydrogen yields from different sugars by batch cultures of metabolically engineered *Escherichia coli* DJT135. *International Journal of Hydrogen Energy*, 34(19), 7979–7982.

Guo, W. Q., Ren, N. Q., Wang, X. J. et al. 2009. Optimization of culture conditions for hydrogen production by *Ethanoligenens harbinense* B49 using response surface methodology. *Bioresource Technology*, 100(3), 1192–1196.

Hafez, H., Nakhla, G., El Naggar, M. H., Elbeshbishy, E., and Baghchehsaraee, B. 2010. Effect of organic loading on a novel hydrogen bioreactor. *International Journal of Hydrogen Energy*, 35(1), 81–92.

Horiuchi, J. I., Kikuchi, S., Kobayashi, M., Kanno, T., and Shimizu, T. 2001. Modeling of pH response in continuous anaerobic acidogenesis by an artificial neural network. *Biochemical Engineering Journal*, 9(3), 199–204.

Hu, B. and Chen, S. 2007. Pretreatment of methanogenic granules for immobilized hydrogen fermentation. *International Journal of Hydrogen Energy*, 32(15), 3266–3273.

Hussy, I., Hawkes, F. R., Dinsdale, R., and Hawkes, D. L. 2005. Continuous fermentative hydrogen production from sucrose and sugarbeet. *International Journal of Hydrogen Energy*, 30(5), 471–483.

Ivanova, G., Rákhely, G., and Kovács, K. L. 2009. Thermophilic biohydrogen production from energy plants by *Caldicellulosiruptor saccharolyticus* and comparison with related studies. *International Journal of Hydrogen Energy*, 34(9), 3659–3670.

Jayasinghearachchi, H. S., Sarma, P. M., and Lal, B. 2012. Biological hydrogen production by extremely thermophilic novel bacterium *Thermoanaerobacter mathranii* A3N isolated from oil producing well. *International Journal of Hydrogen Energy, 37*(7), 5569–5578.

Jo, J. H., Lee, D. S., and Park, J. M. 2008. The effects of pH on carbon material and energy balances in hydrogen-producing *Clostridium tyrobutyricum* JM1. *Bioresource Technology, 99*(17), 8485–8491.

Junghare, M., Subudhi, S., and Lal, B. 2012. Improvement of hydrogen production under decreased partial pressure by newly isolated alkaline tolerant anaerobe, *Clostridium butyricum* TM-9A: Optimization of process parameters. *International Journal of Hydrogen Energy, 37*(4), 3160–3168.

Junyapoon, S., Buala, W., and Phunpruch, S. 2011. Hydrogen production with *Escherichia coli* isolated from municipal sewage sludge. *Thammasat International Journal of Science and Technology, 16*(1), 9–15.

Kamalaskar, L. B., Dhakephalkar, P. K., Meher, K. K., and Ranade, D. R. 2010. High biohydrogen yielding *Clostridium* sp. DMHC-10 isolated from sludge of distillery waste treatment plant. *International Journal of Hydrogen Energy, 35*(19), 10639–10644.

Khanna, N., Dasgupta, C. N., Mishra, P., and Das, D. 2011. Homologous overexpression of [FeFe] hydrogenase in *Enterobacter cloacae* IIT-BT 08 to enhance hydrogen gas production from cheese whey. *International Journal of Hydrogen Energy, 36*(24), 15573–15582.

Kumar, N. and Das, D. 2000. Enhancement of hydrogen production by *Enterobacter cloacae* IIT-BT 08. *Process Biochemistry, 35*(6), 589–593.

Lay, J. J. and Fan, K. S. 2003. Influence of chemical nature of organic wastes on their conversion to hydrogen by heat-shock digested sludge. *International Journal of Hydrogen Energy, 28*(12), 1361–1367.

Lee, K. S., Lo, Y. S., Lo, Y. C., Lin, P. J., and Chang, J. S. 2003. H_2 production with anaerobic sludge using activated-carbon supported packed-bed bioreactors. *Biotechnology Letters, 25*(2), 133–138.

Li, Q. and Liu, C. Z. 2012. Co-culture of *Clostridium thermocellum* and *Clostridium thermosaccharolyticum* for enhancing hydrogen production via thermophilic fermentation of cornstalk waste. *International Journal of Hydrogen Energy, 37*(14), 10648–10654.

Lin, P. Y., Whang, L. M., Wu, Y. R. et al. 2007. Biological hydrogen production of the genus *Clostridium*: Metabolic study and mathematical model simulation. *International Journal of Hydrogen Energy, 32*(12), 1728–1735.

Liu, C. H., Chang, C. Y., Cheng, C. L., Lee, D. J., and Chang, J. S. 2012. Fermentative hydrogen production by *Clostridium butyricum* CGS5 using carbohydrate-rich microalgal biomass as feedstock. *International Journal of Hydrogen Energy, 37*(20), 15458–15464.

Liu, B. F., Ren, N. Q., Xing, D. F. et al. 2009. Hydrogen production by immobilized *R. faecalis* RLD-53 using soluble metabolites from ethanol fermentation bacteria *E. harbinense* B49. *Bioresource Technology, 100*(10), 2719–2723.

Lo, Y. C., Lu, W. C., Chen, C. Y., and Chang, J. S. 2010. Dark fermentative hydrogen production from enzymatic hydrolysate of xylan and pretreated rice straw by *Clostridium butyricum* CGS5. *Bioresource Technology, 101*(15), 5885–5891.

Lu, W., Wen, J., Chen, Y. et al. 2007. Synergistic effect of *Candida maltosa* HY-35 and *Enterobacter aerogenes* W-23 on hydrogen production. *International Journal of Hydrogen Energy, 32*(8), 1059–1066.

Magnusson, L., Cicek, N., Sparling, R., and Levin, D. 2009. Continuous hydrogen production during fermentation of α-cellulose by the thermophillic bacterium *Clostridium thermocellum*. *Biotechnology and Bioengineering, 102*(3), 759–766.

Mars, A. E., Veuskens, T., Budde, M. A. et al. 2010. Biohydrogen production from untreated and hydrolysed potato steam peels by the extreme thermophiles *Caldicellulosiruptor saccharolyticus* and *Thermotoga neapolitana*. *International Journal of Hydrogen Energy*, 35(15), 7730–7737.

Milne, T. A., Elam, C. C., and Evans, R. J. 2002. *Hydrogen from Biomass: State of the Art and Research Challenges*. Golden, CO: National Renewable Energy Laboratory.

Mitchell, R. J., Kim, J. S., Jeon, B. S., and Sang, B. I. 2009. Continuous hydrogen and butyric acid fermentation by immobilized *Clostridium tyrobutyricum* ATCC 25755: Effects of the glucose concentration and hydraulic retention time. *Bioresource Technology*, 100(21), 5352–5355.

Mizuno, O., Ohara, T., Shinya, M., and Noike, T. 2000. Characteristics of hydrogen production from bean curd manufacturing waste by anaerobic microflora. *Water Science and Technology*, 42(3,4), 345–350.

Munro, S. A., Zinder, S. H., and Walker, L. P. 2009. The fermentation stoichiometry of *Thermotoga neapolitana* and influence of temperature, oxygen, and pH on hydrogen production. *Biotechnology Progress*, 25(4), 1035–1042.

Nath, K., Kumar, A., and Das, D. 2006. Effect of some environmental parameters on fermentative hydrogen production by *Enterobacter cloacae* DM11. *Canadian Journal of Microbiology*, 52(6), 525–532.

Ngo, T. A., Kim, M., and Sim, S. J. 2011. Thermophilic hydrogen fermentation using *Thermotoga neapolitana* DSM 4359 by fedbatch culture. *International Journal of Hydrogen Energy*, 36, 14014–14023.

Ngo, T. A., Nguyen, T. H., and Bui, H. T. V. 2012. Thermophilic fermentative hydrogen production from xylose by *Thermotoga neapolitana* DSM 4359. *Renewable Energy*, 37(1), 174–179.

Niu, K., Zhang, X., Tan, W. S., and Zhu, M. L. 2010. Characteristics of fermentative hydrogen production with *Klebsiella pneumoniae* ECU-15 isolated from anaerobic sewage sludge. *International Journal of Hydrogen Energy*, 35(1), 71–80.

Noike, T. and Mizuno, O. 2000. Hydrogen fermentation of organic municipal wastes. *Water Science and Technology*, 42(12), 155–162.

Ntaikou, I., Antonopoulou, G., and Lyberatos, G. 2010. Biohydrogen production from biomass and wastes via dark fermentation: A review. *Waste and Biomass Valorization*, 1(1), 21–39.

O-Thong, S., Prasertsan, P., Karakashev, D., and Angelidaki, I. 2008. Thermophilic fermentative hydrogen production by the newly isolated *Thermoanaerobacterium thermosaccharolyticum* PSU-2. *International Journal of Hydrogen Energy*, 33(4), 1204–1214.

Palazzi, E., Fabiano, B., and Perego, P. 2000. Process development of continuous hydrogen production by *Enterobacter aerogenes* in a packed column reactor. *Bioprocess Engineering*, 22(3), 205–213.

Pan, C. M., Fan, Y. T., Zhao, P., and Hou, H. W. 2008. Fermentative hydrogen production by the newly isolated *Clostridium beijerinckii* Fanp3. *International Journal of Hydrogen Energy*, 33(20), 5383–5391.

Panagiotopoulos, I. A., Bakker, R. R., de Vrije, T., Claassen, P. A. M., and Koukios, E. G. 2012. Dilute-acid pretreatment of barley straw for biological hydrogen production using *Caldicellulosiruptor saccharolyticus*. *International Journal of Hydrogen Energy*, 37(16), 11727–11734.

Panagiotopoulos, I. A., Bakker, R. R., de Vrije, T., Koukios, E. G., and Claassen, P. A. M. 2010. Pretreatment of sweet sorghum bagasse for hydrogen production by

Caldicellulosiruptor saccharolyticus. International Journal of Hydrogen Energy, 35(15), 7738–7747.

Pattra, S., Sangyoka, S., Boonmee, M., and Reungsang, A. 2008. Biohydrogen production from the fermentation of sugarcane bagasse hydrolysate by *Clostridium butyricum. International Journal of Hydrogen Energy, 33*(19), 5256–5265.

Plangklang, P., Reungsang, A., and Pattra, S. 2012. Enhanced biohydrogen production from sugarcane juice by immobilized *Clostridium butyricum* on sugarcane bagasse. *International Journal of Hydrogen Energy, 37*(20), 15525–15532.

Redwood, M. D. and Macaskie, L. E. 2006. A two-stage, two-organism process for biohydrogen from glucose. *International Journal of Hydrogen Energy, 31*(11), 1514–1521.

Redwood, M. D., Orozco, R. L., Majewski, A. J., and Macaskie, L. E. 2012. An integrated biohydrogen refinery: Synergy of photofermentation, extractive fermentation and hydrothermal hydrolysis of food wastes. *Bioresource Technology, 119*, 384–392.

Ren, N. Q., Cao, G. L., Guo, W. Q. et al. 2010. Biological hydrogen production from corn stover by moderately thermophile *Thermoanaerobacterium thermosaccharolyticum* W16. *International Journal of Hydrogen Energy, 35*(7), 2708–2712.

Ren, N., Cao, G., Wang, A., Lee, D. J., Guo, W., and Zhu, Y. 2008. Dark fermentation of xylose and glucose mix using isolated *Thermoanaerobacterium thermosaccharolyticum* W16. *International Journal of Hydrogen Energy, 33*(21), 6124–6132.

Ren, Y., Wang, J., Liu, Z., Ren, Y., and Li, G. 2009. Hydrogen production from the monomeric sugars hydrolyzed from hemicellulose by *Enterobacter aerogenes. Renewable Energy, 34*(12), 2774–2779.

Rosales-Colunga, L. M., Razo-Flores, E., and De León Rodríguez, A. 2012. Fermentation of lactose and its constituent sugars by *Escherichia coli* WDHL: Impact on hydrogen production. *Bioresource Technology, 111*, 180–184.

Seppala, J. J., Puhakka, J. A., Yli-Harja, O., Karp, M. T., and Santala, V. 2011. Fermentative hydrogen production by *Clostridium butyricum* and *Escherichia coli* in pure and cocultures. *International Journal of Hydrogen Energy, 36*, 10701–10708.

Shigeharu, T. and Noriaki, W. 1983. Biological hydrogen production by *Enterobacter aerogenes. Journal of Chemical Engineering of Japan, 16*, 529–530.

Show, K. Y., Lee, D. J., Tay, J. H., Lin, C. Y., and Chang, J. S. 2012. Biohydrogen production: Current perspectives and the way forward. *International Journal of Hydrogen Energy, 37*(20), 15616–15631.

Show, K. Y., Zhang, Z. P., Tay, J. H. et al. 2010. Critical assessment of anaerobic processes for continuous biohydrogen production from organic wastewater. *International Journal of Hydrogen Energy, 35*(24), 13350–13355.

Taguchi, F., Mizukami, N., Hasegawa, K., Saito-Taki, T., and Morimoto, M. 1994. Effect of amylase accumulation on hydrogen production by *Clostridium beijerinckii,* strain AM21B. *Journal of Fermentation and Bioengineering, 77*(5), 565–567.

Taguchi, F., Takiguchi, S., and Morimoto, M. 1992. Efficient hydrogen production from starch by a bacterium isolated from termites. *Journal of Fermentation and Bioengineering, 73*(3), 244–245.

Tanisho, S. and Ishiwata, Y. 1995. Continuous hydrogen production from molasses by fermentation using urethane foam as a support of flocks. *International Journal of Hydrogen Energy, 20*(7), 541–545.

Ueno, Y., Otsuka, S., and Morimoto, M. 1996. Hydrogen production from industrial wastewater by anaerobic microflora in chemostat culture. *Journal of Fermentation and Bioengineering, 82*(2), 194–197.

Van Niel, E. W. J., Budde, M. A. W., De Haas, G. G., Van der Wal, F. J., Claassen, P. A. M., and Stams, A. J. M. 2002. Distinctive properties of high hydrogen producing extreme thermophiles, *Caldicellulosiruptor saccharolyticus* and *Thermotaga elfii*. *International Journal of Hydrogen Energy*, 27(11), 1391–1398.

Wang, X. and Jin, B. 2009. Process optimization of biological hydrogen production from molasses by a newly isolated *Clostridium butyricum* W5. *Journal of Bioscience and Bioengineering*, 107(2), 138–144.

Wang, A., Ren, N., Shi, Y., and Lee, D. 2008. Bioaugmented hydrogen production from microcrystalline cellulose using cocultured *Clostridium acetobutylicum* and *Ethanoigenens harbinense*. *International Journal of Hydrogen Energy*, 33, 912–917.

Xing, D., Ren, N., Wang, A., Li, Q., Feng, Y., and Ma, F. 2008. Continuous hydrogen production of auto-aggregative *Ethanoligenens harbinense* YUAN-3 under non-sterile condition. *International Journal of Hydrogen Energy*, 33, 1489–1495.

Xu, L., Ren, N., Wang, X., and Jia, Y. 2008. Biohydrogen production by *Ethanoligenens harbinense* B49: Nutrient optimization. *International Journal of Hydrogen Energy*, 33(23), 6962–6967.

Yokoi, H., Maki, R., Hirose, J., and Hayashi, S. 2002. Microbial production of hydrogen from starch-manufacturing wastes. *Biomass and Bioenergy*, 22(5), 389–395.

Yokoi, H., Ohkawara, T., Hirose, J., Hayashi, S., and Takasaki, Y. 1995. Characteristics of hydrogen production by aciduric *Enterobacter aerogenes* strain HO-39. *Journal of Fermentation and Bioengineering*, 80(6), 571–574.

Yokoi, H., Tokushige, T., Hirose, J., Hayashi, S., and Takasaki, Y. 1998. H_2 production from starch by a mixed culture of *Clostridium butyricum* and *Enterobacter aerogenes*. *Biotechnology Letters*, 20(2), 143–147.

Yu, H., Zhu, Z., Hu, W., and Zhang, H. 2002. Hydrogen production from rice winery wastewater in an upflow anaerobic reactor by using mixed anaerobic cultures. *International Journal of Hydrogen Energy*, 27(11), 1359–1365.

Zhang, H., Bruns, M. A., and Logan, B. E. 2006. Biological hydrogen production by *Clostridium acetobutylicum* in an unsaturated flow reactor. *Water Research*, 40(4), 728–734.

Zhao, X., Xing, D., Fu, N., Liu, B., and Ren, N. 2011. Hydrogen production by the newly isolated *Clostridium beijerinckii* RZF–1108. *Bioresource technology*, 102(18), 8432–8436.

Zhu, Z., Shi, J., Zhou, Z., Hu, F., and Bao, J. 2010. Photo-fermentation of *Rhodobacter sphaeroides* for hydrogen production using lignocellulose-derived organic acids. *Process Biochemistry*, 45(12), 1894–1898.

Zhu, D., Wang, G., Qiao, H., and Cai, J. 2008. Fermentative hydrogen production by the new marine *Pantoea agglomerans* isolated from the mangrove sludge. *International Journal of Hydrogen Energy*, 33(21), 6116–6123.

Zinchenko, V., Babykin, M., Glaser, V., Mekhedov, S., and Shestakov, S. 1997. Mutation in ntrC gene leading to the derepression of nitrogenase synthesis in *Rhodobacter sphaeroides*. *FEMS Microbiology Letters*, 147(1), 57–61.

5

Molecular Biology of Hydrogenases and Their Accessory Genes

5.1 Introduction

It is well known that the atmosphere of the prebiotic earth was hydrogen rich. Therefore, it is not unreasonable to assume that the early species were capable of metabolizing molecular hydrogen. However, the capacity of these microorganisms to metabolize molecular hydrogen was discovered only at the end of the 19th century. Later, the hydrogen-metabolizing enzyme was identified and established as hydrogenase by Stephenson and Stickland (1931) (Tamagnini et al., 2002). Most of the current researches on hydrogenase focus on prokaryotes, which have helped establish the structural and functional insight into the enzyme. The enzyme catalyzes the simplest reversible reaction (Equation 5.1), the reversible reductive formation of hydrogen from protons and electrons:

$$2H^+ + 2e^- \rightarrow H_2 \tag{5.1}$$

The direction of the reaction mostly depends on the redox potential of the components which interact with the enzyme. In the presence of an electron acceptor, hydrogenase will act as a H_2 uptake enzyme, while in the presence of an electron donor, the enzyme will produce H_2. The hydrogen uptake reaction results in protons and electrons, which are subsequently used to generate ATP and reducing power while the hydrogen-producing reaction involves the reduction of protons by low oxidation electrons generated by fermentation (Nicolet et al., 2000).

Three main approaches have been used to study hydrogenases, namely, biochemical, genetic, and crystallography (Vignais and Billoud, 2007). During the 1960s and 1970s, hydrogenase enzymes from a variety of organisms were purified and characterized. The year 1980 was the age of recombinant DNA technology. Therefore, during the mid-1980s several hydrogenase genes were cloned and sequenced. This allowed classification of hydrogenase based on their sequence analysis and also rationalized some of the earlier researched properties of these enzymes. In the 1990s knowledge of the gene sequence paved the way for molecular engineering of the structural and the related genes involved in the maturation of the hydrogenases. In the same decade, the first of the crystal structure of [NiFe]-hydrogenase was developed from *Desulfovibrio* (*D.*) *vulgaris* (Higuchi et al., 1987). At present, five crystal structures from [NiFe] hydrogenase, two from [NiFeSe] hydrogenase, and two from [FeFe] hydrogenase have been established. [NiFe] hydrogenase include the enzymes from sulfate-reducing bacteria *D. gigas* (Volbeda et al., 1995, 1996), *D. vulgaris* Miyazaki F. (Higuchi et al., 1997, 1999), *D. fructosovorans* (Volbeda et al., 2002, 2005), *D. desulfuricans* (ATCC 27774) (Matias et al., 2001), and photosynthetic bacterium *Allochromatium vinosum* (Ogata et al., 2010). [NiFeSe]-hydrogenase structures from *Desulfomicrobium baculatum* (Garcin et al., 1999) and *D. vulgaris* Hildenborough (Marques et al., 2010), and [FeFe]-hydrogenase structures from anaerobic soil bacterium *Clostridium pasteurianum* (CpI) (Peters et al., 1998) and sulfate-reducing bacterium *D. desulfuricans* (Nicolet et al., 1999).

5.2 Occurrence of Hydrogenase in Nature

Phylogenetic and genome analysis reveals that [NiFe] and [FeFe] hydrogenases are widely distributed in nature. The [FeFe] hydrogenase is only known to occur in obligate anaerobic bacteria, facultative anaerobe, and a small number of green algae. They are found in subcellular organelles of eukaryotes, namely, hydrogenosomes of protozoa and chloroplasts of green algae. However, its presence in archaea is yet to be identified (Vignais et al., 2001; Vignais and Billoud, 2007; Posewitz et al., 2008). The latter domain comprises two kingdoms, the Euryarchaeota (mesophilic and moderately thermophilic methanogens are the best studied Archaea) and the Crenarchaeota (hyperthermophiles from terrestrial volcanic habitats and submarine volcanic habitats). They are known to contain the [Fe-only] hydrogenase. Contrarily, [NiFe] hydrogenases occur predominately in the prokaryotic cyanobacteria and photosynthetic bacteria. Table 5.1 lists the diversity of hydrogenase-containing bacteria. These enzymes are thus present in all major domains of life, and in a considerable variety of physiological contexts, as discussed later in this chapter. The enzyme, depending on its type and location within the cell, may serve any one or more purposes in the organism including reductant generation, energy conservation, hydrogen oxidation, or disposal of excess electrons.

TABLE 5.1

Diversity of Hydrogenase Present in the Various Organisms Found in Nature

Family	Organism	Hyd Diversity	Accession No.
Archaea			
Euryarchaeota			
Archaeo-globaceae	*Archaeoglobus fulgidus* strain VC-16	*vhtG*	AAB89863
		vhtA	AAB89864
	Archaeoglobus fulgidus strain VC-16	*vhuG*	AAB89871
		vhuA	AAB89872
	Azotobacter chroococcum strain MCD1	*hupS*	CAA37133
		hupL	CAA37134
	Azotobacter vinelandii strain OP	*hoxK*	AAA82505
Methano-sarcinaceae	*Methanosarcina barkeri* strain Fusaro	*frhG2*	CAA74092
	(DSM804)	*frhA2*	CAA74090
	Methanosarcina barkeri strain Fusaro	*echC*	CAA76119
	(DSM804)	*echE*	CAA76121
	Methanosarcina mazei strain Go1	*vhoG*	CAA58113
		vhoA	CAA58114
	Methanosarcina mazei strain Go1	*vhtG*	CAA58176
		vhtA	CAA58177
	Methanothermus fervidus	*mvhG*	AAA72831
		mvhA	AAA72832
Thermo-coccaceae	*Pyrococcus abyssi* strain Orsay	*none*	CAB49780
		none	CAB49779
	Pyrococcus abyssi strain Orsay	*hydD2*	CAB49862
		hydA2	CAB49863
	Pyrococcus abyssi strain Orsay	*cooL-like*	CAB49637
		cooH-like	CAB49635
	Pyrococcus furiosus DSM 3638	*hydD*	CAA53036
		hydA (sulfhydro-genase I)	CAA53037
	Pyrococcus furiosus DSM 3638	*shyD2*	AAF61853
		shyA2 (sulfhydro-genase II)	AAF61854
	Pyrococcus horikoshii strain OT3	*PH1292*	BAA30395
		PH1294	BAA30397

continued

TABLE 5.1   (continued)

Diversity of Hydrogenase Present in the Various Organisms Found in Nature

Family	Organism	Hyd Diversity	Accession No.
Bacteria			
Cyanobacteria			
Nostocaceae	*Anabaena* sp. PCC 7120	*hupS*	AAC79877
	Anabaena sp. PCC 7120	*hupL*	AAC79878
		hoxY	
		hoxH	
	Anabaena variabilis ATCC 29413	*hupS*	CAA73658
		hupL	CAA73659
	Anabaena variabilis ATCC 29413	*hoxY*	CAA55876
		hoxH	CAA55878
	Anabaena variabilis strain IAM M58	*hoxY*	BAB39386
	Nostoc muscorum	*hoxY*	Not sequenced
		hoxH	
	Nostoc punctiforme. PCC 73102	*hupS*	AAC16276
		hupL	AAC16277
Proteobacteria			
Alcaligenaceae	*Alcaligenes hydrogenophilus* strain M50	*hoxB*	AAB49360
		hoxC	AAB49361
	Alcaligenes hydrogenophilus strain M50	*hupS*	AAB25779
		hupL	AAB25780
Bacteroidaceae	*Acetomicrobium flavidum* DSM20663	*hydS,* *hydL*	CAA56463 CAA56464
Bradyrhizo-biaceae	*Bradyrhizobium japonicum* strain USDA110	*hupU*	AAA62627
	USDA 110	*hupV*	AAA62628
	Bradyrhizobium japonicum strain	*hupS*	AAA26218
	USDA 122	*hupL*	AAA26219
Campylo-bacteraceae	*Campylobacter jejuni* NCTC 11168	*hydA*	CAB73521
	Campylobacter jejuni NCTC 11168	*hydB* *hydA2*	CAB73520 CAB73823
Desulfovi-brionaceae	*Desulfovibrio baculatus*	*PhsL (hysA)g*	AAA23375
	Desulfovibrio desulfuricans ATCC 27774	*hydA*	

TABLE 5.1   (continued)

Diversity of Hydrogenase Present in the Various Organisms Found in Nature

Family	Organism	Hyd Diversity	Accession No.
		(small subunit)	AAF43137
	Desulfovibrio fructosovorans	(large subunit)	AAF43138
		hynB	AAA23371
	Desulfovibrio gigas	*hynA*	AAA23372
		hynB	AAA23377i
		hynA	AAA23378i
	Desulfovibrio vulgaris Miyazaki F	*hynB*	AAA23369
		hynA	AAA23370
	Desulfotobacterium dehalogenans	*hydA*	AAF13046
	Desulfomicrobium baculatus DSM1743	*hydB*	AAA23376
		PhsS (hysB)g	
Entero-bacteriaceae	*Citrobacter freundii*	*hyaA*	BAA05929
	Escherichia coli strain K12	*hyaA*	AAA23997
		hyaB	AAA23998
		(H$_2$ase 1)	
	Escherichia coli strain K12	*hybO*	AAC76033
		hybC	AAA21591
	Escherichia coli strain K12	*hycG*	CAA35552
		hycE	CAA35550
		(H$_2$ase 3)	
	Escherichia coli strain K12	*hyfI*	AAB88571
		hyfG	AAB88569
		(H$_2$ase 4)	
Clostridiaceae	*Clostridium acetobutylicum* ATCC 824	*hyaB*	BAA05930
		hupS	
Ralstoniaceae	*Ralstonia eutropha* strain H16	*hoxB*	AAB49364
	ATCC 17699	*hoxC*	AAB49365
	Ralstonia eutropha strain H16	*hoxK*	AAA16461
	(ATCC 17699)	*hoxG*	AAA16462
	Ralstonia eutropha strain H16	*hoxY*	AAC06142
	(ATCC 17699)	*hoxH*	
Aquificae			
Aquificaceae	*Aquifex aeolicus*	*mbhS1*	AAC06862
		mbhL1	AAC06861
	Aquifex aeolicus	*mbhS2*	AAC07047
		mbhL2	AAC07046
	Aquifex aeolicus	*mbhS3*	AAC06946
		mbhL3	AAC06045

5.3 Classification of Hydrogenases

Voordouw (1992) proposed the earliest classification scheme. Later Wu and Mandrand-Berthelot (1993) proposed a classification based on the multiple alignments for full-length amino acid. Subsequently, they generated pairwise alignment scores. On the basis of resulting dendograms, 30 hydrogenases were grouped into 6 classes. Recently, Vignais et al. (2001) have provided the most updated classification based on the cluster analysis using both amino acid sequences and segments corresponding to functional domains.

Hydrogenases can be classified according to the metal content of the active site into [NiFe], [FeFe], and [Fe] hydrogenases. The vast majority of known hydrogenases belong to the first two classes, and presently a large number of these enzymes have been characterized genetically and/or biochemically (Vignais et al., 2001). A characteristic feature of all hydrogenases is that the iron atoms are ligated by small inorganic ligands (CO, CN-), which were first detected by an FTIR spectroscopy (Song et al., 2007). The [NiFe] and [FeFe] hydrogenases contain sulfur-bridged bimetallic centers, typically with an open coordination site on one metal. A subgroup of the first class comprises the [Ni-Fe-Se] hydrogenases, in which one of the cysteine ligands of the nickel atom is replaced by a selenocysteine. The third class of enzymes has long been thought to contain no metal. However, it was recently shown that the active site harbors a single iron atom with an unusual coordination sphere. These enzymes are now called [Fe-only]-hydrogenases or iron–sulfur cluster-free hydrogenases. They can activate H_2 only in the presence of a second substrate (methenyl tetrahydromethanopterin). They are quite different from the first two classes and will therefore only briefly be discussed in this chapter.

5.3.1 [Fe-only] Hydrogenases

[Fe-only] hydrogenase (henceforth referred to as [Fe] hydrogenase) is a cytoplasmic homodimeric enzyme (2 × 38 kDa) harboring an iron-containing cofactor. The systematic name for [Fe] hydrogenase is hydrogen-forming methylenetetrahydromethanopterin (methylene-H_4MPT) dehydrogenase, abbreviated as Hmd. It differs from [NiFe] and [FeFe] hydrogenases in that it contains a mononuclear metal center rather than a dinuclear metal center in its active site (Korbas et al., 2006). Moreover, it is devoid of iron–sulfur clusters, that is why the enzyme is also referred to as iron–sulfur-cluster-free hydrogenase. It catalyzes the reversible reduction of methenyltetrahydromethanopterin (methenyl-H_4MPT$^+$) with H_2 to methylene-H_4MPT, a reaction involved in methanogenesis from H_2 and CO_2 in many methanogenic archaea (Thauer et al., 1993). In the reaction, a hydride from H_2 is transferred into the *pro-R* position of the C (14a) methylene group of the reaction product (Schleucher et al., 1999) (Figure 5.1).

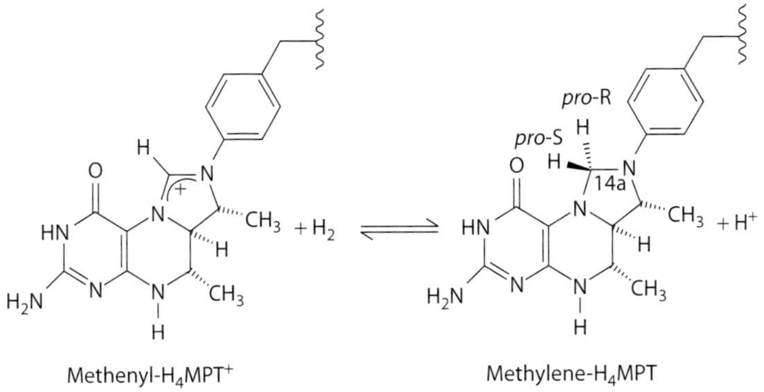

Methenyl-H$_4$MPT$^+$ Methylene-H$_4$MPT

FIGURE 5.1
Reaction catalyzed by [Fe] hydrogenase. N^5,N^{10}-methenyltetrahydromethanopterin (*methenyl-H$_4$MPT$^+$*) reduction with H$_2$ to N^5,N^{10}-methylenetetrahydromethanopterin (*methylene-H$_4$MPT*) and a proton, whereby a hydride is stereospecifically transferred from H$_2$ into the *pro-R* side of methylene-H$_4$MPT. The exchange activities of [Fe] hydrogenase (iron–sulfur-cluster-free hydrogenase) from methanogenic archaea in comparison with the exchange activities of [FeFe] and [NiFe] hydrogenases. (Reprinted from Vogt, S. et al., 2008. *Journal of Biological Inorganic Chemistry*, 13, 97–106. With permission.)

The enzyme harbors an iron-containing cofactor, in which a low-spin iron is complexed by a pyridone, two CO and a cysteine sulfur. In this respect, [Fe] hydrogenase is similar to [NiFe] and [FeFe] hydrogenases, in which a low-spin iron carbonyl complex, albeit in a dinuclear metal center, is also involved in H$_2$ activation. Like the [NiFe] and [FeFe] hydrogenases, [Fe] hydrogenase catalyzes an active exchange of H$_2$ with protons of water; however, this activity is dependent on the presence of the hydride-accepting methenyl-H$_4$MPT$^+$. It has been observed that in its absence the exchange activity is only 0.01% of that in its presence (Vogt et al., 2008). The residual activity has been attributed to the presence of traces of methenyl-H$_4$MPT$^+$ in the enzyme preparations, but it could also reflect a weak binding of H$_2$ to the iron in the absence of methenyl-H$_4$MPT$^+$.

The K_m and V_{max} of the enzyme for H$_2$ and D$_2$ (deuterium) were found to be almost identical, indicating that a step other than the activation of H$_2$ is rate-determining in the reaction shown in Equation 5.2.

$$H_2 + \text{methenyl-H}_4\text{MPT}^+ \rightarrow \text{methylene-H}_4\text{MPT} + H^+ \quad (\Delta G^{\circ\prime} = 5.5 \text{ kJ mol}^{-1}) \quad (5.2)$$

Unlike [FeFe] and [NiFe], [Fe] hydrogenase does not catalyze the reduction of viologen dyes (Equation. 5.3) or other artificial one-electron or two-electron acceptors neither in the absence nor in the presence of methenyl-H$_4$MPT$^+$.

$$H_2 + 2MV \rightarrow 2MV^- + 2\,H^+ \quad (5.3)$$

The presence of an iron carbonyl complex in [Fe], [FeFe], and [NiFe] hydrogenases indicates that the low-spin iron could play a crucial role in H_2 activation in all three enzymes. The low-spin iron in [Fe] hydrogenase could in principle function in the activation of H_2 as a Lewis acid or a Lewis base. As a Lewis acid (electrophile), it will bind H_2 side-on [(η^2-H_2)Fe] by which the pK of H_2 of 35 is significantly lowered (Kubas, 2005). Depending on the decrease in pK, the thus-activated H_2 would exchange more or less rapidly with protons of water, and essentially this exchange is not predicted to require the presence of the hydride acceptor methenyl-H_4MPT$^+$. When the iron functions as a Lewis base (nucleophile), exchange of the bound H_2 with protons of water is only possible in the presence of the hydride acceptor (Berkessel, 2001). These predictions fostered the question whether the residual H_2/H^+ exchange activity shown by [Fe] hydrogenase preparations might not be independent of methenyl-H_4MPT$^+$ after all. However, recent findings suggest that methenyl-H_4MPT$^+$ induces a conformational change within the active site of the enzyme leading to an activation of the iron which catalyzes the H_2/H_2O exchange without the carbocation center of methenyl-H_4MPT$^+$ being involved (Vogt et al., 2008).

5.3.2 [NiFe] Hydrogenase: Structure and Location

The most-studied classes of hydrogenases are the [NiFe] hydrogenases from bacteria. Initial advances in the understanding of the complex enzyme were achieved with the help of spectroscopic analysis and x-ray crystallography studies (Armstrong and Albracht, 2005). The core [NiFe] hydrogenase consists of an $\alpha\beta$ heterodimer with the large α subunit comprising nearly 60 kDa and contains the bimetallic active site whereas the small β subunit comprises the FeS clusters and is ca. 30 kDa. Crystal structures of *Desulfovibrio* reveal that the subunits interact largely through a globular heterodimer. The active site is buried deep inside the large subunit and is coordinated to the protein by four cysteines, two of which are also involved in bridging to the iron ion. Further, infrared spectroscopy and Fourier transformation have also revealed the presence of three diatomic ligands, one CO and two CN$^-$ coordinated to the iron ion (Figure 5.2). The iron ion is largely found to be Fe^{2+} oxidation state and is diamagnetic. Contrarily, the Ni ion exhibits variable oxidation states

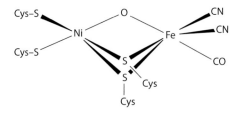

FIGURE 5.2
Active site biochemistry of a standard [NiFe] hydrogenase based on the crystal structure of *D. gigas.*

TABLE 5.2

Classification of [NiFe] Hydrogenases into Subgroups with Respect to Their Function

[NiFe] Group	Function	Organism in Which It Is Found
1	Membrane-bound H_2 uptake hydrogenases	Bacteria, archea
2a	Cyanobacteria uptake hydrogenases	Cyanobacteria, bacteria
2b	H_2-sensing hydrogenases	Bacteria
3a	F_{420}-reducing hydrogenases	F_{420}-reducing archaea
3b	Bifunctional (NADP) hydrogenases	NADP-reducing archaea, bacteria
3c	Methyl-Viologen-reducing hydrogenases	MV-reducing archaea, bacteria
3d	Bidirectional NAD(P)-linked hydrogenases	NAD/NADP-reducing cyanobacteria, bacteria
4	Membrane-bound H_2-evolving hydrogenases	Multimeric H_2-evolving archaea, bacteria

(3^+, 2^+, or 1^+) and is further coordinated to two cysteine-S ligands. The small unit contains the FeS clusters to mediate the electron transfer to and from the active site to the enzyme surface (Vignais et al., 2001). The arrangement of the FeS clusters is found to vary. *D. gigas* is found to contain three FeS clusters, a [4Fe-4S] cluster in close proximity (13 Å) to the active site, a [3Fe-4S] cluster located within a distance of 12 Å, and another located at a distance of 12 Å from the second cluster (Volbeda et al., 1995). This linear arrangement of FeS clusters was proposed to act as a wire to mediate electrons away from the active site to the electron acceptors such as cytochrome *c* located at the enzyme surface. [NiFe] hydrogenases vary in structure and location depending upon their physiological role within the cell. Accordingly, Vignais (2007) classified the [NiFe] hydrogenases into four groups (For more details, consult Vignais and Billoud, 2007, also see Table 5.2).

5.3.2.1 Group 1: [NiFe] Uptake Hydrogenase

These are membrane-bound hydrogenases found mostly in archaea and bacteria (excluding cyanobacteria). They perform respiratory hydrogen oxidation linked to quinone reduction. They link the oxidation of hydrogen to the reduction of anaerobic electron acceptors such as NO_3^-, SO_4^-, fumarate, or CO_2 during anaerobic respiration or to O_2 during aerobic respiration. The energy is recovered in the form of proton motive force. They are linked to the quinone pool of the membrane by a third subunit, cytochrome *b* that has higher redox potential compared to H_2/H^+ couple. Thus, these hydrogenases primarily function to oxidize molecular hydrogen and are called uptake hydrogenases (hup). These uptake hydrogenases are characterized by the presence of a long signal peptide (35–50 amino acid residues) at the N-terminus of their small subunit. The signal consists of the conserved amino acids [DENST] RRxFxK motif, which belongs to the twin arginine translocation (TAT) pathway. It serves as a signal to transfer the completely folded protein to the membrane periplasm.

5.3.2.2 *Group 2: Cyanobacterial Uptake Hydrogenases and Hydrogen Sensors*

This group is further split into two groups: 2a and 2b. Group 2a comprises cyanobacterial uptake hydrogenases mostly found in nitrogen-fixing strains encoded by hup-hydrogen uptake genes (hupSL). These differ from the Group 1 hydrogenase in being cytoplasmic and lacking the characteristic TAT signal of Group 1 hydrogenase. Their role is to detect the presence of H_2 in the environment and to trigger a cascade of cellular reactions controlling the synthesis of respiratory [NiFe] hydrogenases (Vignais and Colbeau, 2004). Thus, these hydrogenases function to provide the reductant to the TCA cycle for CO_2 fixation.

Group 2b included the sensory hydrogenases from bacteria called HupUV or HoxBC. These hydrogenases are characteristically insensitive to oxygen as compared to the other groups of hydrogenases (Volbeda et al., 2002). They function to sense hydrogen concentrations in the regulatory cascade that controls the biosynthesis of proteobacterial uptake hydrogenases.

5.3.2.3 *Group 3: Multimeric Soluble Hydrogenases*

Primarily in this group the dimeric hydrogenase is associated with other subunits able to bind soluble cofactors such as cofactor 420, NAD, or NADP. Besides, Group 3 hydrogenases contain an additional subunit that interacts with the soluble coenzymes, for example, the iron–sulfur flavoprotein FrhB (Mills et al., 2013). These were re-subdivided into four groups; 3a, 3b, 3c, and 3d. They are mostly distributed in archaea and bacteria.

Group 3a comprises the uptake hydrogenases associated with cofactor F_{420}. They are found in the archaea. These hydrogenases catalyze the reversible reduction of the soluble hydride carrier, NAD(P) or F_{420}. F_{420} is a deazaflavin derivative that acts as an important hydride acceptor/donor in the central methanogenic pathway.

Group 3b comprises the uptake hydrogenases associated with cofactor NADP. They are mostly found in bacteria and archaea.

Group 3c comprises the uptake hydrogenases associated with cofactor MV. They are mostly found in bacteria and archaea. Presently, the MV-reducing [NiFe]-hydrogenases have not been extensively characterized, and additional examination is required to better define the biochemical properties and physiological function of these enzymes.

Group 3d cyanobacterial hydrogenase has been purified as a pentameric enzyme and the corresponding structural genes, *hox*, are clustered together with three open reading frames (ORFs) of unknown function(s). Typically, they are composed of two moieties, *hoxY* and *hoxH* genes and the diaphorase moiety, encoded by the *hoxU*, *hoxE*, and *hoxF* genes homologous to some subunits of Complex 1 of mitochondrial and bacterial respiratory chains that contain NAD(P), FMN, and Fe-S-binding sites.

5.3.2.4 *Group 4:* Escherichia coli *Hydrogenase 3*

5.3.2.4.1 Hydrogen Evolving, Energy-Conserving Membrane-Associated Hydrogenases

The multimeric enzymes (six subunits or more) of this group reduce protons from water to dispose of excess-reducing equivalents produced by the anaerobic oxidation of C1 organic compounds of low potential, such as carbon monoxide or formate (Vignais and Colbeau, 2004). Ideal example is *Escherichia coli* hydrogenase 3. It is an H_2-evolving enzyme, which is part of the multisubunit formate–hydrogen lyase complex. Similar enzymes have been identified in various bacteria and archaea including *R. rubrum, C. hydrogenoformans, M. barkeri, M. thermoautotrophicus,* and *P. furiosus.*

5.3.2.5 *Structural Organization of the Genes-Encoding [NiFe] Hydrogenases and Their Physiological Role in the Organism*

Mostly in organisms containing the [NiFe] hydrogenases, except *E. coli,* the genes are clustered in organized operons in the chromosome. As exceptions, in *A. eutrophus* and *R. capsulatus,* the genes are carried in a megaplasmid. The clusters comprise the structural genes generally labeled 'L' for large subunit and 'S' for small subunit and the accessory genes required for maturation and insertion of Ni, CO, and CN^- at the active site of the heterodimer. In the following text the different groups of [NiFe] hydrogenase from *E. coli, R. rubrum,* and cyanobacteria will be discussed to give a deeper perspective of gene organization and the role of the different hydrogenases in the physiological context.

5.3.2.5.1 Operon Arrangement and Function of [NiFe] Hydrogenase in E. coli

Three immunologically and genetically distinct [NiFe] hydrogenases, termed hydrogenase 1, hydrogenase 2, and hydrogenase 3, have been identified and characterized in *Escherichia coli* (Böhm et al., 1990; Menon et al., 1990; Sauter et al., 1992). More recently, the existence of a fourth enzyme has been postulated based on the genetic analysis of the *E. coli* chromosome (Andrews et al., 1997).

The physiological function of hydrogenase 1 is not well defined. It is apparent from its subunit structure and from its overall amino acid sequence similarity with other uptake hydrogenases that it has a role in hydrogen oxidation (Menon et al., 1990). It has been proposed to be involved in recycling the hydrogen produced by hydrogenase 3 based on the fact that its synthesis shows a strong correlation with that of the FHL pathway (Sawers et al., 1985, 1986); however, such an activity has not been demonstrated practically.

Hydrogenase 2 synthesis is known to be induced when cells are grown anaerobically on nonfermentable carbon sources such as hydrogen and fumarate or glycerol and fumarate (Ballantine and Boxer, 1985; Sawers et al., 1985). Thus, it has been suggested that the function of hydrogenase 2 of *E. coli* may be in a respiratory capacity allowing the cells to gain energy from the oxidation of molecular hydrogen. A subsequent mutational analysis of the hyb structural genes encoding the enzyme confirmed this hypothesis (Menon et al., 1994).

Hydrogenase 3 is known to be a part of formate hydrogen lyase complex, which, during fermentation, converts the formate produced by the pyruvate formate-lyase reaction into carbon dioxide and molecular hydrogen. The presence of this hydrogenase thus prevents the accumulation of formate which may be toxic to the cells due to pH imbalance. (Böhm et al., 1990; Rossmann et al., 1991; Sauter et al., 1992). The polycistronic operons encoding hydrogenases 1, 2, and 3 are encoded by *hyaABCDEF, hybABCDEF, and hycABCDEF,* respectively (Menon et al., 1990, 1994). Hyd-3 located on *hyc* operon mediates hydrogen production from formate hydrogen lyase complex (FHL complex), which is active during mixed acid fermentation at slightly acidic pH. Hyd-4 encoded by *hyfABCDEFGHIJR-focB* operon has high homology with *hyc* operon, but *hyf* operon is not expressed in *E. coli* in normal conditions. It can be activated in the presence of effector-independent Fhl A (transcriptional activator of the FHL complex) mutant proteins. The operon encodes a putative 10-subunit hydrogenase complex (hydrogenase-4 [Hyf]); a potential σ 54-dependent transcriptional activator, hyfR (related to FhlA); and a putative formate transporter, FocB (related to FocA) (Skibinski et al., 2002). The role of the four hydrogenases and their positions in the metabolic pathway are highlighted in the Figure 5.3.

5.3.2.5.2 Operon Arrangement and Function of [NiFe] Hydrogenase in R. Rubrum

R. rubrum contains a versatile hydrogen metabolism. Its metabolism involves three enzymes involved in the hydrogen pathway: a CO-dependent enzyme, a formate-dependent enzyme, and a sensory hydrogenase. Biological CO oxidation plays a significant role not only in the generation of H_2 from CO (and H_2O) but also in yielding ATP in *R. rubrum* and related organisms, although the mechanism of the latter is unknown thus far. Genes and enzymes involved in the overall CO oxidation-H_2 production pathway have been elucidated in *R. rubrum* (Figure 5.4). There are two CO-inducible transcripts *cooFSCTJ* and *cooMKLXUH*. The *cooFSCTJ* encodes the CO-oxidizing carbon monooxide dehydrogenase (cooS), an FeS electron transfer protein (cooF) and accessory proteins required for the maturation of CooS (Ensign and Ludden 1991; Kerby et al., 1992). However, the *cooMKLXUH* operon encodes the hydrogenase structural proteins, two of which, *cooL* and *cooH*, resemble the small and large subunits of standard [NiFe] hydrogenase (Fox et al., 1996), while the four additional subunits (*cooMKXU*) bear significant homology to complex I (Fox et al., 1996). Both transcripts are under the exclusive control of *cooA*, a heme-containing CO-sensing transcriptional factor. Due to their characteristic sequence and energy conservation features coupling the oxidation of a carbonyl group (originating from CO, formate, or acetate) to proton reduction, these hydrogenase are classified in the Group 4 energy-converting (or energy-conserving) hydrogenase (Vignais et al., 2001). The second enzyme involved in the hydrogen metabolism of *R. rubrum* is a putative formate hydrogen lyase encoded by *hycBCDEFGI*. It is similar in function to hydrogenase 3 of *E. coli* (for more details, refer to Schwarz et al., 2010).

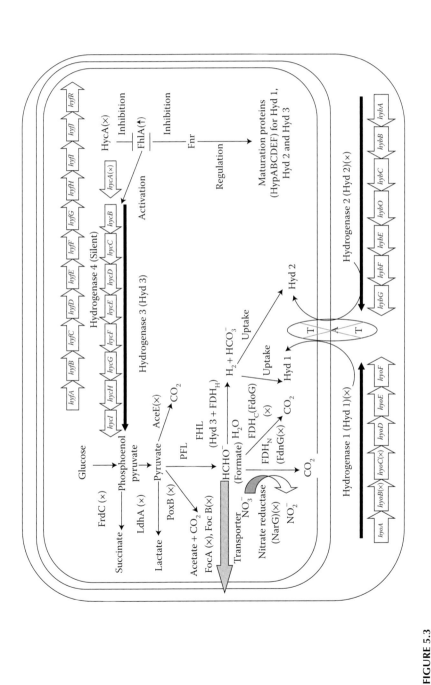

FIGURE 5.3

Role of the four hydrogenases in *E. coli*; their structural operons and their functions. (Reprinted from Sinha, P. and Pandey, A. 2011. An evaluative report and challenges for fermentative biohydrogen production. *International Journal of Hydrogen Energy*, 36(13), 7460–7478. With permission.)

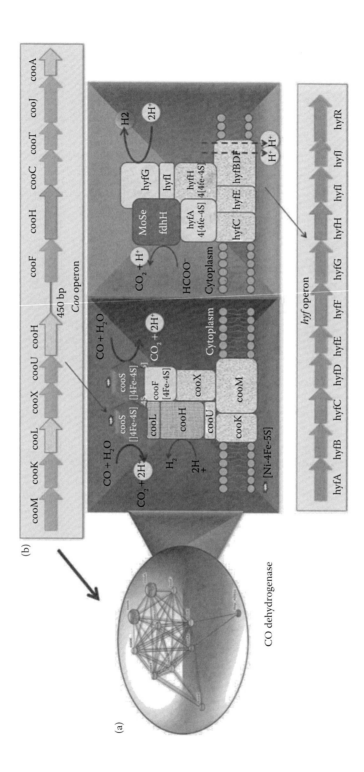

FIGURE 5.4

(a) PDB structure of CO dehydrogenase. (b) Structural operons and structural models of the [NiFe] hydrogensase complex in *R. rubrum*. (With kind permission from Springer Science + Business Media: *Medicine*, Recent advances in phototrophic prokaryotes, 675, 2010, 305-348, Schwarz, C. et al.)

5.3.2.5.3 Operon Arrangement and Function of [NiFe] Hydrogenase in Cyanobacteria

Cyanobacteria comprises uptake and bidirectional hydrogenase (Table 5.3). Uptake hydrogenases are present in only nitrogen-fixing cyanobacteria (Tamagnini et al., 2002). Studies showed that the uptake subunit consists of three genes *hupSLW*, *hupS*, and *hupL* that encode the small and large subunits, respectively. However, *hupW* is an hydrogenase-specific endopeptidase. The ORF is located immediately downstream of *hupSL* and is shown to be cotranscribed in *Gleothece* ATCC 27152 (Oliveira et al., 2004). RT-PCR studies have shown that *hupL* is expressed only under nitrogen-fixing conditions (Carrasco et al., 2005), clearly demonstrating its role in uptake of hydrogen evolved during molecular nitrogen fixation into ammonia.

The bidirectional hydrogenase may be present in both nitrogen-fixing and non-nitrogen-fixing cyanobacteria (Tamagnini et al., 2000, 2002, 2007). The enzyme is a heteropentameric unit encoded by *hoxEFUYH* (hox-hydrogen oxidizing). *hoxEFU* constitutes the diapharose part while *hoxYH* constitutes the hydrogenase part (Schmitz et al., 1995, 2002). The diapharose units (*hoxEFU*) show striking similarity to the NADH dehydrogenase type I (complex I) subunits *NuoE, NuoF,* and *NuoG*. Since the cyanobacterial complex I (NDH-1) and the NADH dehydrogenases in chloroplasts lack homologs to those subunits, it was suggested that the bidirectional hydrogenase could function as the NADH dehydrogenase part of complex I (Figure 5.5) (for detailed review, refer to Tamagnini et al., 2002).

5.3.2.6 Biosynthesis of [NiFe] Hydrogenases

The maturation of the hydrogenase gene cluster involves seven accessory proteins, the products of the 'hyp' for "hydrogenase pleiotropic" genes, namely, *hypA, hypB, hypC, hypD, hypE, hypF,* and an endopeptidase, *hycI*. The endoprotease is necessary for C-terminal peptide cleavage after incorporation of Ni to complete [NiFe] hydrogenase maturation (Casalot and Rousset, 2001). These genes are involved in the biosynthesis of the diatomic ligands (CO, CN^-) that are coordinated to an iron ion, followed by nickel insertion and (c) endoproteolytic activation of the mature complex (Bock et al., 2006). Besides, several chaperones are involved in the transportation of the mature protein to the periplasmic site via the TAT pathway. The most-studied hydrogenase maturation system is that involved in the biosynthesis of *E. coli* hydrogenase 3 and is described here and shown in Figure 5.6.

Synthesis of the [NiFe] active site requires the synthesis and incorporation of the CO and CN^- in the active site. Studies show that CN^- is synthesized by the interaction of two gene products *hypE* and *hypF*, using carbomyl phosphate as precursor to form a thiocarbamate (Brazzolotto et al., 2006). However, the incorporation of CO into the active site is still a debatable question. It is suggested that it may be derived from different metabolic origins (Forzi et al., 2007). Generally, acetate or CO_2 may serve as substrates

TABLE 5.3

Examples of Different Hydrogenase Genes Located in Various Cyanobacteria

(Accession No.)

Strain	Uptake Hydrogenase		Bidirectional Hydrogenase					
	hupS	*hupL*	*hoxE*	*hoxF*	*hoxU*	*hoxY*	*hoxH*	*hoxW*
Filamentous								
N. punctiforme PCC 73102	C651/20, AF030525	C651/21, AF030525	NP[a]	NP[a]	NP[a]	NP[a]	NP[a]	
A. variabilis PCC 7120	all0688, U08013	all068N/C, U08013	alr0751	alr0752	alr0762	alr0764	alr0766	
Unicellular								
Synechocystis PCC 6803	NP[a]	NP[a]	sll1220, X97610	sll1221, X97610	sll1223, X97610	sll1224, X97610	sll1226, X97610	
Synechococcus PCC 6301			Y13471	Y13471	X97797	X97797	X97797	X97797

[a] NP: not present.

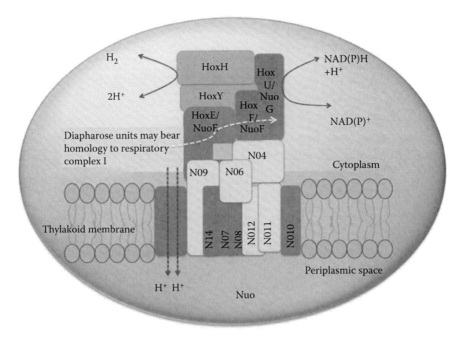

FIGURE 5.5
(See color insert.) Hypothetical model of putative interaction of hox bidirectional hydrogenase with respiratory complex 1.

(Roseboom et al., 2006). According to the current model, biosynthesis of the diatomic ligands and coordination of Fe by CN^- and CO represent independent steps in the assembly of the enzyme (Bock et al., 2006). Once the $Fe(CN)_2(CO)$ complex is established, Ni is introduced into the active site. Studies suggest that under high concentration of Ni, it can be assembled into the active site without the help of accessory proteins. However, under physiological limitations, the nickel requirement is provided by complex of maturation factors, *hypA* and *hypB*. The *hypA* and *hypB* genes are usually clustered together in the genome and are involved in the Ni delivery and incorporation. While *hypA* is a Ni-binding metallo-chaperone, *hypB* provides GTPase activity and provides energy for the insertion of Ni and subsequent release of the maturation factors. Further, *hypC*, *hypD*, and *hypE* are grouped together. The C-terminal region of *hypD* binds a [4Fe-4S] cluster and is able to form a complex with *hypC*. The *hypC*/*hypD* complex is further proposed to mobilize the CN^- for incorporation into the active site. This complex readily interacts with the large subunit of the enzyme. The final step is the maturation that requires the cleavage of the C-terminal peptide of the large subunit apoprotein complex by an endopeptidase. It has been suggested that the C-terminal peptide keeps the apoprotein active site in a particular conformation accessible to metal insertion and may even serve as the site of interaction between

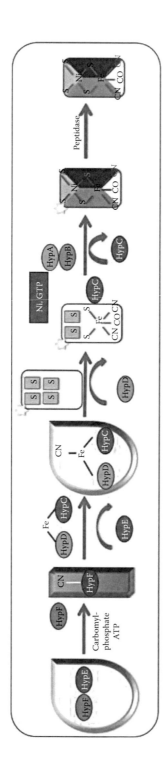

FIGURE 5.6

Biosynthesis of [NiFe] hydrogenase assembly. (Adapted from Posewitz, M. C., Mulder. D. W., and Peters, J. W. 2008. *Current Chemical Biology, 2,* 178–199.)

the large and the small subunit (Sawers et al., 2004). Another *hyp* maturation gene, *hypX*, is present in some organisms. In *R. eutropha*, *hypX* is proposed to be responsible for incorporating an additional CN⁻ ligand in the soluble *R. eutropha* [NiFe] hydrogenase active site (Bleijlevens et al., 2004). Interestingly, there are multiple copies of *hyp* genes in some organisms (Lenz et al., 2005), and it is currently unknown if this represents genetic redundancy or whether the different structural enzymes require unique maturation factors.

5.3.2.7 Transcriptional Regulation of [NiFe] Hydrogenases

The control of hydrogenase synthesis represents a means to quickly and efficiently respond to changes in the environment and in particular to new energy demands. It is exerted at the transcriptional level. Transcriptional control involves one or several two component regulatory systems, which may act positively or negatively (Vignais and Billoud, 2007). In response to a specific signal, a sensor histidine kinase autophosphorylates at a conserved histidine residue and then transphosphorylates the cognate response regulator transcription factor at a conserved aspartate residue that activates or represses gene expression when phosphorylated by the sensor kinase. Hydrogenase responds to several types of signals. In a few special organisms (e.g., methanogens), whose metabolism is strictly adapted to H_2 activation, hydrogenase is synthesized constitutively (Cammack et al., 2001). In other cases, the expression of hydrogenases is dependent on various environmental signals. Some hydrogenases are expressed in the presence of H_2 and are regulated through a H_2-sensing signal-transduction cascade (Lenz et al., 2002) or by other unknown mechanisms (Tamagnini et al., 2002). The other factor responsible for the regulation of several different hydrogenases is the anaerobic environment (Vignais and Colbeau, 2004).

5.3.2.7.1 Aerobic/Anaerobic Control

Under aerobic conditions, hydrogenase activity is suppressed. It has been reported that their transcription is initiated under anaerobiosis in the absence of exogenous electron acceptors. The sensing of low oxygen concentrations in the cell requires the regulatory proteins homologous to *E. coli* fumarate nitrate reduction (FNR) protein. Strains that are unable to synthesize the global transcriptional regulator were also found to be defective in the synthesis of the hydrogenase (Sawers et al., 1985), strongly suggesting that FNR was responsible for anaerobic induction of enzymes. FNR is a cytoplasmic oxygen responsive regulator with a sensory and a regulatory DNA-binding domain. Under anaerobiosis, it activates the genes involved in anaerobic respiration and represses the genes involved in the aerobic pathway (Luchi and Lin, 1995). It was shown by DNA microarray technology that the transcript level of one-third of the genes expressed during aerobic growth were altered when *E. coli* cells were switched to the anaerobic growth state and that the expression of 49% of these genes were either directly or indirectly modulated by FNR (Salmon et al., 2003). The protein is known to bind as a dimer to the FNR consensus sequence TTGAT-N_4-ATCAA (Nakajima et al., 2001). It has been reported that FNR activity depends on the presence of [4Fe-4S]$^{2+}$ cluster. In the presence of oxygen, the cluster is converted to a stabler [2Fe-2S]$^{2+}$ configuration. It has been suggested that the labile nature of the [4Fe-4S]$^{2+}$ cluster makes FNR an oxygen sensor. Accordingly, Wu and Mandrand-Berthelot (1986) showed that the reduced level of hydrogenase (hydrogenase 1 and 2 of *E. coli*) enzymes in FNR mutant was the consequence of impaired nickel transport. In *R. leguminosarum*, FNR-like regulators, *FnrN*, have been identified. These regulators differ from FNR in lacking the N-terminal region for binding to the [4Fe-4S]$^{2+}$ cluster (Patschkowski et al., 1996).

5.3.2.7.2 Nitrogen Stimulation

In *R. leguminosarum* in addition to FnrN hydrogenase, expression is coregulated by nitrogenase *nifA*, which activates hydrogenase by binding to an upstream-activating sequence in the promoter region of the *hupSL* genes (Brito et al., 1997). Higher activity of hox hydrogenase is observed under nitrogen starvation and low-light intensity. It was suggested that bidirectional hydrogenase could act as an alternative electron donor to PSI after inactivation of PSII due to nitrogen starvation. Transcripts of the bidirectional hydrogenase have been shown to be present in NH_4^+-grown filaments and in both vegetative and heterocysts under nitrogen-fixing conditions in *A. variabilis* (Boison et al., 2000). Besides, analysis of the transcription of the *hoxY* and *hoxH* in *G. alpicola*, under combined nitrogen-limiting growth conditions, demonstrated an increase in the enzyme activity, however no regulation was found at the transcriptional level (Sheremetieva et al., 2002). Further, induction of *hupL* transcript has been observed in *Nostoc* strains when shifted from non-nitrogen fixing conditions to nitrogen-fixing conditions (Axelsson et al., 1999).

5.3.2.7.3 LexA and AbrB Regulation

Regulation of bidirectional hydrogenase is under the control of two transcriptional factors: LexA and AbrB (Oliveira and Lindblad, 2005, 2008). It has been shown that LexA and AbrB proteins do not correlate with hydrogenase expression, and hence they may be post-translationally modified. The deletion of *lexA* in *Synechocystis* sp. 6803 indicated their role in the under-inorganic carbon starvation. This differs from their role in bacteria, where LexA together with RecA are known to be a crucial key regulators of the SOS response in *E. coli*. Further, in *Synechocystis* sp. 6803, LexA occupies multiple-binding sites in the *hoxEH* promoter region (Oliveira and Lindblad, 2005), where it appears to activate hydrogenase transcription as confirmed by the hox operon down-regulation in the *lexA* deletion mutant developed by Gutekunst et al. (2005).

5.3.2.7.4 Nickel and Iron-Specific Regulation

[NiFe] hydrogenase is a bimetallic enzyme-containing [FeS] cluster, hence its regulation is also based on the subject of availability of these metals in the growth medium. Axelsson and Lindblad (2002) showed that in heterocystous *N. muscorum* CCAP 1453/12, the addition of external nickel to the growth medium significantly enhanced mRNA abundance of *hoxH*. Gutekunst et al. (2006) showed that the transcript of the bidirectional hydrogenase gene in *Synechocystis* sp. 6803 increased 10 times with 0.22 µM lower concentrations of iron ion as compared to nonstarved cells. However, surprisingly, the hydrogenase activity was found reduced. The authors suggested that the increase in hox transcription, under iron starvation, might be a feedback mechanism to compensate for the lack of functionality of active enzyme (Gutekunst et al., 2006). In *E. coli*, the nickel-specific transporter system is encoded by the *nikABCDE* operon and is a member of the ABC transporter family. It provides Ni^{2+} ions for the anaerobic biosynthesis of hydrogenases. In the presence of excess nickel, expression of the Nik operon is transcriptionally repressed by the Ni-responsive repressor NikR, a direct sensor of Ni^{2+} ions. However, the mechanism of nickel regulation is not fully understood.

5.3.2.7.5 Redox Regulation

Under anaerobiosis, the transcription of hydrogenase 1 and 2 in *E. coli* is under the control of a two component system ArcA and ArcB (aerobic respiratory control). Mutations in *arcA* or *arcB* are known to affect the expression of more than 30 operons. ArcA is a response regulator with a typical N-terminal receiver domain and a C-terminal effector domain containing a helixturnhelix DNA-binding motif (Luchi and Lin, 1995). ArcB is a membrane-bound histidine sensor kinase, which has both transmitter and input domains. The ability of ArcB to auto-phosphorylate and to transfer the phosphorus group to ArcA is regulated by quinones. Oxidized forms of quinones act as direct negative regulators and inhibit autophosphorylation of ArcB during aerobic conditions, thus providing a link between respiratory chain and gene expression (Georgellis et al., 2001). By oxidizing hydrogen and generating low-potential electrons, hydrogenases participate in cellular redox metabolism. In addition to the hydrogen-dependent-positive regulation of the *hupSL* hydrogenase in *R. capsulatus*, the operon is repressed by the RegAB system (Elsen et al., 2000). Therefore, this operon is regulated not only by the availability of hydrogen, but also by the general redox status of the cells.

5.3.2.7.6 Formate Regulation

Optimal expression of the *hyc* operon is regulated by the transcriptional regulator FhlA. Yoshida et al. (2005) showed enhanced biohydrogen production by overexpression of the *fhlA* activator in *E. coli* strains. FhlA shows homology with NtrC family in its central and C-terminal domains, but differs in possessing an extended N-terminal domain lacking the aspartate residue, which is the site for phosphorylation of response regulators. Thus, FhlA is not regulated by phosphorylation but binding an effector molecule formate. FhlA is a homotetramer and activates *hyp, hyc, fhlF*, and *hypF* promoters. Thus, it regulates both the structural genes as well as the accessory genes in *E. coli*.

5.3.2.7.7 Sulfur and Selenium Regulation

The effect of elemental sulfur on gene expression in *P. furiosus* was investigated with using the microarray technique. All the hydrogenases present (hydrogenase 1, hydrogenase 2, and membrane associated) in the organism were found to be downregulated by sulfur deprivation. Hydrogen production was found to be enhanced under conditions of sulfur deprivation. Mechanism of sulfur regulation is still not clearly understood. Melis et al. (2000) showed that under sulfur deprivation, *Chlamydomonas reinhardtii* cells can accumulate starch, downregulate the activity of photosystem II (PSII), consume oxygen under continuous illumination, and, as a result, produce hydrogen for several days.

D. vulgaris contains the periplasmic [FeFe], [NiFe], and [NiFeSe] hydrogenases. They have a similar physiological role in hydrogen oxidation but are differently expressed in response to the availability of different metals in the growth media. Supplementation of selenium in the growth medium leads to a strong repression of [FeFe] and [NiFe] hydrogenases and increase in [NiFeSe] hydrogenases. Supplementation of Ni increases [NiFe] hydrogenases. However, it has been observed that growth under high concentration of hydrogen also induces [NiFe] hydrogenases even without the supplementation of nickel ion (Valente et al., 2006).

5.3.2.7.8 Sigma Factors

The promoters of some hydrogenases contain a consensus sequence -24(GG)/-12(GC) recognized by the product of the *rpoN/ntrA* gene also known as σ^{54} (sigma factors). Gene expression from sigma-dependent promoters is characterized by the requirement of an activator protein, which binds to an activator sequence located some hundred base pairs upstream of the consensus sequence. The interaction between the RNA polymerase holoenzyme bound at the transcription initiation site and the activator is mediated by DNA loop formation. In *E. coli*, the integration host factor (IHF) has been shown to induce DNA bending of nif promoters and facilitate interaction between the transcriptional activator-bound upstream. In *E. coli*, both the *hyc* and *hypA* operon are transcribed by a σ^{54}-dependent promoter under fermentative growth conditions (Vignais and Toussaint, 1994). A second promoter located within *hypA* and FNR dependent is responsible for *hypBCDE* transcription under nonfermentative conditions when the -24/-12 RpoN-dependent promoter upstream of *hypA* is silent (Lutz et al., 1991).

5.3.3 [FeFe]-Hydrogenase: Structure and Location

[FeFe] hydrogenase is the only type of hydrogenase to have been found in eukaryotes, and they are located exclusively in organelles, that is, in chloroplasts or in hydrogenosomes (Vignais and Billoud, 2007). In the green algae, *Chlamydomonas reinhardtii, Scenedesmus obliquus*, and *Chlorella fusca*, the enzyme is located in the chloroplast stroma and is linked via ferredoxin to the photosynthetic electron transport chain, where it functions as an electron "valve" that enables the algae to survive under anaerobic conditions. Hydrogenosomes are peculiar organelles that supply ATP to the cell and make molecular hydrogen. They are found in various unrelated eukaryotes, such as anaerobic flagellates, chytridiomycete fungi, and ciliates. The presence of [FeFe] hydrogenases in these lower eukaryotes has often been deduced from the DNA sequences of complete genes. This is an evidence of the evolutionary relationship of the organisms.

Unlike [NiFe] hydrogenases, [FeFe] hydrogenases are usually monomeric and consist of the catalytic subunit only, although dimeric, trimeric, and tetrameric enzymes are also known (Peters, 1999; Nicolet et al., 2002). Two families of Fe-Hases are recognized: (i) cytoplasmic, soluble, monomeric Fe-Hases, found in strict anaerobes such as *Clostridium pasteurianum* (CpI hydrogenase) and *Megasphaera elsdenii*. They are extremely sensitive to inactivation by O_2 and catalyze both H_2 evolution and uptake; and (ii) periplasmic, heterodimeric Fe-Hases from *Desulfovibrio* spp., that can be purified aerobically and catalyze mainly H_2 oxidation. The location of hydrogenase in a bacterial cell reflects its need within the cell. Periplasmic hydrogenase functions mostly to oxidize hydrogen, which creates a proton gradient across the cell membrane. The resulting electrons are also transferred to the cytoplasm by a redox transmembrane complex 14, where they are used in a stepwise manner to reduce sulfate to sulfide or to generate reducing power for the cell. In contrast, the location of the enzyme in the cytoplasm is justified by its H_2 evolution activity as in fermenting *Clostridia* sp., and ferredoxin transfers two electrons to CpI, which, in turn, uses protons as electron acceptors to generate hydrogen. This reaction rids the cell of an excess of low-potential electrons produced during fermentation in anaerobic organisms (Nicolet et al., 2000).

The [FeFe] enzymes have been relatively less studied as compared to [NiFe] hydrogenase, and till date structures of this enzyme have been elucidated only from *Desulfovibrio desulfuricans* (DdH) and *Clostridium pasteurianum* (CpI) (Nicolet et al., 1999; Peters, 1999). All [FeFe] hydrogenases have been biochemically characterized or identified by the common amino acid sequences they share. The conserved sequence configure a common architecture of various compliments of FeS cluster domains (F cluster) linked to the active site (H cluster). The FeS clusters are typically located at the N terminal of the H cluster and probably function to relay the electrons to and from the active site. They consist of either [4Fe-4S] or [2Fe-2S] cluster-binding domains that resemble ferredoxins (Vignais et al., 2001; Meyer, 2007; Dubini et al., 2009). The protein surface of CpI surrounding the [4Fe-4S] cluster is distinctly positively charged in contrast to the region surrounding the [2Fe-2S], which is distinctly negatively charged, suggesting that the structure of the enzyme may support interactions with multiple external electron transfer partners (Peters et al., 1998; Peters, 1999). The simplest [FeFe] hydrogenases are those which express only the H-cluster without F-cluster domains such as those observed in the chlorophycean algae, including *C.*

reinhardtii, Chlorella fusca, and *Scenedesmus obliquus* (Florin et al., 2001; Happe and Kaminski, 2002; Forestier et al., 2003).

The highly conserved H-cluster-binding domain, the core feature of [FeFe] hydrogenases, can be identified in primary sequences by three distinct binding motifs termed L1 (TSCCPxW), L2 (MPCxxKxxE), and L3 (ExMACxxGCxxGGGxP) (Vignais et al., 2001). The characteristic presence of these motifs distinguishes them from closely related sequences of *hydA* homologs including eukaryotic *narF*-like gene sequences (Balk et al., 2004, 2005; Huang et al., 2007; Boyd et al., 2009, 2010).

5.3.3.1 [FeFe]-Hydrogenase Active Site

The active site of the [FeFe] hydrogenase is the largest domain consisting of nearly 350 amino acid residues and exists as a pseudosymmetric structure in which two β sheets come together to form a cleft in which the H cluster is located. The H cluster comprises a [4Fe-4S] cubane linked to a bimetallic Fe center. The metallic center is coordinated by five diatomic CN^- and CO^- ligands (Pierik et al., 1998) that provide a similar coordination environment as found in [NiFe] hydrogenases (Figure 5.7). With the help of an FTIR and EPR analysis, the role of the ligands is now well understood. Both the CN^- and CO^- ligands act as π acid ligands and form strong metal–metal bonds, undergoing metal to ligand back bonding which helps stabilize the metal in low oxidation states (Mulder et al., 2010). Further, CO is involved in controlling hydrogen catalysis via redox-dependent changes (Liu and Hu, 2002), whereas CN^- is involved in facile electron transfer from the cubane to the 2Fe subcluster. Another characteristic feature of the iron binuclear cluster is the presence of the two terminal nonprotein ligands bound to each Fe atom. However, unlike the [NiFe] metallic center, the Fe_1-Fe_2 atoms at the [FeFe]-hydrogenase active site lack a direct linkage to the protein rather are bridged by sulfur atoms, which are covalently linked to each other through a bridging molecule. In the case of *D. desulfuricans,* this bridging molecule was tentatively identified as propandithiol (Nicolet et al., 1999).

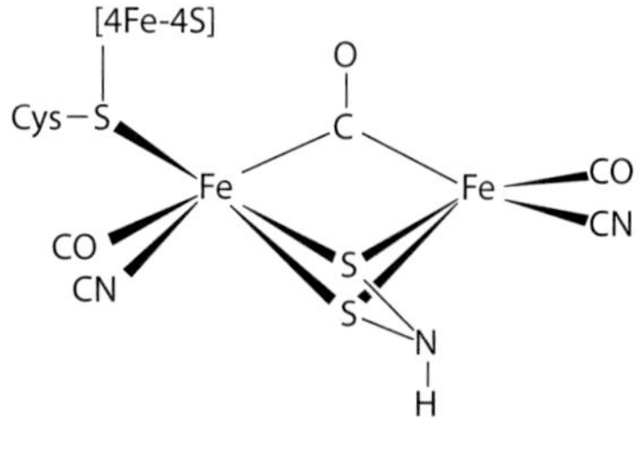

FIGURE 5.7
Active site biochemistry of a standard [FeFe] hydrogenase based on the crystal structure of *C. pasteuranium.*

The CpI and DdH active site structures show different oxidation states (H_{ox} and H_{red}, respectively) of the H cluster and their distal Fe coordination with the [4Fe-4S] subcluster (Nicolet et al., 2000; Mulder et al., 2011). In the CpI structure, which represents the H_{ox} state, the distal Fe is in coordination to two thiolatesulfurs, two CO ligands, a terminal CN^- ligand, and a terminally bound water molecule (Peters et al., 1998). Contrarily in the DdH structure, representing the H_{red} state, the water molecule present in the H_{ox} state is characteristically absent, leaving the distal Fe with an open coordination and a square pyrimadal geometry (Nicolet et al., 1999). These structural observations, related with different oxidation states, indicate the presence of a ligand exchangeable coordination site at the distal Fe of the bimetallic center.

Similar to the [NiFe] active site, the [FeFe] active site is also reversibly inhibited in the presence of CO. X-ray crystal studies of the active site show that CO binding occurs at the ligand exchangeable site, where H_2O molecule is bound at the distal Fe in the H_{ox} state. With these observations, Nicolet et al. (2000) suggest that the distal Fe atom of the bimetallic center may be the region, where hydrogen binding may occur. Further support comes from the observation that a hydrophobic continuous channel leads from the protein surface and ends at the Fe coordination site of the *D. desulfuricans* structure (Hatchikian et al., 1999). This region is highly conserved in the *C. pasteurianum* enzyme and points to a specific access of H_2 to the active site as has been discussed for the [NiFe] hydrogenases (Montet et al., 1997).

With these structural evidence, the mechanism of hydrogen catalysis has been proposed. It has been proposed that in the oxidized state water is present at the ligand exchangeable site. However, when the metal cluster is reduced or hydrogen is available as a coordinating group, water is displaced by the formation of a hydride or by the binding of hydrogen. Several studies have suggested that heterolytic cleavage could take place at the terminal position of the distal Fe site (Nicolet et al., 2000).

5.3.3.2 [FeFe]-Hydrogenase Maturation Machinery

It is only recently that evidence has been obtained that the hydrogenase maturation requires a set of accessory genes. Studies have revealed that the post-translational process of [FeFe] hydrogenase is extremely complex and involves a number of difficult reactions including: (i) the synthesis of CO, CN, and the dithiolate-bridging ligand; (ii) the assembly of the di-iron active site subcluster; (iii) its incorporation into the enzyme already containing the [4Fe–4S] component of the H cluster; and (iv) the assembly and transfer of the accessory [Fe–S] clusters. These reactions have to be stringently controlled because they produce the toxic CN and, possibly CO, as well as a hydrolytically sensitive dithiolate. This suggests that the synthesis of these ligands and their binding to Fe are concerted processes occurring simultaneously (Figure 5.8).

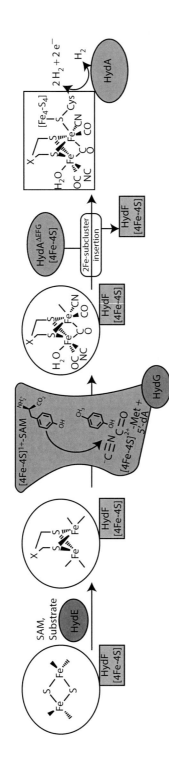

FIGURE 5.8

Hypothetical scheme for [FeFe]-hydrogenase (HydA) H-cluster activation by maturation enzymes *HydE*, *HydF*, and *HydG*. In the molecular representations of 2Fe subcluster precursors and the H cluster, the magenta "X" indicates the dithiolate bridgehead group that is likely to be CH_2, NH/NH_2^+ or O. (Reprinted from Mulder, D. W. et al. 2011. Insights into [FeFe]-hydrogenase structure, mechanism, and maturation. *Structure*, 19(8), 1038–1052. With permission.)

In view of this, studies showed that a deletion mutant of the accessory genes in *C. reinhardtii* was incapable of hydrogen production (Posewitz et al., 2004). Further work by King et al. (2006) showed that the presence of these accessory genes was essential for functional activity of heterologously expressed hydrogenase gene in *E. coli*, which lacks native [FeFe] hydrogenase. These accessory genes have been identified as *hydE, hydF,* and *hydG*. Research in *E. coli* has shown that the expression of hydA1 alone or in combination with only *hydEF* or only *hydG* results in no activity of the hydrogenase gene (King et al., 2006). Further studies revealed that *hydA* expressed in a background devoid of the accessory genes could be activated by addition of a cell extract containing *hydE, hydF,* and *hydG* coexpressed (McGlynn et al., 2007).

The organization of the genes in the chromosome and their spatial location has been found to be different for various organisms (Table 5.4). Thus, the genomic organization of the three hyd genes, *hydE, hydG,* and *hydF*, involved in [FeFe]-hydrogenase maturation varies from one organism to another with three possible patterns: (i) independent genes scattered in the genome (*Clostridium acetobutylicum, Clostridium perfringens*); (ii) fused genes (*C. reinhardtii*); (iii) genes organized in operons (*D. desulfuricans, Bacteroides thetaiotaomicron*). Besides, in a few organisms, other related genes have also been observed. For example, ammonium aspartate lyase and an unidentified ORF, *hydX*, are often found together with the *hydA* genes in *Shewanella oneidensis*. It should be noted that these additional genes are not observed in all organisms containing [FeFe]-hydrogenases, and it is currently unknown whether these gene products participate in [FeFe]-hydogenase maturation in a subset of organisms, or are perhaps associated with some other aspect of H_2 metabolism (Posewitz et al., 2008). In *C. reinhardtii, hydE* and *hydF* are fused and transcribed as one gene. However, for the majority of organisms, *hydE* and *hydF* exist as separately transcribed genes. *HydE* and *HydG* belong to the radical superfamily of proteins and contains a CX_3-CX_2-C Radical-SAM signature motif. Additionally, hydG contains a second-conserved cysteine motif,

TABLE 5.4

Selected Examples of Maturation Gene Organization in Different Microorganisms Containing [FeFe] Hydrogenase

Organism	Gene	Cellular Location	Gene Arrangement	NCBI Accession No.
Desulfovibrio desulfuricans	*hydE, hydF, hydG*	Periplasm	Operon	*hydF*:2276 *hydE*:2277 *hydG*:2278
Clostridium acetobutylicum	*hydE, hydF, hydG*	Cytoplasm	Independent	*hydE*:CAC1631 *hydF*:CAC1651 *hydG*:CAC1356
Chlamydomonas reinhardtii	*hydE, hydF, hydG*	Chloroplast	Fusion	*hydEF*: AY582739 *hydG*:AY582740

CX_2-CX_{22}-C, in the C-terminal portion of the protein (Posewitz et al., 2004; Bock et al., 2006). However, hydF has an N-terminal GTPase domain as well as a putative metal-binding motif, CX-$HX_{(44-53)}$-HCX_2-C (Boichenko et al., 2004).

Preliminary characterization of the reconstituted *hydE* and *hydG* from *T. maritima* showed that *hydE* bound two [4Fe-4S] clusters while *hydG* bound a single [4Fe-4S] cluster. Further, both the enzymes could cleave radical SAM nonproductively. However, reconstituted *hydF* was found to bind a single [4Fe-4S] cluster and was found to hydrolyze GDP to GTP. These preliminary experiments led the chassis for further work, which helped elucidate the mechanism of biosynthesis of the [FeFe] hydrogenase with the help of the accessory proteins. In the first step, the radical SAM proteins *hydE* and/or *hydG* are involved the synthesis of the bridging dithiolate ligand toward the modification of the [2Fe-2S] cluster. Subsequently, either the one or both the enzymes are involved in the generation of a glycyl radical. The interaction of this radical with an Fe (I)-thiolate site results in an exothermic decomposition yielding an Fe(I) radical coordinated to CN^-, CO^-, and H_2O. The GTPase enzyme, hydF, is shown to act as a scaffolding protein in this process with cluster translocation from *hydF* to *hydA* (Mulder et al., 2011). Further, studies suggested that only coexpression of *hydEFG* maturates hydrogenase, and supplementation of single gene products cannot activate the enzyme.

It is interesting to note that the hydrogenases are flexible with the maturation factors. In various studies, unrelated maturation genes have been found to maturate hydrogenases. For example, expression of an active *hydA1* enzyme from *C. reinhardtii* or *S. obliquus* has been shown to be possible using *C. acetobutycicum*, another [FeFe]-hydrogenase-synthesizing organism, as the expression host (Gibral et al., 2005; Kuchenreuther et al., 2010). Further evidence for the lack of selectivity of the *hyd* machinery came from the observation that coexpression of *hydE*, *hydF*, and *hydG* from the bacterium *C. acetobutylicum* with various algal and bacterial [FeFe] hydrogenases in *E. coli* resulted in purified enzymes with specific activities that were not very different from those of their counterparts from native sources (Agapakis et al., 2010).

5.4 Problems Associated with Oxygen Sensitivity of Hydrogenases and Plausible Solutions

As previously discussed, a major bottleneck of biohydrogen production is the oxygen sensitivity associated with these enzymes. While [FeFe] are oxygen intolerant, [NiFe] can tolerate micro-aerobic environment and are only temporarily inhibited by oxygen. In literature, several examples of indigenous oxygen tolerance by organisms known to have [NiFe] hydrogenase have been cited (Table 5.5). These include the well-studied *Ralstonia eutropha* which is shown to possess three indigenous oxygen-tolerant hydrogenases

TABLE 5.5

Characteristic Features of [NiFe] Hydrogenase from Selected Organisms Possessing Indigenous Oxygen Tolerance

Organism	Characteristic Features of [NiFe] Hydrogenase That Impart Oxygen Tolerance	Oxygen Tolerance Potential	Reference
Aquifex aeolicus	*A. aeolicus* has three [NiFe] hydrogenases: hydI-cytochrome *b*, hydII, and hydIII. Among them, hydIII is known to be the most oxygen stable. Stability is attributed to the kinetics of inactivation being much slower as compared to the kinetics of reactivation that is much faster.	50% of the initial activity is still measurable after exposure to O_2 for 24 h at the same temperature.	Guiral et al. (2006)
Rubrivivax gelatinosus	O_2 tolerance is displayed by its CO-linked hydrogenase.	50% of the initial activity is still measurable after exposure to air for 21.	Maness et al. (2002)
Ralstonia eutropha	*R. eutropha* has three hydrogenases: RH (regulatory hydrogenase); SH (soluble hydrogenase), and MBH (membrane bound hydrogenase). Oxygen insensitivity of the SH hydrogenase is related to the unique [NiFe]-active site architecture possessing two extracyanides, one bound to the Ni and the other bound to the Fe.	nd[a]	Happe et al. (2000)
	Oxygen insensitivity of the regulatory hydrogenase (RH) is determined by the shape and size of the intramolecular hydrophobic cavities leading to the active site.		Buhrke et al. (2005)
	Membrane-bound hydrogenases (MBH) possess a novel [3Fe–4S] cluster, which is known to determine oxygen tolerance. Suggestion is that the maturation proteins HoxR and HoxT may play a role in sustaining high oxygen tolerance.		Fritsch et al. (2011)

[a] nd: not determined.

capable of oxidizing hydrogen at atmospheric pO_2. However, to date, it is still not possible to conclusively identify what molecular factors render the enzymes from *Ralstonia* significantly less O_2-sensitive than corresponding enzymes from other species. A number of postulates can be drawn.

5.4.1 Reasons for Oxygen Insensitivity of Hydrogenase

5.4.1.1 Blocking of the Active Site by Partial or Complete Reduction Product of Attacking Oxygen

Compared to the oxygen-sensitive [FeFe], the moderately oxygen-tolerant [NiFe] hydrogenases are known to avoid oxygen toxicity by change in state of the active site when attacked by oxygen. Upon exposure to O_2, the Fe^{2+} atom at the active site during the catalytic cycle remains low spin (S = 0) and does not change the oxidation state. However, Ni at the active site changes the oxidation state (i.e., Ni^{3+}, Ni^{2+}, or even Ni^{1+}) leading to various inactive forms of the enzyme, namely, Ni-A or Ni-B. Depending on the oxidation state of the enzyme, a partial or completely reduced oxygen forms a third bridging ligand between the two metals (Ni and Fe). In the Ni-A state, the oxygenic bridge between Ni and Fe was found to be a per-oxo group or an oxygen atom bound to cysteinyl sulfur (of FeS cluster). Electron deficiency during the oxygen attack leads to the formation of this state, which is also known as the unready state. This state requires long-term (several hours) reactivation to return to its active. Moreover, such activation is known to take place exclusively *in vitro* (Vincent et al., 2007). In the Ni-B, oxygen forms a hydroxyl group at the Ni-Fe active site. Unlike Ni-A state, EPR studies have shown the rapid oxidation of Ni-B state to its active form within several seconds. This process of fast reactivation demands additional electrons that are delivered to the active site at high potential. Once returned to an anaerobic environment, the addition of a reducing agent such as sodium dithionite or H_2 re-reduces the oxo- or hydroxo-group, and hydrogenase activity is restored (Higuchi et al., 1999; Frey, 2002). The development of these inactive forms and their reversibility may help understand some characteristic features associated with the biophotolysis process such as (a) the lack of H_2 evolution in cyanobacterial cultures during photosynthesis, and (b) the rationale for the restoration of hydrogenase activity when cultures are assayed under dark, anaerobic conditions (Ghirardi et al., 2007, 2009). However, very recently, Hamdan et al. (2013) have challenged the concept of development Ni-A and Ni-B state on exposure to oxygen. In their studies, experimental results from both EPR and FTIR have shown that Ni-A and Ni-B states can be formed even anaerobically in comparable proportions. Thus, the earlier basis that a partial or complete reduction product of oxygen bound to the active site rendered it inactive may not be true. Thus, from their studies, Hamdan and group conclude that the attacking oxygen is not incorporated as an active site ligand but, rather, may act as an electron acceptor. They go on to explain that the oxygen tolerant hydrogenases have an unusual [4Fe-3S] cluster, which, they believe, may play a dominant role in developing the oxygen-tolerant behavior of the enzyme.

5.4.1.2 Protective Role of FeS Clusters Surrounding the Active Site

Interestingly the oxygen-tolerant [NiFe] hydrogenases from *Ralstonia-eutoropha* do not generate the Ni-A nor Ni-B inactive states (Bernhard et al., 2001). Instead, their oxygen tolerance is related to the conformational change in the proximal FeS clusters. Recently, Fritsch et al. (2011) described the crystal structure of the oxygen-tolerant [NiFe] hydrogenase (MBH, membrane-bound hydrogenase) from the aerobic hydrogen oxidizer *R. eutropha* H16 at 1.5 A resolution. The crystal structure revealed the presence of a novel [3Fe-4S] cluster coordinated by the sulfur atoms of six cystein residues, two of which are found exclusively in oxygen-tolerant hydrogenases. This cluster adapts a unique, open, noncuboidal conformation with extended distances between the di-iron center. Further, it can adapt three conformational redox states at physiological conditions and is proposed to act as a switch which serves either as an electron acceptor upon H_2 oxidation or as an electron donor. This property to act as an electron donor is vital for oxygen tolerance. Oxygen tolerance implies that upon approaching the catalytic center oxygen must be reduced by the immediate delivery of four electrons and protons to water (Goris et al., 2011). In accordance, the novel [3Fe-4S] cluster upon oxygen attack immediately delivers four electrons and protons to reduce it to water, thereby preventing the formation of the reactive oxygen species that damage or block the active site by formation of Ni-A state. Additionally, formation of water prevents the active site from getting blocked, thereby allowing the hydrogenase activity to continue. To generate further evidence, Lenz et al. (2011) replaced the regular cysteines with glycines in their position. A mutant derivative in which glycine residues replace both cysteines, lost its oxygen tolerance to a remarkable degree. The enhanced oxygen sensitivity of the mutant was accompanied by the conversion of the unique proximal [3Fe-4S] cluster to the standard [4Fe-4S] cluster. Thus, the unorthodox proximal FeS cluster likely plays a key role in developing oxygen sensitivity. Future efforts should concentrate on further understanding of their molecular architecture and further extrapolate the findings in other hydrogenases of choice.

5.4.1.3 Role of Conformation of Gas Channels in Delivering Oxygen Tolerance

Earlier studies showed that the diffusion of gas molecules to the hydrogenase active site is not random. Gas channels have been identified for hydrogen and oxygen diffusion into the active site (Nicolet et al., 1999). However, the conformation of these gas channels is said to play a critical role in determining the oxygen insensitivity. *R. eutropha* hydrogenases contain conserved phenylalanine (Phe) and isoleucine (Ile) positioned at the end of the channel. It was proposed that these residues narrow the gas channels thereby limiting their access to larger molecules like oxygen, however still allowing free diffusion of smaller molecules like hydrogen (Volbeda et al., 2002). Further, Buhrke et al. (2005) developed single and double mutants of the conserved amino acids of *R. eutropha* and found that its oxygen tolerance capacity was significantly compromised. However, to this date, it has not been possible to develop oxygen-tolerant hydrogenase, which indicates that any one but possibly a combination of the factors may account for the oxygen resistivity of *R. eutropha* hydrogenase. However, in literature, studies have been conducted on photosynthetic bacteria and green algal hydrogenases with limited success.

5.4.2 Possible Solutions to Overcome Oxygen Insensitivity of Hydrogenase

5.4.2.1 Change in the Amino Acid Residues of the Gas Channels

It has been suggested that specific gas channels allow hydrogen and oxygen to access the active site. Therefore, it has been postulated that reduction in size of these channels may help overcome oxygen sensitivity. In support of this hypothesis, the oxygen-tolerant hydrogen-sensing hydrogenases from *R. eutropha* contain phenylalanine (Phe) and isoleucine (Ile) positioned at the end of the channel, whereas the wild-type MBH contains luecine (Leu) and Valine (Val) at the same position (Buhrke et al., 2005). However, the O_2 tolerance of *Ralstonia* MBH enzymes cannot be simply linked to single-point mutations in the vicinity of the active site; mutants of *R eutropha* H16 MBH in which the closest varying residues were exchanged for those found in O_2-sensitive hydrogenases did not show significantly enhanced O_2 sensitivity compared to the wild type (Buhrke et al., 2005). Tolerance to O_2 is clearly a complex factor and is determined by a well-adapted spatial and electronic structure of the active site rather than a simple restriction of diffusion of inhibitory gases such as O_2. Further studies to investigate kinetic and thermodynamic details of the reactions of the active site with H_2 and O_2 are required.

5.4.2.2 Overexpression of Oxygen-Tolerant Hydrogenases

Another way to combat the oxygen sensitivity was suggested by the Craig Venter group and NREL, who suggested the overexpression of bidirectional oxygen-tolerant NiFe hydrogenase *hydS* and *hydL* along with their accessory genes of *Thiocapsa roseopersicina* in the cyanobacterium *Synechococcus* PCC7942. Overexpression of an oxygen-tolerant hydrogenase circumvents the need for additional sparging of an inert gas like argon in the culture medium to induce anaerobiosis.

5.4.2.3 Nano-Technology to the Rescue: Creating Anoxic Environments within the Organism to Enhance Hydrogen Production

Using nanotechnology, it may be possible to create a cyanobacteria with a membrane that incorporates an active transport protein system for the facilitated ejection of O_2. This system would rapidly remove the O_2 produced during metabolism. Another possibility to explore may be to create a membrane possessing an O_2-pholbic and an O_2-phillic side. The O_2-philiac side would face the bacteria and the phobic side open to the environment would facilitate the movement of O_2 away from the cells allowing the production of H_2 to continue without being hindered by the O_2.

5.4.2.4 Gene Shuffling for Rapid Generation of Hydrogen

Another approach to circumvent the issue may be obtained by gene shuffling. In an interesting approach, Nagy et al. (2007) showed that the native hydrogenase activity could be increased by gene shuffling. They shuffled the [FeFe]-hydrogenase genes between *C. acetobutylicum* and *C. saccharobutylicum* enzymes, which are 67% identical at the protein level. They showed that this shuffled gene gave 1.3-fold higher production than both the wild-type genes. The group did not show any tests related to increase in oxygen tolerance capacity of the shuffled product. However, construction of such libraries with both [NiFe] hydrogenase and nitrogenase may be attempted in the future to increase oxygen tolerance capacity of the hydrogenases.

In nature, different approaches have evolved to safeguard the oxygen sensitivity of the hydrogenase. This includes spatial and or temporal separation of the enzymes from oxygen. In the laboratory, Melis et al. (2000) found that macronutrient deprivation of sulfur serves a similar purpose by inactivating the oxygen evolving PSII. However, the use of this strategy for cyanobacteria is still debatable.

HYDROGENASE ASSAY

Quantification of hydrogen production and/or consumption can be carried out in vitro by determining the reduction of dyes like Methyl Viologen. The dye acts as a promiscuous electron donor to hydrogenase. It is a quantitative assay, measured using a spectrophotometer or gas chromatography. Native gel assays can also be carried out to determine the activity of the enzyme. Since mostly hydrogenases are highly sensitive to oxygen exposure, the primary requirement of the assay including the preparation of cell extracts is the presence of anaerobic environment. This is a limitation of the assay, because even though ultrasonication and high pressure apparatus have been used to disrupt cells under anaerobic conditions (Nakashimada et al., 2002), it is difficult to exclude air to achieve anaerobic conditions.

H₂ASE ASSAY BY GAS EVOLUTION METHOD

H_2ase activity can be quantified by the amount of H_2 evolved from sodium-dithionite (Na-dithionite)-reduced MV (Figure 5.9). The gas generated by the enzyme is detected by gas chromatograph equipped with a thermal conductivity detector. The assay is performed in a gas-tight screw cap vials containing 2 mL of 50 mM Tris–HCl, pH 8.0, 5 mM MV and 0.1–0.2 mL of the enzyme. The reaction is initiated by the addition of 10 mM Na-dithionite. The assay mixtures must be bubbled with argon to remove traces of dissolved oxygen before addition of the enzyme. The reaction mixture is incubated in a shaker at 37°C for 15 min. One unit of the enzyme is defined as the amount of hydrogenase evolving 1 μmol of H_2/min (Happe and Naber, 1993).

HYDROGENASE ASSAY BY SPECTROPHOTOMETRIC METHOD

H_2ase assay is measured spectrophotometrically with the oxidation of Na-dithionite-reduced MV (MV was turned purple after reduction) (Figure 5.2). The assay mixture (50 mM Tris–HCl, pH 8.0, 5 mM MV, 10 mM Na-dithionite) is taken in a gas-tight anaerobic cuvette (3 mL) and was flushed with argon before addition of the enzyme or crude cell lysate. Activity of the enzyme is monitored by measuring the decrease in absorbance assuming the molar extinction coefficient of MV 13.6 mM/cm at 604 nm. One unit of the enzyme is defined as the amount of H_2ase-catalyzing oxidation of 1 μmol of MV/min.

IN-GEL ASSAY ON NATIVE POLYACRYLAMIDE GEL ELECTROPHORESIS

The H_2ase activity can be detected on the native polyacrylamide gel. The detection of H_2ase activity on native PAGE is termed as activity staining. For activity staining ~2 μg of purified protein is dissolved in 10 μL of native protein-loading buffer (0.5 M Tris–HCl, pH 6.8, 20% glycerol (w/v), 0.2% (w/v) bromophenol blue and loaded onto 12% (w/v) non-denaturing PAGE. After electrophoresis, the lanes are cut and the gel pieces incubated in 50 mM Tris–HCl, pH 8.0 with 10 mM Na-dithionite and bubbled with argon for the next 15 min. Next, methyl viologen (5 mM final concentration) must be added and the solution is incubated at 37°C under nitrogen gas (in anaerobic glove box) until the hydrogenase activity is detected on native PAGE. The stain must be fixed with 2,3,5-triphenyltetrazolium chloride (1 mg/mL) and washed with de-ionized water.

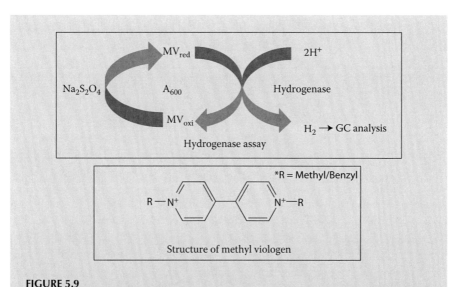

FIGURE 5.9
Mode of electron transfer from reduced methyl viologen (substrate) to the enzyme for hydrogen production. The activity of the enzyme can be determined by the oxidation of MV by measuring decrease in absorbance at 600 nm or by GC.

5.5 Evolutionary Significance of Hydrogenase

In nature, microbial hydrogenase besides hydrogen production are also associated with other functions. Some of these functions are discussed below.

5.5.1 Role of Hydrogenase during Nitrogen Fixation

Nitrogen fixation is catalyzed by the enzyme complex nitrogenase. It involves the reduction of nitrogen to ammonia and also characteristically produces hydrogen (Equation 5.4)

$$8H^+ + 4e^- + N_2 + 16ATP + 2NH_3 + H_2 + 16ADP + 16P_i \qquad (5.4)$$

Besides, in the absence of molecular nitrogen, the nitrogenase solely catalyzes the reduction of protons to molecular hydrogen. Since equivalent amount of energy is required by nitrogenase to reduce proton to molecular hydrogen, hydrogen production by nitrogenase is a waste of energy for the cell. Therefore, uptake hydrogenase recycles the H_2 produced during N_2 fixation, thereby minimizing the loss of energy during nitrogenase catalysis (Figure 5.10). The observation that nitrogen fixation efficiency increases in the presence of hydrogenase led Dixon and Arch (1972) to suggest that

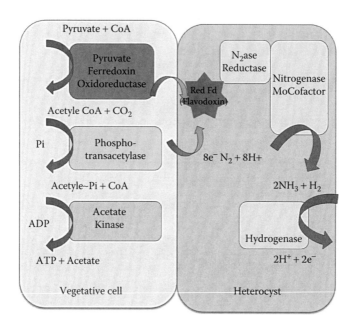

FIGURE 5.10
Schematic representation of relationship between nitrogenase and uptake hydrogenase in heterocystous cyanobacteria.

hydrogenase could support nitrogen fixation in aerobic organisms by (a) preventing reduction of nitrogen reduction by hydrogen generated by nitrogenase and (b) recycling of reducing power back to the cell by breakdown of hydrogen produced by nitrogenase. In accordance, several nitrogenase–hydrogenase interrelationships have been observed in several organisms (Vignais and Billoud, 2007) such as *Rhizobia* (Dixon, 1978), *Azotobacter* (Walker and Yates 1978), cyanobacteria (Bothe et al., 1978, 2010), and photosynthetic bacteria (Meyer et al., 1978).

5.5.2 Role of Hydrogenase during Methanogenesis

The biological sources of methane are wide-ranging, for example, Archaebacteria release the potent greenhouse gas in rice fields, mires, and cows' stomachs. However, the conditions for the occurrence of methonogenesis requires anaerobicity. Methane is produced by three major pathways: (1) reduction of carbon dioxide, (2) fermentation of acetate, and (3) dismutation of methanol or methylamines (Thauer et al., 2010). Hydrogenases play important roles in this metabolic process. The production of methane in these microbes depends on a nickel–iron hydrogenase called *frh*-adding electrons to a coenzyme called F_{420}. This hydrogenase cleaves a hydrogen molecule into two electrons, which are transferred to the F_{420} coenzyme, and two protons. The reduced form of F_{420} is then used for several reactions in the methane production process. This process, which is known as methanogenesis, provides the microbes with energy (for more details, refer to Thauer et al., 2010).

5.5.3 Role of Hydrogenase in Bioremediation

Tetrachloroethene is one of the most commonly present groundwater pollutants. In the past decades, it has accumulated due to the presence of chlorinated compounds such as pesticides, solvents, and cooling agents. Several anaerobic bacteria can use the compound as terminal electron acceptor in a novel kind of respiration known as dehalorespiration. It is widely known that most of the tetrachloroethene-dehalorespiring organisms use hydrogen as electron donor. These organisms have a high affinity for hydrogen and can outcompete methanogens and homoacetogens for this substrate. Examples of bacteria capable of dechlorination process with hydrogen as electron donor include *Dehalobacter restrictus* (Holliger et al., 1998), *Dehalococcoides ethenogenes* (Maymó–Gatell et al., 1997), and so on.

Besides dechlorination, hydrogenases are also known to be involved in reduction of toxic heavy metals from solution by efficient reduction to less soluble metal species (Lloyd and Lovely, 2001). For example, *E. coli* and *D. desulfuricans* reduce Tc (VII) with formate or hydrogen as electron acceptors (Lloyd et al., 1999) (for more details, refer to Vignais and Billoud, 2007).

5.6 Conclusion

Hydrogenase research has been strengthened by the availability of large-scale genome information and x-ray crystallography investigations. Phylogenetic and genome analysis reveal that the [NiFe] and [FeFe] hydrogenases are widely distributed in nature while [Fe-only] hydrogenases have been found restricted to mostly methanogens. These enzymes are complex and require the activity of accessory genes known as maturation machinery. Additionally, the transcription of the enzymes is also under strict regulation. However, hydrogenases are extremely oxygen labile, and, currently, a lot of work is being conducted to overcome this limitation of the enzyme.

Glossary

Arc	Aerobic respiratory control
DCMU	3-(3,4-Dichlorophenyl)-1,1-dimethylurea
EPR	Electron paramagnetic resonance

FHL   Formate hydrogen lyase
FNR   Fumarate nitrate reductase
FTIR   Fourier transform infrared spectroscopy
H$_4$MPT   Tetrahydromethanopterin
Hup   Hydrogenase uptake
SDS-PAGE   Sodium dodecyl sulfate polyacrylamide gel electrophoresis
TAT   Twin arginine translocation

References

Agapakis, C. M. and Silver P. 2010. Modular electron transfer circuits for synthetic biology: Insulation of an engineered biohydrogen pathway. *Bioengineered Bugs*, *1*, 413–418.

Andrews, S. C., Berks, B. C., McClay, J. et al.1997. A 12-cistron *Escherichia coli* operon (hyf) encoding a putative proton-pumping formate hydrogenlyase system. *Microbiology*, *143*, 3633–3647.

Armstrong, F. A. and Albracht, S. P. J. 2005. [NiFe]-hydrogenases: Spectroscopic and electrochemical definition of reactions and intermediates. *Philosophical Transactions. Series A, Mathematical, Physical, and Engineering Sciences*, *363*, 937–954.

Axelsson, R. and Lindblad, P. 2002. Transcriptional regulation of nostoc hydrogenases: Effects of oxygen, hydrogen and nickel. *Applied Environmental Microbiology*, *68*, 444–447.

Axelsson, R., Oxelfelt, F., and Lindblad, P. 1999. Transcriptional regulation of *Nostoc* uptake hydrogenase. *FEMS Microbiology Letters*, *170*, 77–81.

Balk, J., Pierik, A. J., Aguilar Netz, D. J., Mühlenhoff, U., and Lill, R. 2005. Nar1p, a conserved eukaryotic protein with similarity to Fe-only hydrogenases, functions in cytosolic iron-sulphur protein biogenesis. *Biochemical Society Transactions*, *33*, 86–89.

Balk, J., Pierik, A. J., Netz, D. J., Mühlenhoff, U., and Lill, R. 2004. The hydrogenase-like Nar1p is essential for maturation of cytosolic and nuclear iron-sulphur proteins. *EMBO Journal*, *23*, 2105–2115.

Ballantine, S. P. and Boxer, D. H. 1985. Nickel-containing hydrogenase isoenzymes from anaerobically grown *Escherichia coli* K-12. *Journal of Bacteriology*, *163*, 454–459.

Berkessel, A. 2001. Activation of dihydrogen without transition metals. *Current Opinion in Chemical Biology*, *5*, 486–490.

Bernhard, M., Buhrke, T., Bleijlevens, B. et al. 2001. *Journal of Biology Chemistry*, *276*, 15592–15597.

Bleijlevens, B., Buhrke, T., van der Linden, E., Friedrich, B., and Albracht, S. P. 2004. The auxiliary protein HypX provides oxygen tolerance to the soluble [NiFe]-hydrogenase of *Ralstonia eutropha* H16 by way of a cyanide ligand to nickel. *Journal of Biological Chemistry*, *279*, 46686–46691.

Bock, A., King, P. W., Blokesch, M., and Posewitz, M. C. 2006. Maturation of hydrogenases. *Advance Microbiology Physiology*, *51*, 1–71.

Böhm, R., Sauter, M., and Böck, A. 1990. Nucleotide sequence and expression of an operon in *Escherichia coli* coding for formate hydrogenlyase components. *Molecular Microbiology, 4*, 231–243.

Boichenko, V. A., Greenbaum, E., and Seibert, M. 2004. Hydrogen production by photosynthetic microorganisms. In *Photoconversion of Solar Energy: Molecular to Global Photosynthesis*, eds. M. D. Archer and J. Barber, pp. 397–452. London: Imperial College Press.

Boison, G., Bothe, H., and Schmitz, O. 2000. Transcriptional analysis of hydrogenase genes in the cyanobacteria *Anacystis nidulans* and *Anabaena variabilis* monitored by RT-PCR. *Current Microbiology, 40*, 315–321

Bothe, H., Distler, E., and Eisbrenner, G. 1978. Hydrogen metabolism in blue-green algae. *Biochimie, 60*, 277–289.

Bothe, H., Schimtz, O., Yates G. M., and Newton, W. E. 2010. Nitrogen fixation and hydrogen metabolism in cyanobacteria. *Microbiology and Molecular Biology Reviews, 74*, 529.

Boyd, E. S., Hamilton, T. L., Spear, J. R., Lavin, M., and Peters, J. W. 2010. [FeFe]-hydrogenase in Yellowstone National Park: Evidence for dispersal limitation and phylogenetic niche conservatism. *ISME Journal, 4*, 1485–1495.

Boyd, E. S., Spear, J. R., and Peters, J. W. 2009. [FeFe] hydrogenase genetic diversity provides insight into molecular adaptation in a saline microbial mat community. *Applied and Environmental Microbiology, 75*, 4620–4623.

Brazzolotto, X., Rubach, J. K., Gaillard, J., Gambarelli, S., Atta, M., and Fontecave, M. 2006. The [Fe-Fe]-hydrogenase maturation protein HydF from *Thermotoga maritima* is a GTPase with an iron-sulfur cluster. *Journal of Biological Chemistry, 281*, 769–774.

Brito, B., Martinez, M., Fernandez, D. et al. 1997. Hydrogenase genes from *Rhizobium leguminosarum* bv. viciae are controlled by the nitrogen fixation regulatory protein NifA. *Proceedings of National Academy of Sciences, 94*, 6019–6024.

Buhrke, T., Lenz, O., Krauss, N., and Friedrich, B. 2005. Oxygen tolerance of the H_2-sensing [NiFe] hydrogenase from *Ralstonia eutropha* H16 is based on limited access of oxygen to the active site. *Journal of Biological Chemistry, 280*, 23791–23796.

Cammack, R., Frey, M., and Robson, R. 2001. *Hydrogen as a fuel: Learning from Nature.* London and New York: Taylor and Francis.

Carrasco, C. D., Holliday, S. D., Hansel, A., Lindblad, P., and Golden, J. W. 2005. Heterocyst-specific excision of the *Anabaena* sp. strain PCC 7120 hupL element requires xisC. *Journal of Bacteriology, 187*, 6031–6038.

Casalot, L. and Rousset, M. 2001. Maturation of the [NiFe] hydrogenases. *Trends in Microbiology, 9*, 228–237.

Dixon, R. O. 1978. Nitrogenase—Hydrogenase interrelationships. *Rhizobia Biochimie, 60*, 233.

Dixon, R. and Arch, O. 1972. Hydrogenase in legume root nodule bacteroids: Occurrence and properties. *Microbiology, 85*, 193.

Dubini, A., Mus, F., Seibert, M., Grossman, A. R., and Posewitz, M. C. 2009. Flexibility in anaerobic metabolism as revealed in a mutant of *Chlamydomonas reinhardtii* lacking hydrogenase activity. *Journal of Biological Chemistry, 284*, 7201–7213.

Elsen, S., Dischert, W., Colbeau, A. et al. 2000. Expression of uptake hydrogenase and molybdenum nitrogenase in *Rhodobacter capsulatus* is coregulated by the RegB-RegA two-component regulatory system. *Journal of Bacteriology, 182*, 2831–2837.

Ensign, S. A. and Ludden, P. W. 1991. Characterization of the CO oxidation/H_2 evolution system of *Rhodospirillum rubrum* role of a 22-kDa iron-sulfur protein in mediating electron transfer between carbon monoxide dehydrogenase and hydrogenase. *Journal of Biological Chemistry*, 266, 18395–18403.

Florin, L., Tsokoglou, A., and Happe, T. 2001. A novel type of iron hydrogenase in the green alga *Scenedesmus obliquus* is linked to the photosynthetic electron transport chain. *Journal of Biological Chemistry*, 276, 6125–6132.

Forestier, M., King, P., Zhang, L., Posewitz, M., Schwarzer, S., Happe, T. et al. 2003. Expression of two [Fe]-hydrogenases in *Chlamydomonas reinhardtii* under anaerobic conditions. *European Journal of Biochemistry*, 270, 2750–2758.

Forzi, L., Hellwig, P., Thauer, R. K., and Sawers, R. G. 2007. The CO and CN⁻ ligands to the active site Fe in [NiFe]-hydrogenase of *Escherichia coli* have different metabolic origins. *FEMS Letters*, 581, 3317–3321.

Fox, J. D., He, Y., Shelver, D. et al. 1996. Characterization of the region encoding the CO-induced hydrogenase of *Rhodospirillum rubrum*. *Journal of Bacteriology*, 178, 6200–6208.

Frey, M. 2002. Hydrogenases: Hydrogen-activating enzymes. *ChemBioChem*, 3, 153–160.

Fritsch, J., Scheerer, P., Frielingsdorf, S. et al. 2011. The crystal structure of an oxygen-tolerant hydrogenase uncovers a novel iron-sulphur centre. *Nature*, 479, 249–252.

Garcin, E., Vernede, X., Hatchikian, E. C., Volbeda, A., Frey, M., and Fontecilla-Camps, J. C. 1999. The crystal structure of a reduced [Ni–Fe–Se] hydrogenase provides an image of the activated catalytic center. *Journal of Structural Fold and Design*, 7, 557–566.

Georgellis, D., Kwon, O., and Lin, E. C. 2001. Quinones as the redox signal for the arc two-component system of bacteria. *Science*, 292, 2314–2316.

Ghirardi, M. L., Dubini, A., Yu, J., and Maness, P. C. 2009. Photobiological hydrogen-producing systems. *Chemical Society Reviews*, 38, 52–61.

Ghirardi, M. L., Posewitz, M. C., Maness, P. C., Dubini, A, Yu, J. P., and Seibert, M. 2007. Hydrogenases and hydrogen photoproduction in oxygenic photosynthetic organisms. *Annual Review of Plant Biology*, 58, 71–91.

Gibral, L., von Abendroth, G., Winkler, M. et al. 2005. Homologous and heterologous overexpression in *Clostridium acetobutylicum* and characterization of purified clostridial and algal Fe⁻ only hydrogenases with high specific activities. *Applied and Environmental Microbiology*, 71, 2777–2781.

Goris, T., Wait, A. F., Saggu, M. et al. 2011. A unique iron-sulfur cluster is crucial for oxygen tolerance of a [NiFe]-hydrogenase. *Nature Chemical Biology*, 7, 310–318.

Guiral, M., Tron, P., Belle, V. et al. 2006. Hyperthermostable ad oxygen resistant hydrogenases from a hyperthermophilic bacterium *Aquifex aeolicus*: Physicochemical properties. *International Journal of Hydrogen Energy*, 31, 1424–1431.

Gutekunst, K., Hoffmann, D., Lommer, M. et al. 2006. Metal Dependence and Intracellular Regulation of the Bidirectional NiFe- Hydrogenase in *Synechocystis* sp. PCC 6803, *International Journal of Hydrogen Energy*, 31, 1452.

Gutekunst, K., Phunpruch, S., Schwarz, C., Schuchardt, S., Schulz-Friedrich, R., and Appel, J. 2005. LexA regulates the bidirectional hydrogenase in the cyanobacterium *Synechocystis* sp. PCC 6803 as a transcription activator. *Molecular Microbiology*, 58, 810–823.

Hamdan, A. A., Burlat, B., Gutiérrez-Sanz, O. et al. 2013. O_2-independent formation of the inactive states of NiFe hydrogenase. *Nature Chemical Biology*, 9, 15–17.

Happe, T. and Kaminski, A. 2002. Differential regulation of the Fe-hydrogenase during anaerobic adaptation in the green alga *Chlamydomonas reinhardtii*. *European Journal of Biochemistry*, 269, 1022–1032.

Happe, T. and Naber, J. D. 1993. Isolation, characterization and N-terminal amino acid sequence of hydrogenase from the green alga *Chlamydomonas reinhardtii*. *European Journal of Biochemistry*, 214, 475–481.

Happe, R. P., Roseboom, W., Egert, G. et al. 2000. Unusual FTIR and EPR properties of the H_2-activating site of the cytoplasmic NAD-reducing hydrogenase from *Ralstonia eutropha*. *FEBS Letters*, 466, 259–263.

Hatchikian, E. C., Magro, V., Forget, N., Nicolet, Y., and Fontecilla-Camps, J. C. J. 1999. Carboxy-terminal processing of the large subunit of [Fe] hydrogenase from *Desulfovibrio desulfuricans* ATCC 7757. *Bacteriology*, 181, 2947.

Higuchi, Y., Ogata, H., Miki, K., Yasuoka, N., and Yagi, T. 1999. Removal of the bridging ligand atom at the Ni-Fe active site of [NiFe] hydrogenase upon reduction with H_2, as revealed by X-ray structure analysis at 1.4 Å resolution. *Structure*, 7, 549–556.

Higuchi, Y., Yagi, T., and Yasuoka, N. 1997. Unusual ligand structure in Ni-Fe active center and an additional Mg site in hydrogenase revealed by high resolution X-ray structure analysis. *Structure*, 5, 1671–1680.

Higuchi, Y., Yasuoka, N., Kakudo, M., Katsube, Y., Yagi, T., and Inokuchi, H. J. 1987. Isolation and crystallization of high molecular weight cytochrome from *Desulfovibrio vulgaris Hildenborough*. *Journal of Biological Chemistry*, 262, 2823.

Holliger, C., D. Hahn, H. Harmsen, W. et al. 1998. *Dehalobacter restrictus* gen. nov. and sp. nov., a strictly anaerobic bacterium that reductively dechlorinates tetra- and trichloroethene in an anaerobic respiration. *Archives of Microbiology*, 169, 313–321.

Huang, J., Song, D., Flores, A. et al. 2007. IOP1, a novel hydrogenase-like protein that modulates hypoxia-inducible factor-1alpha activity. *Journal of Biochemistry*, 401, 341–352.

Kerby, R. L., Hong, S. S., Ensign, S. A. et al. 1992. Genetic and physiological characterization of the *Rhodospirillum rubrum* carbon monoxide dehydrogenase system. *Journal of Bacteriology*, 174, 5284–5294.

King, P. W., Posewitz, M. C., Ghirardi, M. L., and Seibert, M. 2006. Functional studies of [FeFe] hydrogenase maturation in an *Escherichia coli* biosynthetic system. *Journal of Bacteriology*, 188, 62163–2172.

Korbas, M., Vogt, S., Meyer-Klaucke, W. et al. 2006. The iron-sulfur cluster-free hydrogenase (Hmd) is a metalloenzyme with a novel iron binding motif. *Journal of Biological Chemistry*, 281, 30804–30813.

Kubas, G. J. 2005. Catalytic processes involving dihydrogen complexes and other sigma-bond complexes. *Catalysis Letters*, 104, 79–101.

Kuchenreuther, J. M., Grady-Smith, C. S., Bingham, A. S., George, S. J., Cramer, S. P., and Swartz, J. R. 2010. High-yield expression of heterologous [FeFe] hydrogenases in *Escherichia coli*. *PloS one*, 5(11), e15491.

Lenz, O., Bernhard, M., Buhrke, T., Schwartz, E., and Friedrich, B. 2002. The hydrogen-sensing apparatus in *Ralstonia eutropha*. *Journal of Molecular Microbiology Biotechnology*, 4, 255–262.

Lenz, O., Gleiche, A., Strack, A., and Friedrich, B. 2005. Requirements for heterologous production of a complex metalloenzyme, the membrane-bound [NiFe] hydrogenase. *Journal of Bacteriology*, 187, 6590–6595.

Lenz, O., Ludwig, M., Schubert, T. et al. 2011. H_2 conversion in the presence of O_2 as performed by the membrane-bound [NiFe]-hydrogenase of *Ralstonia eutropha*. *European Journal of Chemistry and Physics, 11*, 1107–1119.

Lloyd, J. R. and Lovley, D. R. 2001. Microbial detoxification of metals and radionuclides. *Current Opinion in Biotechnology, 12*, 248.

Lloyd, J. R., Thomas, G. H., Finlay, J. A., Cole, J. A., and Macaskie, L. E. 1999. Microbial reduction of technetium by Escherichia coli and *Desulfovibrio desulfuricans*: Enhancement via the use of high-activity strains and effect of process parameters. *Biotechnology and Bioengineering, 66*, 122.

Liu, Z. P. and Hu, P. 2002. A density functional theory study on the active center of Fe-only hydrogenase: Characterization and electronic structure of the redox states. *Journal of American Chemical Society, 124*, 5175–5182.

Luchi, S. and Lin, E. C. C. 1995. Signal transduction in the Arc system for control of operons encoding aerobic respiratory enzymes. In *Two-Component Signal Transduction*, eds. J. A. Hoch and T. Silhavy, pp. 223–231. Washington, DC: ASM Press.

Lutz, S., Jacobi, A., Schlensog, V., Bohm, R., Sawers, G., and Bock, A. 1991. Molecular characterization of an operon (hyp) necessary for the activity of the three hydrogenase isoenzymes in *Escherichia coli. Molecular Microbiology, 5*, 123–135.

Maness, P., Smolinski, S., Dillon, A. C., Heben, M. J., and Weaver, P. F. 2002. Characterization of the oxygen tolerance of a hydrogenase linked to a carbon monoxide oxidation pathway in *Rubrivivax gelatinosus. Applied and Environmental Microbiology, 68*, 2633–2636.

Marques, M. C., Coelho, R., De Lacey, A. L., Pereira, I. A., and Matias, P. M. 2010. The three-dimensional structure of [NiFeSe] hydrogenase from *Desulfovibrio vulgaris Hildenborough*: A hydrogenase without a bridging ligand in the active site in its oxidized, "as-isolated" state. *Journal of Molecular Biology, 396*, 893–907.

Matias, P. M., Soares, C. M., Saraiva, L. M. et al. 2001. [NiFe] hydrogenase from *Desulfovibrio desulfuricans* ATCC 27774, gene sequencing, three-dimensional structure determination and refinement at 1.8 A and modelling studies of its interaction with the tetrahaem cyto-chrome c3. *Journal of Biological Inorganic Chemistry, 6*, 63–81.

McGlynn, S. E., Ruebush, S. S., Naumov, A. et al. 2007. *Journal of Biological and Inorganic Chemistry, 12*, 443–447.

Melis, A., Zhang, L., Forestier, M., Ghirardi, M. L., and Seibert, M. 2000. Sustained photobiological hydrogen gas production upon reversible inactivation of oxygen evolution in the green alga *Chlamydomonas reinhardtii. Plant Physiology, 122*, 127–136.

Menon, N. K., Chatelus, C. Y., Dervartanian, M. et al. 1994. Cloning, sequencing. and mutational analyses of the hyb operon encoding *Escherichia coli* hydrogenase 2. *Journal of Bacteriology, 176*, 4416–4423.

Menon, N. K., Robbins, J., Peck, H. D., Jr., Chatelus, C. Y., Choi, E. S., and Przybyla, A. E. 1990. Cloning and sequencing of a putative *Escherichia coli* [NiFe] hydrogenase-1 operon containing six open reading frames. *Journal of Bacteriology, 172*, 1969–1977.

Meyer, J. 2007. [FeFe] hydrogenases and their evolution: A genomic perspective. *Cell Molecular Life Sciences, 64*, 1063–1084.

Meyer, J., Kelley, B. C., and Vignais, P. M. 1978. Effect of light nitrogenase function and synthesis in *Rhodopseudomonas capsulata. Journal of Bacteriology, 136*, 201–208.

Mills, D. J., Vitt, S., Strauss, M., Shima, S., and Vonck, J. 2013. De novo modeling of the F420-reducing [NiFe]-hydrogenase from a methanogenic archaeon by cryo-electron microscopy. eLife 2013;2:e00218 http://dx.doi.org/10.7554/eLife.00218.

Montet, Y., Amara, P., Volbeda, A. et al. 1997. Gas access to the active site of Ni-Fe hydrogenases probed by X-ray crystallography and molecular dynamics. *Nature Structural and Molecular Biology*, 4, 523–526.

Mulder, D. W., Boyd, E. S., Sarma, R. et al. 2010. Stepwise [FeFe]-hydrogenase H-cluster assembly revealed in the structure of HydA(DeltaEFG. *Nature*, 465, 248–251.

Mulder, D. W., Shepard, E. M., Meuser, J. E. et al. 2011. Insights into [FeFe]-hydrogenase structure, mechanism, and maturation. *Structure*, 19(8), 1038–1052.

Maymó-Gatell, X., Chien, Y., Gossett, J. M., and Zinder, S. H. 1997. Isolation of a bacterium that reductively dechlorinates tetrachloroethene to ethene. *Science*, 276, 1568–1571.

Nagy, L. E., Meuser, J. E., Plummer, S. et al. 2007. Application of gene-shuffling for the rapid generation of novel [FeFe]-hydrogenase libraries. *Biotechnological Letters*, 29, 421–430.

Nakajima, Y., Tsuzuki, M., and Ueda, R. 2001. Improved productivity by reduction of the content of light-harvesting pigment in *Chlamydomonas perigranulata*. *Journal of Applied Phycology*, 13, 95–101.

Nakashimada, Y., Rachman, M. A., Kakizono, T., and Nishio, N. 2002. Hydrogen production of *Enterobacter aerogenes* altered by extracellular and intracellular redox states. *International Journal of Hydrogen Energy*, 27, 1399–1405.

Nicolet, Y., Cavazza, C., and Fontecilla-Camps, J. C. 2002. Fe-only hydrogenases: Structure, function and evolution. *Journal of Inorganic Biochemistry*, 91, 1.

Nicolet, Y., Lemon, B. J., Fontecilla-Camps, J. C., and Peters, J. W. 2000. A novel FeS cluster in Fe-only hydrogenases. *Trends in Biochemistry*, 25, 138–143.

Nicolet, Y., Piras, C., Legrand, P., Hatchikian, C. E., and Fontecilla-Camps, J. C. 1999. *Desulfovibrio desulfuricans* iron hydrogenase: The structure shows unusual coordination to an active site Fe binuclear center. *Structural Fold and Design*, 7, 13–23.

Ogata, H., Kellers, P., and Lubitz, W. 2010. The crystal structure of the [NiFe] hydrogenase from the photosynthetic bacterium *Allochromatium vinosum*: Characterization of the oxidized enzyme Ni-A state. *Journal of Molecular Biology*, 402, 428–444.

Oliveira, P., Leita, E., Tamagnini, P., Moradas-Ferreira, P., and Oxelfelt, F. 2004. Characterization and transcriptional analysis of hupSL Win *Gloeothece* sp. ATCC 27152, an uptake hydrogenase from a unicellular cyanobacterium. *Microbiology*, 150, 3647–3655.

Oliveira, P. and Lindblad, P. 2005. LexA, a transcription regulator binding in the promoter region of the bidirectional hydrogenase in the cyanobacterium *Synechocystis* sp. PCC 6803. *FEMS Microbiology Letters*, 251, 59–66.

Oliveira, P. and Lindblad, P. 2008. An AbrB-like protein regulates the expression of the bidirectional hydrogenase in *Synechocystis* sp. strain PCC 6803. *Journal of Bacteriology*, 190, 1011–1019.

Peters, J. W. 1999. Structure and mechanism of iron-only hydrogenases. *Current Opinion in Structural Biology*, 9, 670–676.

Peters, J. W., Lanzilotta, W. N., Lemon, B. J., and Seefeldt, L. C. 1998. X-ray crystal structure of the Fe-only hydrogenase (CpI) from *Clostridium pasteurianum* to 1.8 angstrom resolution. *Science*, 282, 1853–1858.

Pierik, A. J, Hulstein, M., Hagen, W. R., and Albracht, S. P. J. 1998. A low-spin iron with CN and CO as intrinsic ligands forms the core of the active site in [Fe]-hydrogenases. *European Journal of Biochemistry*, 258, 572–578.

Patschkowski, T., Schlüter, A. and Priefer, U.B. 1996. *Rhizobium leguminosarum* bv. viciae contains a second fnr/fixK-like gene and an unusual fixL homologue. *Molecular Microbiology, 21,* 267–280.

Posewitz, M. C., King, P. W., Smolinski, S. L., Zhang, L., Seibert, M., and Ghirardi, M. L. 2004. Discovery of two novel radical S-adenosylmethionine proteins required for the assembly of an active [Fe] hydrogenase. *Journal of Biological Chemistry, 279,* 25711–25720.

Posewitz, M. C., Mulder. D. W., and Peters, J. W. 2008. New frontiers in hydrogenase structure and biosynthesis. *Current Chemical Biology, 2,* 178–199.

Roseboom, W., De Lacey, A. L., Fernandez, V. M., Hatchikian, E. C., and Albracht, S. P. 2006. The active site of the [FeFe]-hydrogenase from *Desulfovibrio desulfuricans*. II. Redox properties, light sensitivity and CO-ligand exchange as observed by infrared spectroscopy. *Journal of Biological and Inorganic Chemistry, 11,* 102–118.

Rossmann, R., Sawers, G., and Bock, A. 1991. Mechanism of regulation of the formate-hydrogenlyase pathway by oxygen, nitrate and pH: Definition of the formate regulon. *Molecular Microbiology, 5,* 2807–2814.

Salmon, K., Hung, S. P., Mekjian, Baldi, P., Hatfield, G. W., and Gunsalus, R. P. 2003. Global Gene Expression Profiling in *Escherichia coli* K12: The effects of oxygen availability and FNR. *Journal of Biological Chemistry, 278,* 29837–29855.

Sauter, M., Bohm, R., and Bock, A. 1992. Mutational analysis of the operon (hyc) determining hydrogenase 3 formation in *Escherichia coli*. *Molecular Microbiology, 6,* 1523–1532.

Sawers, R. G., Ballantine, S. P., and Boxer, D. H. 1985. Differential expression of hydrogenase isoenzymes in *Escherichia coli* K-12: Evidence for a third isoenzyme. *Journal of Bacteriology, 164,* 1324–1331.

Sawers, R. G., Blokesch M., and A. Böck. 2004. Anaerobic formate and hydrogen metabolism. September 2004, posting date. In *R. Curtiss III EcoSal—Escherichia coli and Salmonella: Cellular and Molecular Biology*. Washington, DC: ASM Press.

Sawers, R. G. and Boxer, D. H. 1986. Purification and properties of membrane-bound hydrogenase isoenzyme 1 from anaerobically grown *Escherichia coli*. *European Journal of Biochemistry, 156,* 265–275.

Schleucher, J., Vanderveer, P., Markley, J. L., and Sharkey, T. D. 1999. Intramolecular deuterium distributions reveal disequilibrium of chloroplast phosphoglucose isomerase. *Plant Cell Environment, 22,* 525–534.

Schmitz, O., Boison, G., Hilscher, R. et al. 1995. Molecular biological analysis of a bidirectional hydrogenase from cyanobacteria. *European Journal of Biochemistry, 233,* 266–276.

Schmitz, O., Boison, G., Salzmann, H. et al. 2002. HoxE—A subunit specific for the pentameric bidirectional hydrogenase complex (HoxEFUYH) of cyanobacteria. *Biochemistry and Biophysical Acta, 1554,* 66–74.

Schwarz, C., Poss, Z., Hoffmann, D., and Appel, J. 2010. Recent advances in phototrophic prokaryotes. In *Medicine 675*, eds. P. C. Hallenbeck, pp. 305–348. New York, NY: Springer New York.

Sheremetieva, M. E., Troshina, O. Y., Serebryakova, L. T., and Lindblad, P. 2002. Identification of hox genes and analysis of their transcription in the unicellular cyanobacterium *Gloeocapsa alpicola* CALU 743 growing under nitrate-limiting conditions. *FEMS Microbiology Letters, 214,* 229–233.

Sinha, P. and Pandey, A. 2011. An evaluative report and challenges for fermentative biohydrogen production. *International Journal of Hydrogen Energy, 36*(13), 7460–7478.

Skibinski, D. A., Golby, P., Chang, Y. S. et al. 2002. Regulation of the hydrogenase-4 operon of *Escherichia coli* by the sigma (54)-dependent transcriptional activators FhlA and HyfR. *Journal of Bacteriology, 184*, 6642–6653.

Song, L. C., Yang, Z. Y., Hua, Y. J., Wang, H. T., Liu, Y., and Hu, Q. M. 2007. *Organometallics, 26*, 2106–2110.

Stephenson, M. and Stickland, L. H. 1931. Hydrogenlyases. *Journal of Biochemistry, 26*, 712–724.

Tamagnini, P., Axelsson, R., Lindberg, P. et al. 2002. Hydrogenases and hydrogen metabolism of cyanobacteria. *Microbiology and Molecular Biology Reviews, 66*, 1–20.

Tamagnini, P., Costa, J. L., Almeida, L., Oliveira, M. J., Salema, R., and Lindblad, P. 2000. Diversity of cyanobacterial hydrogenases, a molecular approach. *Current Microbiology, 40*, 356–361.

Tamagnini, P., Leitão, E., Oliveira, P. et al. 2007. Cyanobacterial hydrogenases: Diversity, regulation and applications. *FEMS Microbiology Reviews, 31*, 692–720.

Thauer, R. K., Hedderich, R., and Fischer, R. 1993. Biochemistry, reactions and enzymes involved in methanogenesis from CO, and H. In *Methanogenesis*, eds. J. G. Ferry, pp. 209–252. New York and London: Chapman and Hall.

Thauer, R. K., Kaster, A. K., Goenrich, M., Schick, M., Hiromoto, T., and Shima, S. 2010. Hydrogenases from methanogenic archaea, nickel, a novel cofactor and H_2 storage. *Annual Review Of Biochemistry, 79*, 507–536.

Valente, F. A., Almeida, A. C., Pacheco, I., Carita, J., Saraiva, L. M., and Pereira, I. A. C. 2006. Selenium is involved in regulation of periplasmic hydrogenase gene expression in *Desulfovibrio vulgaris* Hildenborough. *Bacteriology, 188*, 3228–3235.

Vignais, P. M. 2007. Hydrogenases and H^+ reduction in primary energy conservation. Results and problems in cell differentiation, March. doi:10.1007/400.

Vignais, P. M. and Billoud, B. 2007. Occurrence, classification, and biological function of hydrogenases: An overview. *Chemical Reviews, 107*, 4206–4272.

Vignais, P. M., Billoud, B., and Meyer, J. 2001. Classification and phylogeny of hydrogenases 1. *FEMS Microbiology Reviews, 25*, 455–501.

Vignais, P. M. and Colbeau, A. 2004. Molecular biology of microbial hydrogenases. *Current Issues in Molecular Biology, 6*, 159–188.

Vignais, P. M. and Toussaint, B. 1994. Molecular biology of membrane-bound H_2 uptake hydrogenases. *Archives of Microbiol*ogy, *1161*, 1–10.

Vincent, K. A., Parkin, A., and Armstrong, F. A. 2007. Investigating and exploiting the electrocatalytic properties of hydrogenases. *Chemical Reviews, 107*, 4366–4413.

Vogt, S., Lyon, E. J., Shima, S., and Thauer, R. K. 2008. The exchange activities of [Fe]-hydrogenase (iron-sulfurcluster-free hydrogenase) from methanogenic archaea in comparison with the exchange activities of [FeFe] and [NiFe] hydrogenases. *Journal of Biological Inorganic Chemistry, 13*, 97–106.

Volbeda, A., Charon, M. H., Piras, C., Hatchikian, E. C., Frey, M., and Fontecilla-Camps, J. C. 1995. Crystal structure of the nickel-iron hydrogenase from *Desulfovibrio gigas. Nature, 373*, 580–587.

Volbeda, A., Garcin, E., Piras, C. et al. 1996. Structure of the [NiFe] hydrogenase active site: Evidence for biologically uncommon Fe ligands. *Journal of American and Chemical Society, 118*, 12989–12996.

Volbeda, A., Martin, L., Cavazza, C. et al. 2005. Structural differences between the ready and unready oxidized states of [NiFe] hydrogenases. *Journal of Biological Inorganic Chemistry, 10,* 239–249.

Volbeda, A., Montet, Y., Vernède, X., Hatchikian, E. C., and Fontecilla-Camps, J. C. 2002. High resolution crystallographic analysis of *Desulfovibrio fructosovorans* [NiFe] hydrogenase. *International Journal of Hydrogen Energy, 27,* 1449–1461.

Voordouw, G. 1992. Evolution of hydrogenase genes. *Advances Inorganic Chemistry, 38,* 397–422.

Walker, C. C. and Yates, M. G. 1978. The hydrogen cycle in nitrogen-fixing *Azotobacter chroococcum. Biochemistry, 60,* 225.

Wu, L. F. and Mandrand-Berthelot, M. A. 1986. Genetic and physiological characterization of new *E. coli* mutants impaired in hydrogenase activity. *Biochemistry, 68,* 167–179.

Wu, L. F. and Mandrand-Berthelot, M. A. 1993. Microbial hydrogenases—Primary structure, classification, signatures and phylogeny. *FEMS Microbiology Review, 104,* 243–270.

Yoshida, A., Nishimura, T., Kawaguchi, H., Inui, M., and Yukawa, H. 2005. Enhanced hydrogen production from formic acid by formate hydrogen lyase-overexpressing *Escherichia coli* strains. *Applied and Environmental Microbiology, 71,* 6.

FIGURE 3.10
Pilot plant of 0.8 m³ capacity installed at IIT, Kharagpur, India.

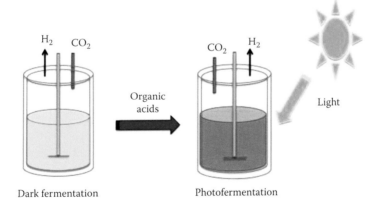

FIGURE 3.11
Integrated dark fermentation and photofermentation to increase the overall yield of the process.

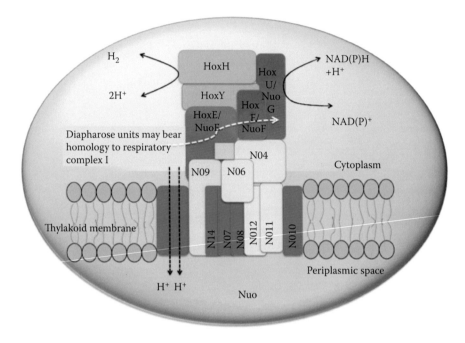

FIGURE 5.5
Hypothetical model of putative interaction of hox bidirectional hydrogenase with respiratory complex 1.

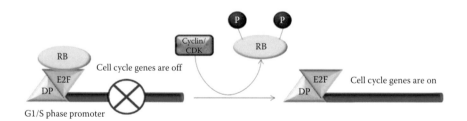

FIGURE 6.2
Schematic diagram represents the cyclin (CYC)/retinoblastoma (RB) pathway, showing E2F transcription factors, inactive when bound by RB protein. The phosphorylation of RB by cyclin dissociates RB from E2F factors, triggering the passage of cells from G1- to S-phase in the cell cycle.

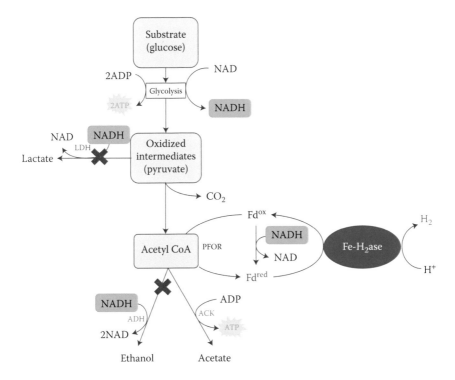

FIGURE 6.3
Schematic diagram represents the redirection of metabolic pathways toward hydrogenase by inhibiting the alcohol and acid production pathways for enhanced hydrogen production.

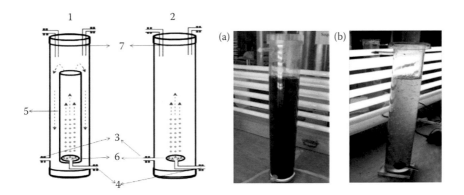

FIGURE 8.1
Schematic diagram and photographs of (a) airlift and (b) bubble column PBRs for algal biomass production using *C. sorokiniana* (1: airlift PBR, 2: bubble column PBR, 3: sampling port, 4: gas inlet, 5: inert draft tube, 6: sparger, 7: gas outlet).

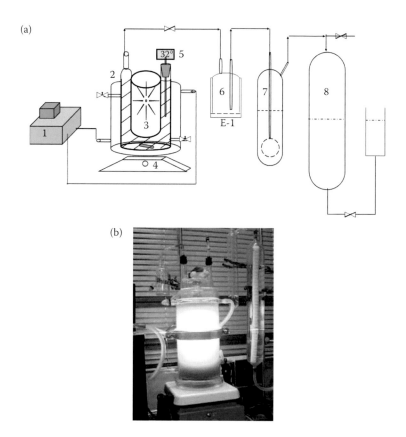

FIGURE 8.7
(a) Schematic diagram of triple-jacketed reactor (1: Temperature control water reservoir, 2: Triple-jacketed PBR, 3: Light source, 4: Magnetic stirrer, 5: Temperature monitoring, 6: Gas trap, 7: CO_2 absorber, 8: Gas collector). (b) Photograph of the triple-jacketed PBR using *Rhodobacter sphaeroides* O.U.001.

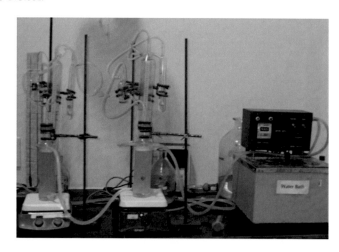

FIGURE 10.3
Batch experimental set up with double-jacketed reactor for the biohydrogen production.

6

Improvement of Hydrogen Production through Molecular Approaches and Metabolic Engineering

6.1 Introduction

Microorganisms possess several unique genetic and metabolic attributes in relevance to biohydrogen production. The knowledge of molecular basis of H_2 production and application of genetic engineering could improve energy production phenotypes up to commercial levels. Different potential avenues of genetic manipulation would be the best way to create a platform for the economically feasible H_2 production. Significant advances in the development of genetic manipulation tools have recently been achieved with some bacterial, cyanobacterial, and microalgal systems, and they are being used to manipulate the central carbon metabolism skeleton, to improve their enzymatic activities, to redirect the metabolic pathways, and to introduce foreign genes within these organisms. Different strategies for genetic modification are given in Table 6.1.

6.2 Molecular Approaches

Fermentative bacteria and microalgae are an attractive resource of renewable energy. Different approaches are available to increase the growth rate and hydrogen production including suitable strain selection, physicochemical parameter optimization, bioreactor design, and genetic engineering approaches.

6.2.1 Improvement of Biomass Production

Low-biomass yields are identified as a key driver of the high cost of H_2 production. In bacterial system, generation time can be very fast like within around

TABLE 6.1

Table Represents the Genetic and Metabolic Engineering Strategies, Genetic Modification in Different Microorganisms for Overcoming the Bottlenecks of Hydrogen Production

Strategy	Procedure Adopted	Genetic Modification	Functions of the Gene	Organism	Observation	Reference
Improvement of biomass production	Targeting CO_2-concentrating mechanisms (CCMs)	Induction of *ccm1*	Master regulator of CCM	*Chlamydomonas reinhardtii*		Fett and Coleman (1994)
		Activation of *lciA* and *lciB*	Probable involvement in inorganic carbon transport	*Chlamydomonas reinhardtii*		Miura et al. (2004)
	Triggering the passage of cells from G1- to S-phase in the cell cycle	*mat3*	Retinoblastoma homolog of cyclin (CYC)/retinoblastoma (RB) pathway	*Chlamydomonas reinhardtii*		Umen et al. (2001)
Enhancement of the uptake of external substrate	Improvement of the supply of external glucose into the cell	*HUP1* gene integration from *Chlorella kessleri*	Hexose sympoter	*Chlamydomonas reinhardtii* strain stm6	H_2 production increased about 150% by supplying 1 mM glucose	Doebbe et al. (2007)
Improvement of photoconversion efficiency	Decrease in core antennal (LH1) content and increases in peripheral antennal (LH2) content			*Rhodobacter sphaeroides* (P3 mutant)	Produced 1.5fold higher rate of H_2 at light intensities at 800 and 850 nm compared to the wild-type	Vasilyeva et al. (1999)

Goal	Method	Description	Organism	Result	Reference
Decrease in bacteriochlorophyll and carotenoids in the chromatophores	Irradication with UV light (254 nm) for 10 s		*Rhodobacter sphaeroides* (MTP4 mutant)	Produced 50% more H$_2$ than wild type	Kondo et al. (2002)
Deficient in phycobilisomes	Deletion of *apcE* and *apcAB* operons	Deficient of phycocyanin and phycobilisome	*Synechocystis* PCC 6803 (Olive and PAL mutant)	200 mL H$_2$/h/L culture	Bernát et al. (2009)
Truncated chlorophyll antenna	Interruption of *tla1* by insertional mutagenesis	Responsible for the regulation of the Chl antenna size in green algae	*Chlamydomonas reinhardtii*	Chlorophyll antenna size of PSI and PSII being about 50% and 65% of that of the wild type.	Polle et al. (2003)
Deficient in phycocyanin	Single-base substitution upstream of the *cpc* operon by chemical mutagenesis	Coding for phycocyanin	*Synechocystis* PCC 6714 (PD1 mutant)	In PD 1 mutant expression of *cpc* were 1/10 to 1/6 than wild type	Nakajima et al. (2001)
Reduced levels of LHCI and LHCII	Downregulation of the entire LHC gene family by RNAi technology		*Chlamydomonas reinhardtii* (*Stm3LR3* mutant)	The mutant *Stm3LR3* reduced levels of LHCI and LHCII	Mussgnug et al. (2007)
Reduction of LHC antenna size	Transformation of a permanently active variant NAB1	NAB1 is an LHC translation repressor	*Chlamydomonas reinhardtii* (*Stm6Glc4T7* mutant)		Beckmann et al. (2009)
Improvement of tolerance to oxygen	Random and site-directed mutagenesis		*Chlamydomonas reinhardtii*	Threefold more oxygen tolerance than wild type	Ghirardi et al. (2000)
Improvement of hydrogenase					

continued

TABLE 6.1 (continued)

Table Represents the Genetic and Metabolic Engineering Strategies, Genetic Modification in Different Microorganisms for Overcoming the Bottlenecks of Hydrogen Production

Strategy	Procedure Adopted	Genetic Modification	Functions of the Gene	Organism	Observation	Reference
Improvement of nitrogenase	Under air H_2 production	Site-directed mutagenesis of *nifD*	Encodes for MoFe protein of nitrogenase	*Anabaena* sp. PCC 7120 (Q193 K, H197T and R284H NifD mutants)	Under air mutants approached or exceeded their H_2 production rates under Argon	Masukawa et al. (2010)
	Blocking of the CO_2 fixation pathway	Combination of mutations in *nifA*, *draT*, and *cbbM*	*nifA*—Transcriptional activator of the nitrogen fixation	*Rhodospirillum rubrum*	Mutants are able to produce H_2 over a much longer time frame than the wild type	Wang et al. (2010)
	Reduction of ammonium ion-mediated repression of nitrogenase	Knocking out *glnB* and *glnK*	Play key roles in repressing the nitrogenase expression in the presence of ammonium ion	*Rhodobacter sphaeroides*	More H_2 accumulation by the mutant in the presence of ammonium ion	Kim et al. (2008)
	Reduction of N_2 fixation activity	*nifV1* and *nifV2*	Homocitrate synthase genes	*Nostoc* PCC 7120	Prolong the high-hydrogen production period.	Sakurai et al. (2008)

Overexpression of enzymes	Overexpression of hydrogenase	Electrotransformation of 2327-bp DNA (ORF of hydA + putative promoter) through pJIR418 shuttle vector	[FeFe] hydrogenase	*Clostridium tyrobutyricum* JM1	1.7-fold and 1.5-fold increase in hydrogenase activity and H_2 yield with the shift of metabolic pathway	Jo et al. (2010)
	Overexpression of hydrogenase	Transformation of *hydA* ORF by pGEX-Kan-*hydA* vector	[FeFe] hydrogenase	*Enterobacter cloacae* IIT-BT 08	The hydrogen yield and rate of production were found to be 1.2-fold and 1.6-fold higher, respectively, compared to the wild type strain	Khanna et al. (2011)
Introduction of foreign hydrogenase	Introduce from *Clostridium acetobutylicum*	Fe-only hydrogenase cloned in pRKhydA vector and mobilized from *E. coli* S17-1 to *Rhodospirillum rubrum* by conjugation	Fe-only hydrogenase	*Rhodospirillum rubrum*	Fe-only hydrogenase requires pyruvate as an electron donor	Kim et al. (2008)
	Introduce from *Clostridium pasteurianum* ATCC6013	By direct electroporation creating "psuedo-transformants"		*Synechococcus elongatus*	Hydrogenase activity was 11 nmol H_2 evolved with reduced Methyl Viologen/mg chlorophyll	Miyake and Asada (1997)

continued

TABLE 6.1 (continued)

Table Represents the Genetic and Metabolic Engineering Strategies, Genetic Modification in Different Microorganisms for Overcoming the Bottlenecks of Hydrogen Production

Strategy	Procedure Adopted	Genetic Modification	Functions of the Gene	Organism	Observation	Reference
	Introduce from *Clostridium pasteurianum*	Hydrogenase I in pKE4-9 vector which carried the chloramphenicol-resistant gene and a strong promoter	[FeFe]-hydrogenase	*Synechococcus* PCC7942		Asada et al. (2000)
	Introduce from *Clostridium acetobutylicum*	*hydA*	[FeFe]-hydrogenase	*Synechococcus elongatus* sp. 7942	H_2 evolution at a rate500-fold greater than that supported by the endogenous [NiFe]-hydrogenase	Ducat et al. (2011)
	Introduce from *Shewanella oneidensis* MR-1	Conjugal mobilization of plasmids from *E. coli* into the Hup-74 mutant AMC414	[FeFe]-hydrogenase	*Anabaena* sp. strain 7120	Heterocysts protected the [FeFe]-hydrogenase against inactivation by O_2	Gärtner et al. (2012)
Deletion of hydrogen uptake genes	Deletion	*hupL* gene	Encoding the large subunit of uptake hydrogenase	*Rhodospirillum rubrum*	H_2 yield of *R. rubrum* UR801 under continuous light is 5700 mL of H_2/L	Ruiyan et al. (2006)

	Suicide vector for site-directed mutagenesis	*hupSL*	Uptake hydrogenase	*Rhodobacter sphaeroides*	Kars et al. (2008)
		hypF		*Thiocapsa roseopersicina*	Fodor et al. (2001)
	Disruption	*hupS*		*Anabaena siamensis* TISTR 8012	Khetkorn et al. (2012)
	Single deletion	double mutant of Δ*hox* and Δ*hup*.			Masukawa et al. (2002)
Generation of anaerobic condition	Attenuated photosynthesis/respiration ratio (P/R ratio)	*apr* mutant		*Chlamydomonas reinhardtii*	Melis (2007)
	Double amino acid substitution to generate D1 mutant	Leucine residue L159 in the D1 protein was replaced by isoleucine, and the asparagine N230 was replaced by tyrosine (L159I-N230Y)		*Chlamydomonas reinhardtii*	Torzillo et al. (2009)
ATP synthase modification	Synthetic proton channel targeted in the thylakoid membranes, under the control of an anaerobic hydrogenase-driven promoter			*Chlamydomonas reinhardtii*	Lee and Greenbaum (2003)
	Mutation	AtpE	ATP synthase subunits		Robertson et al. (1990)

continued

TABLE 6.1 (continued)

Table Represents the Genetic and Metabolic Engineering Strategies, Genetic Modification in Different Microorganisms for Overcoming the Bottlenecks of Hydrogen Production

Strategy	Procedure Adopted	Genetic Modification	Functions of the Gene	Organism	Observation	Reference
Linking of hydrogenase to cyanobacterial photosystems	Development of PsaE-Hyd chimera	Fusion of PSI subunit PsaE (*Thermosynechococcus elongates*) and O$_2$-tolerant [NiFe]-hydrogenase (*Ralstonia eutropha* H16)	[NiFe]-hydrogenase	*Synechocystis* PCC 6803		Ihara et al. (2006)
	Development of cyc3-PSI chimera	Fusion of [FeFe]-hydrogenase (*Clostridium acetobutylicum*) and PSI (*Synechococcus* sp. PCC 7002 PS) using 1,6-hexanedithiol molecular wire	[FeFe]-hydrogenase			Lubner et al. (2010)
Engineering of heterocyst frequency	Overexpression	*hetR*	Serine-type protease crucial to the heterocyst differentiation		Multiple heterocysts	Buikema and Haselkorn (2001)
Redirection of the electron pull	Defected in both alcohol and organic acid formation pathways	Double mutants by chemical mutagenesis		*Enterobacter cloacae* IIT-BT08		Kumar et al. (2001)

Upregulation of formate hydrogen lyase (FHL) complex and disruption of the pathways competing for pyruvate		*Escherichia coli*		Yoshida et al. (2005)
Gene silencing	NAD(P) H-dehydrogenase	*Synechocystis* M55		Cournac et al. (2004)
Reduction of plastoquinon (PQ) by Ndh		Cyanobacteria and green algae	Maintain the NAD/NADH balance and ATP supply	Hemschemeier and Happe (2005)
Random gene insertion		*Chlamydomonas reinhardtii* strain Stm6	Modification of respiratory metabolism	Kruse et al. (2005)

few minutes as, for example, under optimal growth conditions the doubling time for *E. coli* is around 15–20 min. Whereas, doubling time of microalga *Chlamydomonas* sp. at optimal physicochemical condition is around 5–8 h and it takes 8–10 days to reach the stationary phase (Lee and Fiehn, 2008). Biomass productivity needs to be improved for cost-effective H_2 production.

6.2.1.1 CO_2-Concentrating Mechanisms (CCMs)

Photosynthetic acclimation due to limitation of CO_2 is associated with control of genetic and physiological responses through a signal transduction pathway, followed by integrated monitoring of the environmental changes. During photosynthesis, CO_2 is converted into sugar by the carboxylase activity of the enzyme ribulose 1,5-bisphosphate carboxylase oxygenase (RuBisCO). Glycolate 2-phosphate produced by oxygenase activity of RuBisCO has no use to cell, and its synthesis consumes significant amount of cellular energy and also releases previously fixed CO_2 by carboxylase activity of RuBisCO (Figure 6.1). Oxygenase activity of RuBisCO inhibits biomass formation of around 50% (Giordano et al., 2005). To overcome the low affinity of RuBisCO for CO_2, most algae and cyanobacteria have different CO_2-concentrating mechanisms (CCMs). Many species of eukaryotic algae and cyanobacteria are able to grow efficiently at low-inorganic carbon (Ci) concentrations by inducing the expression of CCMs (Aizawa and Miyachi, 1986; Coleman, 1991) leading to

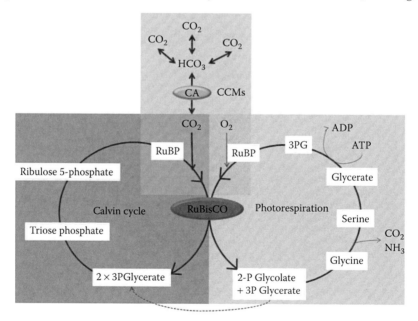

FIGURE 6.1
Schematic diagram represents two different activities of RuBisCO (carboxylase and oxygenase) and CO_2-concentrating mechanisms (CCMs).

the accumulation of a large intracellular Ci pool from which CO_2 is obtained for fixation by RuBisCO into specialized structures like the cyanobacterial carboxysome or the pyrenoid in eukaryotic algae. Several CO_2-responsive genes have been isolated and well studied from cyanobacteria than from green algae (Price et al., 2008). However, it has been found that the *ccm1* is the master regulator of CCM in *Chlamydomonas reinhardtii*. The expression of the enzyme carbonic anhydrase (CA) has been found to be associated with induction of the CCMs. CA catalyzes the interconversion of CO_2 and HCO_3^- and is an important component in the intracellular mobilization of the HCO_3^- pool, by catalyzing the production of CO_2 for RuBisCO (Fett and Coleman, 1994). Among low-CO_2 inducible genes, two novel genes, *lciA* and *lciB*, were identified, which may be involved in inorganic carbon transport (Miura et al., 2004). Genetic engineering research at the Iowa State University reveals that the algae biomass can be boosted by 50–80% by expressing or activating two genes—*lciA* and *lciB* that promote photosynthesis, capture more CO_2, and increase the biomass productivity. However, when algae live in an environment with enough CO_2, these two genes are shut down. The researchers found that expressing them, even in carbon-rich environments, significantly increases growth. This has been first tested in *Chlamydomonas reinhardtii*. Each gene individually yielded a 10–15% increase in biomass. Expressing them together boosted it 50–80% (www.oilgae.com).

6.2.1.2 Cell Cycle

Cell cycle plays a crucial role in the growth and development of multicellular as well as unicellular algae (Dewitte and Murray, 2003). The key players of the cyclin (CYC)/retinoblastoma (RB) pathway are conserved between mammals and plants, albeit under substantially different regulatory mechanisms (Dewitte and Murray, 2003) (Figure 6.2). In unicellular algae, the cyclin (CYC)/retinoblastoma (RB) pathway is believed to be involved in controlling both the commitment of cells to the mitotic cell cycle and decisions involving cell growth and cell division (Umen and Goodenough, 2001; Fang and

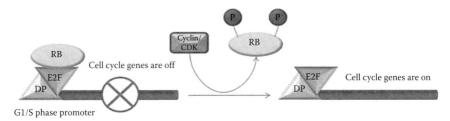

FIGURE 6.2
(**See color insert.**) Schematic diagram represents the cyclin (CYC)/retinoblastoma (RB) pathway, showing E2F transcription factors, inactive when bound by RB protein. The phosphorylation of RB by cyclin dissociates RB from E2F factors, triggering the passage of cells from G1- to S-phase in the cell cycle.

Umen, 2008). The key genes for growth and cell division are regulated by E2F transcription factors, which are inactive when bound by Retinoblastoma protein (RB). The phosphorylation of RB is initiated by Cyclin (CYC)-containing cyclin-dependent kinases (CDKs), resulting in the dissociation of RB from E2F factors (RB-suppressor), triggering the passage of cells from G1- to S-phase in the cell cycle. Umen identified an RB homolog encoded by the *mat3* gene in *C. reinhardtii* and later discovered algal counterparts of other players in the RB pathway in humans and mice. To analyze their function in *Chlamydomonas*, the Salk team isolated cells with mutations in individual members of the RB-signaling pathway (Umen and Goodenough, 2001). Reported suppressors of *mat3* are *dp1* and *e2f1*. Four other weaker suppressors of *mat3* are also reported *smt7-1, smt14-1, smt15-1,* and *smt16-1*. Like *dp1* and *e2f1*, these mutants suppressed *mat3* by affecting size checkpoint control. It has been observed that cell-cycle regulatory function and genetic wiring of the RB pathway are conserved in *Chlamydomonas*, making it attractive model algae for further analyses (Fang and Umen, 2008).

6.2.2 Enhancing the Uptake of External Substrate

In fermentative hydrogen production, external substrates are the sole source of electrons. By overexpressing the external substrate transporter proteins, it might be possible to increase the hydrogen production. There are some reports about modification in transporter proteins. Species of *Chlorella* has a hexose uptake protein that is involved in transferring external carbohydrate to the cell. Recently, the HUP1 (hexose uptake protein) from *Chlorella kessleri* was introduced into *C. reinhardtii* to increase the supply of external glucose into the cell. Hydrogen production capacity was increased about 150% by supplying 1 mM glucose to a strain of Stm6, where HUP1 has been inserted (Doebbe et al., 2007).

6.2.3 Improvement of Photoconversion Efficiency

Photosynthetic bacteria, cyanobacteria, and green algae, all of them convert light energy into chemical energy. Thus the photosynthetic efficiency is important for the supply of energy. Dense culture of them imposes a technical barrier called "self shading." Such cells adopted to have large antenna complex to acclimatize in low-light environment (Polle et al., 2003). Under bright light, they absorb more photons than can be utilized by photosynthesis. This over-excitation causes dissipation of excess (~80%) photons as fluorescence or heat, lowering the light conversion efficiency and cellular productivity. In photosynthetic bacteria and cyanbacteria, H_2 is catalyzed mostly by nitrogenase with expanse of 4 mol of ATP for generation of 1 mol H_2. Thereby, formation of ATP by conversion of light energy is important factor for H_2 production. Photosynthetic microorganisms capture light energy by the photosynthetic pigments present in light-harvesting complex (LHC). The up- and downregulation of LHC proteins are constantly adjusted to prevent photodamage and fulfill the energy demand by the cell (Adir et al., 2003). A mechanism called "energy-dependent nonphotochemical quenching" (NPQ) of chlorophyll fluorescence protects dissipation of a large proportion of light energy as heat or fluorescence and reduces oxidative damage (Müller et al., 2001; Polle et al., 2003) (Box 6.1). In the photobioreactor, photosynthetic yields are known to drop significantly along the reactor walls due to heat dissipation from high-light intensity. Limitation of light is a major bottleneck, because it cannot be stored and ideally should be distributed uniformly in the photobioreactor. Under bright light, the cells at the surface absorb most of the incoming light, however, heat dissipates most of it, as they absorb more photons than can be utilized by photosynthesis. Photosynthetic efficiency can be increased by modifying the light-harvesting antenna complexes responsible for capturing the solar energy. Reduction of pigments can reduce self shading and loss of energy by fluorescence and heat.

Photofermentative bacteria *Rhodobacter* is well explored for hydrogen production. It has two antenna complexes LH1 and LH2. Mutant *Rhodobacter* (P3 mutant) with 2.7-fold decreases in core antennal (LH1) content and 1.6-fold increases in peripheral antennal (LH2) content has given accelerated H_2 production compared to wild type (Vasilyeva et al., 1999). Using a *Rhodobacter sphaeroides* mutant MTP4 within the plate-type reactor, 50% more hydrogen is produced than its wild-type counterpart *Rhodobacter sphaeroides* RV (Kondo et al., 2002).

BOX 6.1 NONPHOTOCHEMICAL QUENCHING (NPQ)

The photosynthetic organisms protect themselves from the adverse effect of high-light intensity by a mechanism called nonphotochemical quenching (NPQ). Light is absorbed by the light-harvesting complex (LHC), which consists of several chlorophyll (Chl) molecule and transfers to the reaction center (RC). In excess, light chlorophyll turns to singlet excited state from the ground state. The excitation energy harmlessly dissipates through heat and florescence by molecular vibration.

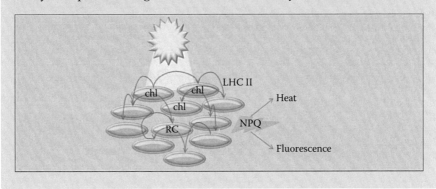

In view of this, Myers (1957) and later Radmer and Kok (1977) suggested the need of developing microalgae with truncated antenna size to increase the low-light conversion efficiency. Algae are able to acclimatize sunlight to some extent by modulating their photosynthetic apparatus in size and composition in order to avoid photo-oxidative damage (Escoubas et al., 1995; Durnford and Falkowski, 1997; Yang et al., 2001). Genetic manipulation of Chlorophyll antenna size in algal chloroplasts could increase the overall photosynthetic conversion efficiency of cyanobacteria and green algae. The photosynthetic apparatus of cyanobacteria, in addition to chlorophyll in the photosystems, possesses phycobilisomes with the pigments phycocyanin and phycoerythrin primarily attached to the photosystem II (PSII) dimers. Presence of these pigments extends the spectrum of visible light usable for photosynthesis (Rascher et al., 2003). However, cyanobacteria also suffer from the light limitation, and an initiative is taken by a Japanese group to develop truncated photosystems to enhance the biohydrogen potential (Nakajima et al., 2001). Using chemical mutagen, they developed phycocyanin-deficient PD1 mutant of *Synechocystis* PCC 6714. The mutant showed 1.5 times the photosynthetic productivity than wild type under high-light intensity. Furthermore, a single mutation in the *cpc* operon coding for phycocyanin has been identified. However, the potential of this mutant as a hydrogen producer has not been reported. Cyanobacteria deficient in phycobilisomes expected to produce enhanced rate of H_2. Deletion of *apcE* and *apcAB* operons helped to improve the H_2 production rate in *Synechocystis* PCC6803 (Bernát et al., 2009).

The similar trial has been taken for green algae also. At least 20 members of a large extended family of chlorophyll-binding protein subunits have been identified in LHC complex of *C. reinhardtii*. Nine of these proteins are predicted to be predominantly associated with photosystem (PS) II and therefore are being referred to as LHCII proteins (Elrad and Grossman, 2004). The regulation of LHCII expression occurs at many levels including transcription initiation (Maxwell et al., 1995; Millar et al., 1995), and post-transcriptional regulation (Flachmann and Kühlbrandt, 1995; Lindahl et al., 1995; Durnford et al., 2003). Greenbaum has reported maximum light conversion efficiency of 80% for green algae (Greenbaum, 1988). The recent research has mostly focused on improving the light absorption capacity. Reduction in pigment content can lead to better penetration of light inside the reactor and reduce the wastage of light energy by fluorescence and heat. Polle et al. (2003) recently developed the chlorophyll antenna-truncated *C. reinhardtii*, which showed higher hydrogen production as compared to the wild type. Recently, the RNA interference or RNAi technology has also been used to down regulate the entire family of light-harvesting complexes (LHC) in *C. reinhardtii* strain Stm3LR3. This resulted in a chlorophyll content of around 32% compared to the control strain. This mutant showed an increase in photosynthetic conversion efficiency when cultivated under high light (Mussgnug et al., 2007). Most likely, this strategy can be used more effectively than random mutagenesis to develop varied LHC mutants and should be stabler in the long term. NAB1 has been identified as LHC translation repressor. Transformation of a permanently active repressor variant NAB1 leading to a reduction of LHC antenna size by 10–17%. *C. reinhardtii* strain Stm6Glc4T7 showed a ~50% increase of photosynthetic efficiency (PSII) at saturating light intensity compared to the parental strain (Beckmann et al., 2009).

6.2.4 Improvement of Hydrogen-Producing Enzymes

Hydrogen production is catalyzed mostly by hydrogenase. But, in diazo-trophs bacteria and cyanobacteria, H_2 production mostly coupled with nitrogen fixation and was catalyzed by the enzyme nitrogenase. Both the enzymes are characterized and studied well. Phylogenetically different enzymes, their corresponding genes, maturation systems, catalytic domains, and mechanism of actions are important to better understand the enzyme characteristics and their limitations. These information help researchers to improve hydrogen production by doing modification in molecular level. Details of these enzymes are discussed in Chapter 5. In this chapter, we have focused the limitation of those enzymes and how to improve using molec-ular approaches. If we look into the limitation of the hydrogen-producing enzyme, major is the sensitivity to oxygen.

6.2.4.1 Improvement of Hydrogenase

Hydrogenase enzymes are ubiquitous in bacteria and archaea but also found in microalgae. They are able to catalyze both H_2 oxidation and H_2 evolu-tion. Both photosynthetic and fermentative bacteria are anoxygenic during hydrogen production. However, cyanobacteria and green algae produce O_2 due to their PSII activity, which inhibits the hydrogenase. According to the metal content of their active sites, they are classified as nickel-iron (NiFe) and di-iron (FeFe) hydrogenases as discussed in Chapter 5. Among them [Fe-Fe]-hydrogenases are found to be more biased to produce H_2 than oth-ers. X-ray absorption spectroscopy shows that reaction with oxygen results in destruction of the [4Fe-4S] domain of the active site H cluster while leaving the di-iron domain (2FeH) essentially intact (Stripp et al., 2009). However, it is also observed that [FeFe]-hydrogenases are more sensitive to oxygen than [NiFe]-hydrogenases. Algal hydrogenase is the smallest and the simplest and more sensitive to oxygen inactivation, and it consists of only the H cluster. Although additional F-cluster domain present in bac-terial hydrogenases might provide additional protection to the active site from oxygen inactivation, till date only a few hydrogenases are known to display tolerance to oxygen. Hydrogenases from some extremophiles are known to be naturally tolerant to oxygen. *Aquifex aeolicus* hydrogenase is a good candidate for biotechnological applications due to their high resistance to aerobic and thermal inactivation (Guiral et al., 2006). When the crystal structure of an O_2-tolerant [NiFe]-hydrogenase from the aerobic H_2 oxidizer *Ralstonia eutropha* has been investigated, it has been found that the heterodimeric enzyme consists of a large subunit harboring the catalytic center in the H_2-reduced state and a small subunit containing an electron relay consisting of three different iron–sulphur clusters. This cofactor operates as an electronic switch. It serves as an electron acceptor in the course of H_2 oxidation and as an electron-delivering device upon O_2 attacks at the active site (Fritsch et al., 2011). Significant research has been performed to increase the oxygen tolerance of hydrogen-producing enzymes, hydrogenases in particular (Xu et al., 2005). By introducing random and site-directed mutagenesis in *Chlamydomonas*, strains with 10-fold more oxygen tolerance have been obtained (Ghirardi et al., 2000).

6.2.4.2 Improvement of Nitrogenase

In photosynthetic bacteria and N_2-fixing cyanobacteria, enzyme nitrogenase is involved in scavenging of excess protons by generating H_2 though it is a highly energy-consuming process under nitrogen-starved conditions. Filamentous cyanobacteria can produce H_2 through nitrogenase that reduces dinitrogen to ammonia and, in this process, it evolves H_2. It may be of particular interest to improve the nitrogenase-based system for H_2 evolution for the following reasons:

1. Nitrogenase is also sensitive to oxygen, but less than hydrogenase. Hydrogen production by nitrogenase is mostly inhibited by the ammonium ions rather than O_2.
2. In cyanobacteria, nitrogenase is separated from oxygen through special (heterocyst) or temporal (light/dark) separation.
3. Biochemistry of nitrogenase enzyme is well understood and protein crystal structures are available (Uzumaki et al., 2004).
4. Presence of nitrogenase in heterocysts naturally protects them from O_2 evolution during photosynthesis by PSII.
5. Some organisms are capable of simultaneously producing fixing carbon by photosynthesis and evolving nitrogenase-mediated hydrogen.
6. The enzyme is able to generate hydrogen under high-partial pressure of the same.

Most of these traits are absent in hydrogenase-mediated H_2 production, hence this enzyme appears as a suitable candidate for genetic engineering to enhance the hydrogen yields from cyanobacteria. The nitrogenase system is the primary means of H_2 production under photosynthetic and nitrogen-limiting conditions in many photosynthetic bacteria. The efficiency of this biological H_2 production largely depends on the nitrogenase enzyme and the availability of ATP and electrons in the cell. Although replacement of N_2 by Argon is effective for increasing H_2 production, this approach increases the operational cost for large-scale generation of H_2 (Masukawa et al., 2010). Mutagenesis offers an alternative mechanism to overcome N_2 competition. One approach to increase H_2 production by nitrogenase is to enhance the electron flux to proton reduction and away from N_2 reduction. In view of this recently, Masukawa et al. (2010) engineered nitrogenase in an effort to increase hydrogen production from *Anabena* sp. PCC 7120. They modified six amino acid residues within 5 Å of the FeMo-co to direct the electron flow selectively toward proton reduction in a nitrogen environment. Several variants examined under an N_2 atmosphere significantly increased their *in vivo* rates of H_2 production, approximating rates equivalent to those under an argon atmosphere, and accumulated high levels of H_2 compared to the reference strains. These results demonstrate the feasibility of engineering cyanobacterial strains for enhanced photobiological production of H_2 in an aerobic, nitrogen-containing environment. Blockage of the CO_2 fixation pathway in *R. rubrum* induced nitrogenase activity even in the presence of ammonium, presumably to remove excess reductant in the cell. *cbbM* mutants grew poorly in malate medium under anaerobic conditions. However, the introduction of constitutively active NifA (NifA*), the transcriptional activator of the nitrogen fixation (*nif*) genes, allows *cbbM* mutants to dissipate the excess reductant through the nitrogenase system and improves their growth. The deletion of *cbbM* alters the post-translational regulation of nitrogenase activity, resulting in partially active nitrogenase in the presence of ammonium. The combination of mutations in *nifA*, *draT*, and *cbbM* greatly increased H_2 production of *R. rubrum*, especially in the presence of excess of ammonium. Furthermore, these mutants are able to produce H_2 over a much longer time frame than the wild type, increasing the potential of these recombinant strains for the biological production of H_2 (Wang et al., 2010). By knocking out *glnB* and *glnK*, the problem of repression of nitrogenase by ammonium ions has been overcome in *Rhodobacter sphaeroides* (Kim et al., 2008). Double mutation in hydrogenase and homocitrate synthase gene could improve the nitrogenase-based hydrogen production activity in cyanobacteria. *Nostoc* PCC 7120 has two homocitrate synthase genes, *nifV1* and *nifV2*, and homocitrate bound to Fe-Mo cofactor of dinitrogenase is considered to be important for efficient nitrogen fixation, but not for H_2 production. Mutants disrupted in one of the two genes and the both of them with Δ*hupL* as the parent were developed. It was found that modulation of homocitrate synthase activity is effective in prolonging the high H_2 production period (Sakurai et al., 2008). Skizim (2012) calculated the ratio of hydrogen obtained from nitrogenase (light induced) as compared to that from hydrogenase (dark induced) in *Cyanothece* sp. Miami BG 043511. Such comparable rates would be interesting to interpret to decide which would be more prudent for future engineering studies. Besides, it may be interesting to create a mutant library of nitrogenase by random mutagenesis and gene shuffling. However, such a system also requires the design of high-throughput screening processes. Besides, it may be worthwhile to understand the biochemical requisites for efficient H_2 production by the molybdenum nitrogenase as a basis for its re-engineering.

6.2.4.3 Overexpression of Enzymes

The production of high amounts of enzymes could possibly increase the product. Multiple copies of that gene could produce high amount of enzyme. With this assumption, different enzymes are overexpressed within the cell by transforming with copies of that particular gene (homologous overexpression). The H_2-evolving [FeFe]-hydrogenase in *Clostridium tyrobutyricum* JM1 was isolated to elucidate molecular characterization and modular structure of the hydrogenase. Furthermore, homologous overexpression of the hydrogenase gene was for the first time performed to enhance hydrogen production. Homologous overexpression of the [FeFe] hydrogenase gene resulted in a 1.7-fold and 1.5-fold increase in hydrogenase activity and hydrogen yield concomitant with the shift of metabolic pathway (Jo et al., 2010). The influence of increase in intracellular [FeFe] hydrogenase levels, in *Enterobacter cloacae* IIT-BT 08, on the formation of molecular hydrogen has been studied. Homologous overexpression of the [FeFe] hydrogenase gene increased the hydrogenase activity by 1.3-fold as compared to the wild type (Khanna et al., 2011).

6.2.5 Introduction of Foreign Hydrogenase

The expression of efficient foreign enzymes within the cell could overcome the bottlenecks and be more favorable for improvement of the expected product. However, the technique is more complicated than homologous overexpression due to the transformation of foreign gene into the cell. Scientists have tried heterologous overexpression of the efficient and O_2-tolerant enzymes into the cell. Clostridial *hydA* has also been cloned into *R. rubrum*, and the native hydrogenase of *R. rubrum hydC* has been overexpressed. In both cases, it was observed that pyruvate was the electron donor for H_2 production (Kim et al., 2008).

Hydrogen production from cyanobacteria is well known. However, its low rate of production and yield have been the point of interest for several decades. Considerable engineering of both strain and enzyme are required to counter the problem to enable higher-end production. Cyanobacteria produce hydrogen through [NiFe] hydrogenase and or nitrogenase. Nitrogenases have relatively low-turnover numbers, require two molecule of ATP for every electron used for reduction and produce 1 mol of ATP for every electron reduced. However, [NiFe] hydrogenase may be bidirectional or uptake with low rate of hydrogen evolution. In contrast, [FeFe] hydrogenases can produce high-rate hydrogen with no ATP requirements. Moreover, their conserved structure and relatively simple maturation pathway make [FeFe] hydrogenases excellent enzymatic modules for recombinant expression in synthetic systems (Agapakis et al., 2010). Thus, an institutive strategy to increase the hydrogen yield would be to heterologously express [FeFe] hydrogenase from bacteria into cyanobacteria. An efficient [FeFe]-hydrogenase (*hydA*) from *Clostridium pasteurianum* was introduced and expressed in the cyanobacteria *Synechococcus elongatus* and succeeded to achieve enhanced hydrogen production without co-expression of maturation proteins (Asada et al., 2000; King et al., 2006). Most of the work in the beginning concentrated on the hydrogenase from *Clostridium* sp. because the gene had been well characterized, sequenced, and crystallized (Gorwa et al., 1996; Peters et al., 1998; Kaji et al., 1999). Also, different species of the organism were known to degrade and ferment various biomass polymers such as polysaccharides and proteins to obtain energy and reducing powers such as the proton/electron and reduced compounds in cells. Since anaerobic bacteria possess a mechanism to remove excessive-reducing powers as hydrogen gas using hydrogenase, it is possible that they produce huge amounts of hydrogen gas during growth on biomass materials. Moreover, the clostridial hydrogenases, namely, [Fe]-hydrogenase I of *Clostridium pasteurianum* (Peters et al., 1998), hydrogenase A of *Clostridium perfringens* (Kaji et al., 1999), and hydrogenase A of *Clostridium acetobutylicum* P262 (Gorwa et al., 1996), were sequenced and well characterized. However, later, attention was also focused on HydA1 from *Chlamydomonas reinhardtii*. It was considered a promising model for better understanding of [FeFe]-hydrogenase structure and function as it is one of the smallest known member (48 kDa) of this enzyme family, due to lack of the FeS-cluster containing F-domain (Happe and Naber, 1993). Moreover, [Fe]-hydrogenases from alga were unique since it coupled directly to water oxidation through photosystem II and the photosynthetic electron transport chain, providing the means to generate H_2 using sunlight. Recently, Hopkins et al. (2011) introduced the concept of "minimal hydrogenase" where they have successfully shown functional activity of heterodimeric form that contains only two of the four subunits found in the native heterotetrameric enzyme. It is noteworthy that irrespective of the type of hydrogenase, for the purpose of overexpression, the gene must be placed under the control of a synthetic constitutive or inducible promoter with a strong ribosome-binding site. Such a system can bypass the strict cellular regulations on different levels (transcription and translation) and may result in significant increase in the amount of hydrogen produced. However, successful heterologous expression of [FeFe] hydrogenase in cyanobacteria still remains a challenge and till date very few reports are available in literature. Limited success rate has been attributed to varied limitations which include lack of sufficient knowledge of hydrogenase gene, hydrogenase crystal study, process of maturation of the gene, and so on. In view of these limitations, the earliest work in this direction focused on *in vitro* studies to increase the photosynthetic hydrogen efficiency. In a classic study, Adams et al. (1980) *in vitro* tried to couple the photosynthetic system with hydrogenase using a mediator. However, they were not very successful in their attempts since the *in vitro* systems are unstable due to sensitivity of hydrogenase against O_2 (Rao et al., 1976). Later, among the premier *in vivo* trials were conducted by Asada's group in Japan, where they demonstrated coupling cyanobacterial ferredoxin and clostridial hydrogenase. The study showed that the cyanobacterial ferredoxin could act as an electron carrier between the photosynthetic system and the clostridial hydrogenase *in vivo*. However, in their study, they introduced the clostridial hydrogenase protein into *Synechococcus elongatus*, by direct electroporation creating "psuedotransformants." Interestingly, the cells with clostridial hydrogenase showed simultaneous light-dependent evolution of H_2 and O_2. They reported 11 nmol H_2 evolved with reduced methyl viologen per mg chlorophyll. Though the yield was low, these findings were significant as they demonstrated successful *in vivo* coupling of cyanobacterial photosynthesis and clostridial hydrogenase in cells and thus paved the way for future genetic engineering of cyanobacteria (Miyake and Asada, 1997). Later, the same group attempted to genetic engineer *Synechococcus* sp. PCC7942 by heterologous expression of clostridial hydrogenase gene in pKE4-9 vector, which carried the chloramphenicol-resistant gene and a strong promoter. The clostridial hydrogenase gene was cloned upstream of the cat gene in pKE4-9 and transformed into PCC7942. However, in their studies though the cat protein was expressed, the clostridial hydrogenase was not detected by Western blot analysis. Interestingly, the hydrogenase mRNA was detected in PCC7942, and it was co-transcribed with cat (Asada et al., 2000). The study bought two points into perspective: (i) role of accessory proteins associated with the maturation of the hydrogenase genes and (ii) the need for codon optimization of the ribosome-binding sequence along with the promoter for successful heterologous overexpression. It is only recently that Ducat et al. (2011) made a significant breakthrough by successful overexpression of clostridial hydrogenase in unicellular cyanobacterium, *Synechococcus elongatus* sp. 7942. In their studies, they were able to overcome both the limitations of Asada's group. They used codon-optimized synthetic genes and well-characterized *hydEFG* accessory proteins to maturate the clostridium hydrogenase. They demonstrated that the heterologously expressed hydrogenase was functional *in vitro* and *in vivo*, and that the *in vivo* hydrogenase activity was connected to the light-dependent reactions of the electron transport chain. Following this report Gärtner et al. (2012) showed successful expression of [FeFe]-hydrogenase in heterocysts of *Anabena* sp. strain 7120 under the control of a *hetR* promoter. This was a significant step forward in photobiological hydrogen research as they were able to show for the first time the spatial separation of H_2 production and photosynthetic O_2 evolution. For a long time, it had been postulated that since heterocysts maintain a microanoxic environment, due to the presence only PSI and a specialized cell envelope, it may be a suitable organelle to express the oxygen sensitive hydrogenase. Moreover, the electron transport chain that shuttles low-potential PSI electrons to nitrogenase via ferredoxin may also be suitable for electron transport via hydrogenase. Realization of this technique may go a long way in obtaining sustainable energy from photosynthetic organisms, as it allows for the twin activities of hydrogen production along with photosynthesis. However, it also imposes the challenge of economic way of separation of oxygen from hydrogen. A question that is often raised during such heterologous trials with [FeFe]-hydrogenase is the need for simultaneous expression of two or more gene products (*hydE, hydF,* and *hydG*) which are considered essential for hydrogenase assembly. Along this line, Posewitz et al. (2004) established that functional expression of [FeFe] hydrogenase requires the assistance of the accessory genes when expressed in hosts, which lacked the native FeFe hydrogenase. Asada et al. (2000) reported *in vivo/in vitro* functional expression of clostridial hydrogenase in *Synechococcus* PCC7942 without a maturation operon. Interestingly, however, the same construct was found to be inactive in *E. coli*. To this end, Asada et al. postulated that the [NiFe]-hydrogenase assembly present only in *Synechococcus* PCC7942 may have been flexible enough to also maturate the foreign hydrogenase. However, it is not clearly understood why it is necessary to express the [FeFe]-maturation operon in some organisms and not in others. In stark contrast to this is the [NiFe]-hydrogenase, where heterologous production is more difficult due to the strict host specificity of the maturation machinery. For example, the [NiFe]-hydrogenase from *Desulfovibrio gigas* is poorly matured when produced in the closely related organism *Desulfovibrio fructosovorans* (Rousset et al., 1998). To be properly matured and functional, the structural subunits need the assistance of the parent maturation genes (Rousset et al., 1998; Lenz et al., 2005). Also, to produce the membrane bound hydrogenase enzyme from *Ralstonia eutropha* in *Pseudomonas stutzeri*, the *R. eutropha* maturation machinery was found to be necessary (Lenz et al., 2005). Given the complex nature of the different metallo-enzyme, this observation highlights the difficulty of hydrogenase expression and activation in different microorganisms. Next, the process is riddled by the question whether a foreign hydrogenase can efficiently transfer to or receive electrons from the endogenous redox partner of the host. Hydrogenases transfer electrons to or from an acceptor/donor such as ferredoxins, directly or via accessory subunits. It is not certain that heterologously produced hydrogenase will optimally interact with the redox partner or the accessory subunits of the host. Therefore, transferring and expressing foreign hydrogenases into chosen hosts appeared to be difficult, and homologous production of modified hydrogenases seems to be a more suitable alternative. Ferredoxins are small and soluble redox proteins that are found in bacteria and photosynthetic eukaryotes (Matsubara et al., 1979). The average redox potential is -420 mV vs. NHE. They are characteristically acidic, and electrostatic forces have been shown to be important for interactions between ferredoxins and oxido-reductases like hydrogenase (Agapakis and Silver, 2010). However, contrasting reports are available in literature based on this hypothesis. However, Moulis and Davasse (1995) reported that such electro-static forces were not necessary for foreign clostridial hydrogenase in *E. coli*. To further investigate such discrepancies in literature, recently Agapakis et al. (2010) carried out detailed investigations regarding the linking of fer-redoxins to hydrogenase to enhance hydrogen production potential of the organism. They also showed that co-expression of ferredoxin enhances hydrogen production in a range of organisms. In another study, a complex consisting of ferredoxin I fused to the [FeFe]-hydrogenase (HydA) from *C. reinhardtii* was shown to catalyze light-dependent H_2 production *in vitro* when incubated with purified thylakoids (Yacoby et al., 2011). The group showed that upon addition of competing FNR and $NADP^+$, ~90% of the elec-trons passed from PSI to hydrogenase. This highlighted the importance of interaction of ferredoxin to hydrogenase for efficient transfer of electron. In addition, in *Synechococcus elongatus*, different ferredoxins showed different abilities to interact with a heterologous [FeFe]-hydrogenase and with PSI, and to accept electrons from glycogen reserves or to donate electrons into the plastoquinone pool (Ducat et al., 2011). In view of these findings, it may be reasonable to suggest that if ferredoxins with different affinities for FNR and hydrogenase could be found or developed, it is possible that they could be concurrently expressed at different levels to strategically divide reduc-tant between the carbon dioxide fixation and H_2 production (Kontur et al., 2011). In a contrasting report, recently Agapakis et al. (2010) observed very low-hydrogenase activity with *S. oneidensis* MR-1 dimeric [FeFe] hydroge-nase and different ferredoxin. However, there is evidence that some dimeric hydrogenases preferentially receive electrons from cytochrome *c* instead of ferridoxins. Two *Synechocystis* sp. PCC 6803 strains reported in literature appear to be attractive targets for heterologous overexpression of foreign hydrogenase. These include the PSII mutant strain developed by Howitt et al. (2001), which is devoid of photosynthetic evolution and respiratory oxygen consumption and another strain developed by Pinto et al. (2012), where they created a *hoxYH* mutant as a photoautotrophic chassis for hydro-gen production. Further, in spite of the advances made, we still know little about the regulation of [FeFe]-hydrogenase gene transcription and matura-tion. Further expansion of our knowledge base in these areas may go a long way in resolving the current issues. Moreover, presently, only four crystal structures of [NiFe] hydrogenase from *Desulfovibrio* species and two of [FeFe] hydrogenase one each from *Clostridium pasteurianum* (Peters et al., 1998) and *Desulfovibrio desulfuricans* (Nicolet et al., 1999) have been solved. However, no algal hydrogenases have been crystallized yet.

6.2.6 Deletion of Hydrogen Uptake Genes

The produced hydrogen could be re-oxidized again by another enzyme known as uptake hydrogenase. Hydrogen uptake activity is another obsta-cle for H_2 production. Structurally, uptake [NiFe]-hydrogenase contains two subunits, HupL and HupS, respectively, encoded by the *hupL* and *hupS* genes (Shestakov and Mikheeva, 2006). The active site is encoded by the large subunit (*hupL*), whereas the small subunit contains the [FeS] clusters that transport electron from the active site to the electron transport chain. A discernible strategy to enhance hydrogen production from microorganisms would be to disrupt the uptake of hydrogenase. Initially, this was achieved by chemical mutagenesis (Mikheeva et al., 1995; Kumar and Kumar, 1991), however, later with the advent of advanced molecular tools, targeted knockouts and site-directed mutagenesis studies were carried out.

Inactivation of hydrogen uptake activity in *T. roseopersicina* (*hypF*-deficient mutant) under nitrogen-fixing condition caused a significant increase in the hydrogen evolution capacity. This mutant is therefore a promising candidate for use in practical biohydrogen-producing systems (Fodor et al., 2001). Mutant deficient of uptake hydrogenase activity dramatically increases the net production of hydrogen. Ruiyan and co-researchers reported that a *Rhodospirillum rubrum* mutant deleted of *hupL* gene encoding the large subunit of uptake hydrogenase produced increased amount of H_2 (Ruiyan et al., 2006). To date, significant research has been performed to inhibit the uptake hydrogenase activity using different approaches. Gokhan and co-researchers used a suicide vector for site-directed mutagenesis of uptake hydrogenase (*hupSL*) in *Rhodobacter sphaeroides*. They obtained 20% more H_2 production than wild type (Kars et al., 2008).

Unlike bidirectional hox hydrogenase, uptake hydrogenase is found only in the heterocysts of nitrogen-fixing cyanobacteria where they function to recycle the hydrogen produced by the nitrogenase, in keeping with the view of conservation of energy. [NiFe]-uptake hydrogenase is known to be the smallest hydrogenase found in cyanobacteria probably because it only catalyzes the oxidation of hydrogen (Srirangan et al., 2011). Mostly, the researchers have focused on the knockout of *hupL* as it encodes the active site. However, in an interesting study, Khetkorn et al. (2012) disrupted the *hupS* structural gene of *Anabaena siamensis* TISTR 8012 by Neomycin resistance cassette and demonstrated enhanced hydrogen production. It may be interesting to note that while the approach of deleting the uptake hydrogenase results in significantly higher-hydrogen production, however, hydrogen consumption may still be carried out by the cell via bidirectional [NiFe]-hydrogenase (Tamagnini et al., 2007) as it is also known to uptake hydrogen due to its low K_m for hydrogen. In view of this, Masukawa et al. (2002) developed single deletions of Δhox and Δhup genes and compared it to the double mutant of Δhox and Δhup. In the double mutant, they observed lower nitrogenase activity as compared to the wild type. Since hydrogen consumption serves as a physiological means to balance cell's redox potential by replacing the reducing power lost during nitrogen fixation, deletion of both the processes (*hup* and *hox*) was found to impair nitrogen assimilation capacity. Thus, it appears that hydrogen consumption is physiologically inevitable, particularly for nitrogen-fixing cyanobacteria.

6.2.7 Other Approaches

6.2.7.1 Generation of Anaerobic Condition

The extreme sensitivity of hydrogenases to oxygen presents a major challenge for exploiting these organisms to achieve sustainable hydrogen production. There is no need to generate anoxic condition for anaerobic bacteria to produce H_2. For O_2-evolving cyanobacteria and green algae, H_2 production can be obtained only by generating anaerobic conditions. Different trial has been taken for generating the anaerobic condition that includes sparging with inert gas inside the reactor or depleting the media of sulfur, which stalls the PSII activity (Melis, 2007). Induction of artificial anaerobic conditions, however, adds significant cost to hydrogen production in large scale. Scientists have also developed different genetic modification strategies to achieve the anaerobic condition in light. Under normal growth condition, photosynthetic rate is about four- to sevenfold higher than the respiration rate. By using attenuated photosynthesis/respiration ratio (P/R ratio) mutants (*apr* mutants) of *C. reinhardtii,* the P/R ratio drops below one, thereby establishing anaerobic conditions which mimic the physiological status of sulfur-deprived cells (Melis, 2007). So by using the *apr* mutant, continuous hydrogen production can be obtained. The sulfur-deprivation method that may cause cell growth inhibition and death can be avoided (Rühle et al., 2008). It has also been reported that a sulfur-deprived *Chlamydomonas reinhardtii* D1 mutant that carried a double amino acid substitution is superior to the wild type for hydrogen production. The leucine residue L159 in the D1 protein was replaced by isoleucine, and the asparagine N230 was replaced by tyrosine (L159I-N230Y). This strain is very efficient for prolonged H_2 production and also having lower chlorophyll content and higher-respiration rate that contribute in net yield (Torzillo et al., 2009).

6.2.7.2 ATP Synthase Modification for Enhanced Hydrogen Production

An anaerobic environment may not always be sufficient for enzyme activation, as there are other competitors present for capturing electrons. It has been previously shown that in algae hydrogen production is limited by buildup of proton barrier as evidenced by enhanced hydrogen production by uncoupling of protons (Lien and Pietro, 1981). Electron transport from water to ferredoxin is accompanied by the formation of a proton gradient that drives the production of ATP, which is essential for CO_2 fixation. In a nitrogenase-based system, ATP is required for nitrogen fixation/hydrogen production. However, during photoproduction of molecular H_2 catalyzed by hydrogenase, ATP demands drop, impeding the efficient dissipation of the proton gradient and impairing efficient electron transport for H_2 production. In view of this, Lee and Greenbaum (2003) while working *with C. reinhardtii* suggested the introduction of a synthetic proton channel targeted in the thylakoid membranes, under the control of an anaerobic hydrogenase-driven promoter. In another approach, Robertson's group developed *C. reinhardtii* mutants, which produced lower amounts of ATP. They developed mutants were defective in one of the ATP synthase subunits, AtpE (Robertson et al., 1990). Physiological studies with the mutants showed that though they grew at a relatively slower rate, they produced higher amounts of H_2 notably at higher-light intensity.

6.2.7.3 Linking of Hydrogenase to Cyanobacterial Photosystems

Although heterologous hydrogenase overexpression in cyanobacteria has come a long way, the photon conversion efficiency is relatively low because the Fd-mediated electron transfer rate from PSI to hydrogenase has to compete with the electron transfer channel to the carbon dioxide fixation pathway. Some electron flow through the carbon fixation pathway is necessary for cell survival and thus limits the maximal electrons that can be channeled through hydrogenase (Mcintee-Chmielewski, 2010). Moreover, technically, direct-coupled complex must have electron flow comparable to that of ferredoxin and must be capable of self-assembly *in vivo*. Application of proximity-based electron transfer technique may be considered an ideal solution to circumvent the problem. In view of this, direct linking of an oxygen-tolerant hydrogenase to PSI was suggested. It was considered an ideal system since it would allow hydrogen production coupled with photosynthesis. This would circumvent the technical barrier of spatial and/or temporal separation of hydrogen production from photosynthesis. Technically, it was feasible since the terminal-bound FeS clusters FA and FB of PSI have redox potentials between −440 and −480 mV, which is in the range of the H^+/H_2 couple (−420 mV). Thus, although hydrogenase and PSI are not physiological redox partners, an electron transfer between both components is thermodynamically feasible (Mcintee-Chmielewski, 2010). Ihara et al. (2006) for the first time showed the feasibility of the process although, *in vitro*. They fused the PSI subunit PsaE from the cyanobacterium *Thermosynechococcus elongatus* to an O_2-tolerant [NiFe]-hydrogenase from *Ralstonia eutropha* H16 to develop PsaE-Hyd complex. In *in vitro* trials, the PsaE-hyd chimera was shown to associate with PSI of PsaE-deficient *Synechocystis* PCC 6803 spontaneously and to accept electrons from PSI directly. They showed light-driven hydrogen production at a rate of 0.088 μmol H_2 h^{-1}. However, this was found to be substantially lower than the theoretically predicted value of 1.2 μmol H_2 h^{-1}. Moreover, H_2 production activity was totally suppressed in the presence of *in vivo* physiological partners such as ferredoxin (Fd) and ferredoxin-NADP$^+$-reductase (FNR), severely limiting the *in vivo* application of the process. The authors attributed the results to low-efficiency electron transfer between the chimera and PSI due to relative change in the optimal distance for high-rate electron transfer due to conformational changes in the chimera protein. To overcome the technical difficulties faced by them, the same group subsequently improved the system by direct coupling of hydrogenase to PSI crosslinked with cytochrome *c3* (*cyc3*) on its Fd-binding site. They developed the *cyc3*-PSI chimera because cyc3 could mediate electron transfer from PSI and simultaneously hinder the association between Fd and PSI, a major limitation of the first strategy. *In vitro* experiments showed sevenfold enhanced hydrogen production from the *cyc3*/PSI/hyd complex. Moreover, they could demonstrate successful hydrogen production in the presence of FNR enzymes. These results suggest that the *cyc3*/ PSI/hyd complex may produce H_2 *in vivo*. However, this system is yet to be tested *in vivo*. Another strategy to accomplish the same was demonstrated by coupling to the hydrogenase using molecular wires. The wire may be aliphatic chain of hydrocarbons such as 1,6 hexanedithial or aromatic chain of hydrocarbons such as 1,4 benzenedithial. Aromatic wires show improved performance over aliphatic chain wires. Using such approaches, *Clostridium acetobutylicum* [FeFe]-hydrogenase was fused with PSI of *Synechococcus* sp. PCC 7002 PS using 1,6-hexanedithiol molecular wire (Lubner et al., 2010). While this complex also catalyzed light-dependent H_2 production *in vitro*, unlike genetic engineering strategies, it may not be suitable for *in vivo* implementation.

6.2.7.4 Engineering of Heterocyst Frequency

Metabolic uncoupling by upregulating heterocyst differentiation is hypothesized to further enhance hydrogen production. In free living filamentous *Nostoc punctiforme*, the heterocyst is found to be around 8%. However, under nitrogen fixing, symbiotic association with higher plants, the heterocyst frequency is increased to 30–35% with a parallel increase in ammonium production. Filamentous cyanobacteria are known to produce H_2 at higher rates as compared to the single-celled cyanobacteria. Thus, it is postulated that since H_2 production in filamentous cyanobacteria is associated with nitrogenase, increase in heterocyst frequency may enhance hydrogen production. Moreover, increased hydrogen production by enhanced heterocyst frequency also uncouples metabolic hydrogen from microbial growth. In this context, Jeffries et al. (1978) physiologically showed continuous production of hydrogen under nitrogen-deficient conditions by *Anabaena cylindrica*. Recently, the genetic basis of heterocyst development has been elucidated (Ehira et al., 2003; Sato et al., 2004; Campbell et al., 2007). HetR, a serine-type protease encoded by *hetR*, is expressed early during heterocyst differentiation and is crucial to the differentiation process. PatA is also known to be closely associated with heterocyst differentiation. This role may be in the form of post-translational modification of HetR, because PatA is a member of the response regulator family of proteins. Studies have shown that mutation in the *hetR* inhibits early steps in the formation of heterocysts while overexpression of *hetR* gives rise to multiple heterocysts (Buikema and Haselkorn, 2001). However, a practical study with such strains depicting genetically engineered high-heterocyst frequency and enhanced H_2 production is yet to be reported.

6.3 Metabolic Engineering

Different groups of microorganism mostly have different molecular mechanisms for generating H_2. Many proteins are involved for H_2 generation. Some of them are well characterized at protein as well as at gene level. Researchers are targeting the genes involved directly or indirectly in H_2 generation for maximization of hydrogen production. Many genetic and metabolic engineering works have been done in different H_2-producing organisms to enhance the yield. The enzymes involved in the production of H_2 have been characterized including their types, maturation systems, different domains, and their functions. Three different enzyme hydrogenase, nitrogenise, and formate hydrogen lyase can catalyze the reduction of protons to hydrogen. The ultimate goal of biohydrogen research is to produce as much hydrogen as possible. Metabolic modifications can be a very effective and promising method to optimize and redirect the flow of reducing equivalents (electrons) to the enzyme and also the accumulation of protons (H^+), to achieve a satisfactory yield of hydrogen (H_2) in large scale.

6.3.1 Proteomic Analysis

Proteomic analysis during hydrogen production can depict the metabolic state of the organism and many previously uncharacterized proteins that may play a role in hydrogenase activity. Those proteins could provide suitable targets for metabolic engineering.

Scientists from Manitoba Centre for Proteomics and Systems Biology, University of Manitoba, Canada have done differential shotgun proteome analysis of a mesophilic, hydrogen, and ethanol-producing cellulolytic bacterium, *Clostridium termitidis*, which was performed using two growth substrates. Total proteins were extracted and, approximately, 1370 proteins were identified (Ramachandran et al., 2012). A quantitative comparative proteomic analysis of *Clostridium thermocellum* (Gram-positive thermophilic anaerobic bacterium) using a metabolic 15N isotope labeling method in conjunction with nano-LC-ESI-MS/MS (liquid chromatography–electrospray ionization–tandem mass spectrometry) identifies proteins and biochemical pathways that are differentially utilized by the organism after growth on cellobiose or cellulose. In total, 1255 proteins were identified in the study, and 129 of those were able to have their relative abundance per cell compared in at least one cellular compartment in response to the substrate provided. It has the ability to convert directly cellulosic biomass into useful products such as ethanol and hydrogen. It can divert carbon into alternative pathways for the purpose of producing biosynthetic intermediates necessary to respond to growth on cellulose (Burton and Martin, 2012).

There are also reports of proteomic analysis of hydrogen production from cyanobacteria and green algae. Cultures of the cyanobacterial genus *Cyanothece* have been shown to produce high levels of biohydrogen. Quantitative proteome analysis has been done for *Cyanothece* ATCC 51142 and PCC 7822 grown under eight different nutritional conditions. According to the analysis, it has been found that nitrogenase expression was limited to N_2-fixing conditions, and in the absence of glycerol, nitrogenase gene expression was linked to the dark period. However, in the presence of glycerol, expression of nitrogenase, in light, correlated with cytochrome c oxidase (Cox), glycogen phosphorylase (Glp), and glycolytic and pentose-phosphate pathway (PPP) enzymes. This indicated that nitrogenase expression in the light was facilitated via higher respiration and glycogen breakdown. Key enzymes of the Calvin cycle were inhibited in *Cyanothece* ATCC 51142 in the presence of glycerol under H_2-producing conditions, suggesting a competition between these sources of carbon. However, in *Cyanothece* PCC 7822, the Calvin cycle still played a role in cofactor recycling during H_2 production (Aryal et al., 2013).

The green alga *Chlamydomonas reinhardtii* is a model organism to study H_2 metabolism in photosynthetic eukaryotes. The molecular mechanism of proteomic changes of different phases of sulfur-depleted H_2 photoproduction process has been investigated by 2-DE coupled with MALDI-TOF and MALDI-TOF/TOF-MS. Eighty-two unique gene products have been identified. Major changes included photosynthetic machinery, protein biosynthetic apparatus, molecular chaperones, and 20S proteasomal components. A number of proteins related to sulfate, nitrogen assimilation, acetate assimilation, and antioxidative reactions were also changed significantly. Other proteins showing alteration were proteins involved in cell wall and flagella metabolisms. In addition, among these differentially expressed proteins, 11 were found to be predicted proteins without functional annotation in the *Chlamydomonas* genome database. Proteomic analysis provides new insight into molecular basis of H_2 photoproduction in *Chlamydomonas* under sulfur depletion (Chen et al., 2003).

6.3.2 Redirecting the Electron Pull toward Hydrogen Production

Mostly, the microbial communities produce hydrogen by fermentative metabolism of organic substrates. Fermentative hydrogen production depends on the substrate conversion efficiency. All the metabolic engineering has been done for better conversion of substrate to hydrogen. Theoretically, 1 mol of glucose can produce 12 mol of hydrogen in the absence of oxygen. But this reaction thermodynamically is less favorable ($\delta G = -9.5$ kcal/mol glucose under ambient condition), and no ATP has been formed. Most of the electron flasks have been redirected toward formation of acetate, yielding −51.6 kcal/mol (Thauer et al., 1977), theoretically sufficient for the production of several ATPs. Strategy has been taken by the researchers to redirect all electron pull toward proton reduction. Reducing equivalents (NADH) generated during fermentative oxidation are utilized for some other acids and alcohol formation (Figure 6.3). Tanisho et al. reported that these excess reducing equivalents could be disposed off via proton reduction mediated by electrons carrier and hydrogenase leading to the formation of H_2 in *E. aerogenes* (Tanisho et al., 1989). Redirection of metabolic pathway toward hydrogen generation has been achieved in *Enterobacter cloacae* IIT-BT08 by blocking alcohol and some acid formations. Double mutants, with defects in both alcohol and organic acid formation pathways, had higher H_2 yields (3.4 mol/mol-glucose) than the wild-type strain (2.1 mol/mol-glucose) (Kumar et al., 2001). Double mutant is obtained by using chemical mutagen.

Metabolic pathway redirection has been done in *E. coli* also by upregulating formate hydrogen lyase (FHL) complex and disrupting the pathways competing for pyruvate (Yoshida et al., 2005). Many different genes are

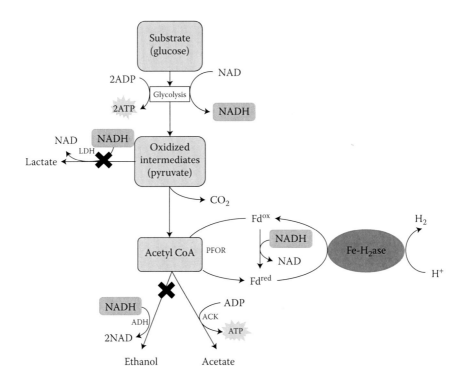

FIGURE 6.3
(**See color insert.**) Schematic diagram represents the redirection of metabolic pathways toward hydrogenase by inhibiting the alcohol and acid production pathways for enhanced hydrogen production.

involved in fermentative hydrogen metabolism including the key proteins and accessory proteins. Modification in accessory genes can also help to increase hydrogen production (Fan et al., 2009). By improving the synthesis of D-malic enzyme and using D-malate as the sole carbon source in strain IR4 of *Rhodobacter capsulatus,* a 50% higher-hydrogen production has been demonstrated (Vignais et al., 2006). Along with external electron donors like H_2O for cyanobacteria and green algae, and sulfur containing inorganic compounds or organic acids for photosynthetic bacteria, reducing equivalents can also be derived from either reserve carbon source produced during biomass formation or from an external organic source. In cyanobacteria and green algae, this alternative metabolic process can produce hydrogen via fermentative reactions and/or reducing plastoquinon (PQ) by Ndh to maintain the NAD/NADH balance and ATP supply (Gfeller and Gibbs, 1984; Hemschemeier and Happe, 2005). *Synechocystis* mutant M55 was developed that was defective in the *ndhB* gene encoding type 1 NADPH-dehydrogenase complex (NDH-1). This mutant was deficient in the NADP(H) oxidation. This generated large reserves (10 times more as compared to the wild type) of reduced NAD(P) pool that helped in sustained hydrogen production (Cournac et al., 2004). In another similar study by Cooley and Vermas (2001), NAD(P) concentration was shown to increase by 50–100% in the NDH-1 deletion mutant, however, the group did not estimate its potential for hydrogen production. NDH complex comprises four subunits, namely, NDH1, NDH2, NDH3, and NDH4. Cournac et al. (2004) and Cooley and Vermas (2001) investigated the role of only NDH1. In future, it may be interesting to study the effects of deletion of all or different subunits on the enhancement of hydrogen production from Synechocystis by effectively blocking the electron transport chain and increasing the reduced NAD(P) pool.

Another point of discussion is whether the recombinant organism is producing optimal amounts of hydrogen or whether the electron supply to hydrogenase is limited by other competing pathways. Central to the understanding of this question is to determine (i) the pathways that contribute electrons to hydrogensase and (ii) the competitive pathways biasing the electron flux away from hydrogen. The competing pathways mainly include cyclic electron flow (CEF), carbon dioxide fixation, proton buildup and production of secondary metabolites. It has previously been evidenced in algal research, which contains native [FeFe] hydrogenase that hydrogen production is greatly enhanced in the presence of proton uncouplers (Lien and Pietro, 1981). After random gene insertion in *Chlamydomoans reinhardtii*, a strain named Stm6 with modified respiratory metabolism was isolated. This strain is able to accumulate large amount of starch in the cells, and has low-dissolved oxygen concentration. It can produce 5–13 times more hydrogen than the wild type (Kruse et al., 2005). Furthermore, hydrogen production is greatly enhanced in a RuBisCO mutant that was incapable of carbon dioxide fixation (Hemscheimeier et al., 2008). Recently, Yacoby et al. (2011) provided evidence in *C. reinhardtii* that during anaerobic hydrogen production, a considerable amount of electron flow from ferredoxin was diverted toward FNR (ferredoxin: NADP⁺ oxidoreductase) rather than toward native [FeFe] hydrogenase (Figure 6.3). The findings highlight the role of competing pathways in decreasing photobiological hydrogen yields from wild type. Although, such studies have not been carried out in cyanobacteria, for attaining efficient conversion of solar energy into photobiological hydrogen production, it will be interesting to develop similar mutants and further to express in them the foreign hydrogenase. Among the few studies published using cyanobacteria as model organisms, McNeeley et al. (2010) enhanced hydrogen by developing an IdhA mutant in *Synechococcus* sp. strain PCC 7002 responsible for NADH-dependent reduction of pyruvate to D-lactate, considered the major fermentative reductant sink for this organism. They showed up to a fivefold increase in hydrogen production due to increase electron flux toward NADPH-dependent [NiFe] hydrogenase. Supplying a carbon source like acetate or glucose externally, hydrogen production can be increased more than when using inorganic media. Metabolic engineering by genetic modification could be a stable solution for efficient hydrogen production. This approach could rationalize different other problems and parameters for producing hydrogen in a large scale with a pilot reactor.

6.4 Conclusion

Microalgae are an extremely diverse group of organisms, many of which possess novel metabolic features that can be exploited for the production of hydrogen. In conclusion, it can be said that there is a scope to improve the H_2 production by genetic modification and synthetic biology by improving the biomass production, photoconversion efficiency and H_2-producing enzymes. Research has also been focused on uptake of external substrate, overexpression of enzymes, introduction of foreign hydrogenase, deletion of hydrogen uptake genes, generation of anaerobic condition, ATP synthase modification, linking of hydrogenase to cyanobacterial photosystems, and engineering of heterocyst frequency. The improvement of genetic tools, genome sequences, and high-throughput analytical techniques allows scientists to analyze and manipulate metabolic pathways and redirect the electron pull toward H_2 production. In future, genetically modified microorganisms are assumed to be the source of H_2 energy. The application of these modern metabolic engineering tools in photosynthetic microalgae has the potential to create important sources of hydrogen.

Glossary

2D	Two-dimensional gel electrophoresis
ATP	Adenosine triphosphate
CA	Carbonic anhydrase
CCMs	CO_2-concentrating mechanisms
CDKs	Cyclin-dependent kinases
CEF	Cyclic electron flow
Ci	Inorganic carbon
Cox	Cytochrome-c-oxidase
CYC	Cyclin
Fd	Ferredoxin
FeMo-co	Iron-molybdenum cofactor
FeS-cluster	Iron-sulfur-cluster
FHL	Formate hydrogen lyase
FNR	Ferredoxin-$NADP^+$-reductase
G1-phase	Gap-1 phase
Glp	Glycogen phosphorylase
K_m	Michaelis-Menten constant
LH1	Light-harvesting complex 1
LH2	Light-harvesting complex 2
LHC	Light-harvesting complex
MALDI-TOF	Matrix-assisted laser desorption/ionization–time of flight
MALDI-TOF/TOF-MS	Matrix-assisted laser desorption/ionization–time of flight–mass spectrometry
mRNA	Messenger ribonucleic acid
NAB1	NGFI-A binding protein 1
NAD	Nicotinamide adenine dinucleotide
NADH	Nicotinamide adenine dinucleotide (reduced form)
NADPH	Nicotinamide adenine dinucleotide phosphate (reduced form)
Nano-LC-ESI-MS/MS	Nanoliquid chromatography–electrospray ionization–mass spectrometry
NDH-1	NADPH-dehydrogenase complex (Type 1)
NHE	Na^+ (sodium)/H^+ (hydrogen) exchanger
NPQ	Nonphotochemical quenching
ORF	Open-reading frame
P/R ratio	Photosynthesis/respiration ratio
PPP	Pentose-phosphate pathway
PQ	Plastoquinon
PSI	Photosystem I
PSII	Photosystem II
RB	Retinoblastoma

RNAi · · · · · · · · · · · Ribonucleic acid-interference technology
RuBisCO · · · · · · · · · Ribulose-1,5-bisphosphate carboxylase/oxygenase
S-phase · · · · · · · · · · Synthesis phase
δG · · · · · · · · · · · · · Gibbs-free energy

References

Adams, M. W., Mortenson, L. E., and Chen, J. S. 1980. Hydrogenase. *Biochimica et Biophysica Acta (BBA)—Reviews on Bioenergetics*, 594(2), 105–176.

Adir, N., Zer, H., Shochat, S., and Ohad, I. 2003. Photoinhibition—A historical perspective. *Photosynthesis Research*, 76, 343–370.

Agapakis, C. M., Ducat, D. C., Boyle, P. M., Wintermute, E. H., Way, J. C., and Silver, P. A. 2010. Insulation of a synthetic hydrogen metabolism circuit in bacteria. *Journal of Biological Engineering*, 4(1), 3.

Agapakis, C. M. and Silver, P. A. 2010. Modular electron transfer circuits for synthetic biology: Insulation of an engineered biohydrogen pathway. *Bioengineered Bugs*, 1(6), 413–418.

Aizawa, K. and Miyachi, S. 1986. Carbonic anhydrase and CO_2 concentrating mechanisms in microalgae and cyanobacteria. *FEMS Microbiology Letters*, 39(3), 215–233.

Aryal, U. K., Callister, S. J., Mishra, S. et al. 2013. Proteome analyses of strains ATCC 51142 and PCC 7822 of the diazotrophic cyanobacterium *Cyanothece* sp. under culture conditions resulting in enhanced H_2 production. *Applied and Environmental Microbiology*, 79(4), 1070–1077.

Asada, Y., Koike, Y., Schnackenberg, J., Miyake, M., Uemura, I., and Miyake, J. 2000. Heterologous expression of clostridial hydrogenase in the cyanobacterium *Synechococcus* PCC7942. *Biochimica et Biophysica Acta (BBA)-Gene Structure and Expression*, 1490(3), 269–278.

Beckmann, J., Lehr, F., Finazzi, G. et al. 2009. Improvement of light to biomass conversion by de-regulation of light-harvesting protein translation in *Chlamydomonas reinhardtii*. *Journal of biotechnology*, 142(1), 70–77.

Bernát, G., Waschewski, N., and Rögner, M. 2009. Towards efficient hydrogen production: The impact of antenna size and external factors on electron transport dynamics in *Synechocystis* PCC 6803. *Photosynthesis Research*, 99, 205–216.

Buikema, W. J. and Haselkorn, R. 2001. Expression of the Anabaena hetR gene from a copper-regulated promoter leads to heterocyst differentiation under repressing conditions. *Proceedings of the National Academy of Sciences*, 98(5), 2729–2734.

Burton, E. and Martin, V. J. 2012. Proteomic analysis of *Clostridium thermocellum* ATCC 27405 reveals the upregulation of an alternative transhydrogenase-malate pathway and nitrogen assimilation in cells grown on cellulose. *Canadian Journal of Microbiology*, 58(12), 1378–1388.

Campbell, E. L., Summers, M. L., Christman, H., Martin, M. E., and Meeks, J. C. 2007. Global gene expression patterns of *Nostoc punctiforme* in steady-state dinitrogen-grown heterocyst-containing cultures and at single time points during the differentiation of akinetes and hormogonia. *Journal of Bacteriology, 189*(14), 5247–5256.

Chen, H. C., Yokthongwattana, K., Newton, A. J., and Melis, A. 2003. SulP, a nuclear gene encoding a putative chloroplast-targeted sulfate permease in *Chlamydomonas reinhardtii. Planta, 218*(1), 98–106.

Coleman, J. R. 1991. The molecular and biochemical analyses of CO_2-concentrating mechanisms in cyanobacteria and microalgae. *Plant, Cell and Environment, 14*(8), 861–867.

Cooley, J. W. and Vermaas, W. F. 2001. Succinate dehydrogenase and other respiratory pathways in thylakoid membranes of *Synechocystis* sp. strain PCC 6803: Capacity comparisons and physiological function. *Journal of Bacteriology, 183*(14), 4251–4258.

Cournac, L., Guedeney, G., Peltier, G., and Vignais, P. M. 2004. Sustained photoevolution of molecular hydrogen in a mutant of *Synechocystis* sp. strain PCC 6803 deficient in the type I NADPH-dehydrogenase complex. *Journal of Bacteriology, 186*(6), 1737–1746.

Dewitte, W. and Murray, J. A. 2003. The plant cell cycle. *Annual Review of Plant Biology, 54*(1), 235–264.

Doebbe, A., Rupprecht, J., Beckmann, J. et al. 2007. Functional integration of the HUP1 hexose symporter gene into the genome of *C. reinhardtii*: Impacts on biological H_2 production. *Journal of Biotechnology, 131*(1), 27–33.

Ducat, D. C., Sachdeva, G., and Silver, P. A. 2011. Rewiring hydrogenase-dependent redox circuits in cyanobacteria. *Proceedings of the National Academy of Sciences, 108*(10), 3941–3946.

Durnford, D. G. and Falkowski, P. G. 1997. Chloroplast redox regulation of nuclear gene transcription during photoacclimation. *Photosynthesis Research, 53*(2–3), 229–241.

Durnford, D. G., Price, J. A., McKim, S. M., and Sarchfield, M. L. 2003. Light-harvesting complex gene expression is controlled by both transcriptional and post-transcriptional mechanisms during photoacclimation in *Chlamydomonas reinhardtii. Physiologia Plantarum, 118*(2), 193–205.

Ehira, S., Ohmori, M., and Sato, N. 2003. Genome-wide expression analysis of the responses to nitrogen deprivation in the heterocyst-forming cyanobacterium *Anabaena* sp. strain PCC 7120. *DNA Research, 10*(3), 97–113.

Elrad, D. and Grossman, A. R. 2004. A genome's-eye view of the light-harvesting polypeptides of *Chlamydomonas reinhardtii. Current Genetics, 45*(2), 61–75.

Escoubas, J. M., Lomas, M., LaRoche, J., and Falkowski, P. G. 1995. Light intensity regulation of cab gene transcription is signaled by the redox state of the plastoquinone pool. *Proceedings of the National Academy of Sciences, 92*(22), 10237–10241.

Fan, Z., Yuan, L., and Chatterjee, R. 2009. Increased hydrogen production by genetic engineering of *Escherichia coli. PloS One, 4*(2), e4432.

Fang, S. C. and Umen, J. G. 2008. A suppressor screen in *Chlamydomonas* identifies novel components of the retinoblastoma tumor suppressor pathway. *Genetics, 178*(3), 1295–1310.

Fett, J. P. and Coleman, J. R. 1994. Regulation of periplasmic carbonic anhydrase expression in *Chlamydomonas reinhardtii* by acetate and pH. *Plant Physiology, 106*(1), 103–108.

Flachmann, R. and Kühlbrandt, W. 1995. Accumulation of plant antenna complexes is regulated by post-transcriptional mechanisms in tobacco. *The Plant Cell Online, 7*(2), 149–160.

Fodor, B., Rákhely, G., Kovács, Á. T., and Kovács, K. L. 2001. Transposon mutagenesis in purple sulfur photosynthetic bacteria: Identification of *hypF*, encoding a protein capable of processing [NiFe] hydrogenases in α, β, and γ subdivisions of the proteobacteria. *Applied and Environmental Microbiology, 67*(6), 2476–2483.

Fritsch, J., Scheerer, P., Frielingsdorf, S. et al. 2011. The crystal structure of an oxygen-tolerant hydrogenase uncovers a novel iron-sulphur centre. *Nature, 479*(7372), 249–252.

Gärtner, K., Lechno-Yossef, S., Cornish, A. J., Wolk, C. P., and Hegg, E. L. 2012. Expression of Shewanella oneidensis MR-1 [FeFe]-hydrogenase genes in *Anabaena* sp. strain PCC 7120. *Applied and Environmental Microbiology, 78*(24), 8579–8586.

Gfeller, R. P. and Gibbs, M. 1984. Fermentative metabolism of *Chlamydomonas reinhardtii* I. Analysis of fermentative products from starch in dark and light. *Plant Physiology, 75*(1), 212–218.

Ghirardi, M. L., Zhang, L., Lee, J. W. et al. 2000. Microalgae: A green source of renewable H_2. *Trends in Biotechnology, 18*(12), 506–511.

Giordano, M., Beardall, J., and Raven, J. A. 2005. CO_2 concentrating mechanisms in algae: Mechanisms, environmental modulation, and evolution. *Annual Review of Plant Biology, 56*, 99–131.

Gorwa, M. F., Croux, C., and Soucaille, P. 1996. Molecular characterization and transcriptional analysis of the putative hydrogenase gene of *Clostridium acetobutylicum* ATCC 824. *Journal of Bacteriology, 178*(9), 2668–2675.

Greenbaum, E. 1988. Energetic efficiency of hydrogen photoevolution by algal water splitting. *Biophysical Journal, 54*(2), 365–368.

Guiral, M., Tron, P., Belle, V. et al. 2006. Hyperthermostable and oxygen resistant hydrogenases from a hyperthermophilic bacterium *Aquifex aeolicus*: Physicochemical properties. *International Journal of Hydrogen Energy, 31*(11), 1424–1431.

Happe, T. and Naber, J. D. 1993. Isolation, characterization and N-terminal amino acid sequence of hydrogenase from the green alga *Chlamydomonas reinhardtii*. *European Journal of Biochemistry, 214*(2), 475–481.

Hemschemeier, A., Fouchard, S., Cournac, L., Peltier, G., and Happe, T. 2008. Hydrogen production by *Chlamydomonas reinhardtii*: An elaborate interplay of electron sources and sinks. *Planta, 227*(2), 397–407.

Hemschemeier, A. and Happe, T. 2005. The exceptional photofermentative hydrogen metabolism of the green alga *Chlamydomonas reinhardtii*. *Biochemical Society Transactions, 33*(1), 39–41.

Hopkins, R. C., Sun, J., Jenney, Jr., F. E., Chandrayan, S. K., McTernan, P. M., and Adams, M. W. 2011. Homologous expression of a subcomplex of *Pyrococcus furiosus* hydrogenase that interacts with pyruvate ferredoxin oxidoreductase. *PloS One, 6*(10), e26569.

Howitt, C. A., Cooley, J. W., Wiskich, J. T., and Vermaas, W. F. 2001. A strain of *Synechocystis* sp. PCC 6803 without photosynthetic oxygen evolution and respiratory oxygen consumption: Implications for the study of cyclic photosynthetic electron transport. *Planta, 214*(1), 46–56.

Ihara, M., Nishihara, H., Yoon, K. S. et al. 2006. Light-driven hydrogen production by a hybrid complex of a [NiFe]-hydrogenase and the cyanobacterial photosystem I. *Photochemistry and photobiology, 82*(3), 676–682.

Jeffries, T. W., Timourian, H., and Ward, R. L. 1978. Hydrogen production by *Anabaena cylindrica*: Effects of varying ammonium and ferric ions, pH, and light. *Applied and Environmental Microbiology, 35*(4), 704–710.

Jo, J. H., Jeon, C. O., Lee, S. Y., Lee, D. S., and Park, J. M. 2010. Molecular character-ization and homologous overexpression of [FeFe]-hydrogenase in *Clostridium tyrobutyricum* JM1. *International Journal of Hydrogen Energy*, 35(3), 1065–1073.

Kaji, M., Taniguchi, Y., Matsushita, O. et al. 1999. The *hydA* gene encoding the H₂-evolving hydrogenase of *Clostridium perfringens*: Molecular characterization and expression of the gene. *FEMS Microbiology Letters*, 181(2), 329–336.

Kars, G., Gündüz, U., Rakhely, G., Yücel, M., Eroğlu, İ., and Kovacs, K. L. 2008. Improved hydrogen production by uptake hydrogenase deficient mutant strain of *Rhodobacter sphaeroides* O.U.001. *International Journal of Hydrogen Energy*, 33(12), 3056–3060.

Khanna, N., Dasgupta, C. N., Mishra, P., and Das, D. 2011. Homologous overex-pression of [FeFe] hydrogenase in *Enterobacter cloacae* IIT-BT 08 to enhance hydrogen gas production from cheese whey. *International Journal of Hydrogen Energy*, 36(24), 15573–15582.

Khetkorn, W., Lindblad, P., and Incharoensakdi, A. 2012. Inactivation of uptake hydrogenase leads to enhanced and sustained hydrogen production with high nitrogenase activity under high light exposure in the cyanobacterium *Anabaena siamensis* TISTR 8012. *Journal of Biological Engineering*, 6(1), 19.

Kim, E. J., Lee, M. K., Kim, M. S., and Lee, J. K. 2008. Molecular hydrogen produc-tion by nitrogenase of *Rhodobacter sphaeroides* and by Fe-only hydrogenase of *Rhodospirillum rubrum*. *International Journal of Hydrogen Energy*, 33(5), 1516–1521.

King, P. W., Posewitz, M. C., Ghirardi, M. L., and Seibert, M. 2006. Functional stud-ies of [FeFe] hydrogenase maturation in an *Escherichia coli* biosynthetic sys-tem. *Journal of Bacteriology*, 188(6), 2163–2172.

Kondo, T., Arakawa, M., Hirai, T., Wakayama, T., Hara, M., and Miyake, J. 2002. Enhancement of hydrogen production by a photosynthetic bacterium mutant with reduced pigment. *Journal of Bioscience and Bioengineering*, 93(2), 145–150.

Kontur, W. S., Noguera, D. R., and Donohue, T. J. 2011. Maximizing reductant flow into microbial H₂ production. *Current Opinion in Biotechnology*, 23(3), 382–389.

Kruse, O., Rupprecht, J., Bader, K. P. et al. 2005. Improved photobiological H₂ pro-duction in engineered green algal cells. *Journal of Biological Chemistry*, 280(40), 34170–34177.

Kumar, N., Ghosh, A., and Das, D. 2001. Redirection of biochemical pathways for the enhancement of H₂ production by *Enterobacter cloacae*. *Biotechnology Letters*, 23(7), 537–541.

Kumar, D. and Kumar, H. D. 1991. Effect of monochromatic lights on nitrogen fixation and hydrogen evolution in the isolated heterocysts of *Anabaena* sp. strain CA. *International Journal of Hydrogen Energy*, 16(6), 397–401.

Lee, D. Y. and Fiehn, O. 2008. High quality metabolomic data for *Chlamydomonas rein-hardtii*. *Plant Methods*, 4(1), 7.

Lee, J. W. and Greenbaum, E. 2003. A new oxygen sensitivity and its potential appli-cation in photosynthetic H₂ production. *Applied Biochemistry and Biotechnology*, 106, 303–313.

Lenz, O., Gleiche, A., Strack, A., and Friedrich, B. 2005. Requirements for heterolo-gous production of a complex metalloenzyme: The membrane-bound [NiFe] hydrogenase. *Journal of Bacteriology*, 187(18), 6590–6595.

Lien, S. and Pietro, A. S. 1981. Effect of uncouplers on anaerobic adaptation of hydrogenase activity in *C. reinhardtii*. *Biochemical and Biophysical Research Communications*, 103(1), 139–147.

Lindahl, M., Yang, D. H., and Andersson, B. 1995. Regulatory proteolysis of the major light-harvesting chlorophyll a/b protein of photosystem II by a light-induced membrane-associated enzymic system. *European Journal of Biochemistry*, 231(2), 503–509.

Lubner, C. E., Knörzer, P., Silva, P. J. et al. 2010. Wiring an [FeFe]-hydrogenase with photosystem I for light-induced hydrogen production. *Biochemistry*, 49(48), 10264–10266.

Masukawa, H., Inoue, K., Sakurai, H., Wolk, C. P., and Hausinger, R. P. 2010. Site-directed mutagenesis of the *Anabaena* sp. strain PCC 7120 nitrogenase active site to increase photobiological hydrogen production. *Applied and Environmental Microbiology*, 76(20), 6741–6750.

Masukawa, H., Mochimaru, M., and Sakurai, H. 2002. Disruption of the uptake hydrogenase gene, but not of the bidirectional hydrogenase gene, leads to enhanced photobiological hydrogen production by the nitrogen-fixing cyano-bacterium *Anabaena* sp. PCC 7120. *Applied Microbiology and Biotechnology*, 58(5), 618–624.

Matsubara, H., Hase, T., Wakabayashi, S., and Wada, K. 1979. Gene duplications during the evolution of chloroplast-type ferredoxin. In *Proceedings for the Evolution of Proteins Molecules*, eds. H. Matsubara and T. Yamanaka, pp. 209–219. Tokyo: Tokyo University Press.

Maxwell, D. P., Falk, S., and Huner, N. P. 1995. Photosystem II excitation pressure and development of resistance to photoinhibition (I. light-harvesting complex II abundance and zeaxanthin content in *Chlorella vulgaris*. *Plant Physiology*, 107(3), 687–694.

Mcintee-Chmielewski N. 2010. Photobiological hydrogen production by a direct coupled photosystem I/hydrogenase complex. *MMG 445 Basic Biotechnology eJournal*, 6(1).

McNeely, K., Xu, Y., Bennette, N., Bryant, D. A., and Dismukes, G. C. 2010. Redirecting reductant flux into hydrogen production via metabolic engineering of fer-mentative carbon metabolism in a cyanobacterium. *Applied and Environmental Microbiology*, 76(15), 5032–5038.

Melis, A. 2007. Photosynthetic H_2 metabolism in *Chlamydomonas reinhardtii* (unicellular green algae). *Planta*, 226, 1075–1086.

Mikheeva, L. E., Schmitz, O., Shestakov, S. V., and Bothe, H. 1995. Mutants of the cyano-bacterium *Anabaena variabilis* altered in hydrogenase activities. *Z. Naturforsch*, 50, 505–510.

Millar, A. J., Straume, M., Chory, J., Chua, N. H., and Kay, S. A. 1995. The regulation of circadian period by phototransduction pathways in Arabidopsis. *Science*, 267(5201), 1163–1166.

Miura, K., Yamano, T., Yoshioka, S. et al. 2004. Expression profiling-based identification of CO_2-responsive genes regulated by CCM1 controlling a carbon-concentrating mechanism in *Chlamydomonas reinhardtii*. *Plant physiology*, 135(3), 1595–1607.

Miyake, M. and Asada, Y. 1997. Direct electroporation of clostridial hydrogenase into cyanobacterial cells. *Biotechnology Techniques*, 11(11), 787–790.

Moulis, J. M. and Davasse, V. 1995. Probing the role of electrostatic forces in the interaction of *Clostridium pasteurianum* ferredoxin with its redox partners. *Biochemistry*, 34(51), 16781–16788.

Müller, P., Li, X. P., and Niyogi, K. K. 2001. Non-photochemical quenching. A response to excess light energy. *Plant Physiology*, 125(4), 1558–1566.

Mussgnug, J. H., Thomas-Hall, S., Rupprecht, J. et al. 2007. Engineering photosynthetic light capture: Impacts on improved solar energy to biomass conversion. *Plant Biotechnology Journal*, 5(6), 802–814.

Myers, J. 1957. Algal culture. In *Encyclopedia of chemical technology. Interscience*, eds. R. E. Kirk and D. E. Othmer, pp. 649–668. New York: Interscience Encyclopedia, Inc.

Nakajima, Y., Fujiwara, S., Sawai, H., Imashimizu, M., and Tsuzuki, M. 2001. A phycocyanin-deficient mutant of *Synechocystis* PCC 6714 with a single-base substitution upstream of the *cpc* operon. *Plant and Cell Physiology*, 42(9), 992–998.

Nicolet, Y., Piras, C., Legrand, P., Hatchikian, C. E., and Fontecilla-Camps, J. C. 1999. *Desulfovibrio desulfuricans* iron hydrogenase: The structure shows unusual coordination to an active site Fe binuclear center. *Structure*, 7, 13–23.

Peters, J. W., Lanzilotta, W. N., Lemon, B. J., and Seefeldt, L. C. 1998. X-ray crystal structure of the Fe-only hydrogenase (CpI) from *Clostridium pasteurianum* to 1.8 angstrom resolution. *Science*, 282, 1853–1858.

Pinto, F., van Elburg, K. A., Pacheco, C. C. et al. 2012. Construction of a chassis for hydrogen production: Physiological and molecular characterization of a *Synechocystis* sp. PCC 6803 mutant lacking a functional bidirectional hydrogenase. *Microbiology*, 158(2), 448–464.

Polle, J. E. W., Kanakagiri, S., and Melis, A. 2003. tla1, a DNA insertional transformant of the green alga *Chlamydomonas reinhardtii* with a truncated light-harvesting chlorophyll antenna size. *Planta*, 217, 49–59.

Posewitz, M. C., King, P. W., Smolinski, S. L., Zhang, L., Seibert, M., and Ghirardi, M. L. 2004. Discovery of two novel radical S-adenosylmethionine proteins required for the assembly of an active [Fe] hydrogenase. *Journal of Biological Chemistry*, 279(24), 25711–25720.

Price, G. D., Badger, M. R., Woodger, F. J., and Long, B. M. 2008. Advances in understanding the cyanobacterial CO_2-concentrating-mechanism (CCM): Functional components, Ci transporters, diversity, genetic regulation and prospects for engineering into plants. *Journal of Experimental Botany*, 59(7), 1441–1461.

Radmer, R. and Kok, B. 1977. Photosynthesis: Limited yields, unlimited dreams. *Bioscience*, 29, 599–605.

Ramachandran, U., Shamshurin, D., Spicer, V. et al. 2012. Total proteome analysis of anaerobic, cellulolytic and biofuels producing bacteria *Clostridium termitidis*. Conference paper hosted by Oak Ridge National Laboratory and NERL.

Rao, K. K., Rosa, L., and Hall, D. O. 1976. Prolonged production of hydrogen gas by a chloroplast biocatalytic system. *Biochemical and Biophysical Research Communications*, 68(1), 21–28.

Rascher, U., Lakatos, M., Büdel, B., and Lüttge, U. 2003. Photosynthetic field capacity of cyanobacteria of a tropical inselberg of the Guiana Highlands. *European Journal of Phycology*, 38(3), 247–256.

Robertson, D., Boynton, J. E., and Gillham, N. W. 1990. Cotranscription of the wild-type chloroplast atpE gene encoding the CF1/CF0 epsilon subunit with the 3' half of the rps7 gene in *Chlamydomonas reinhardtii* and characterization of frameshift mutations in *atpE*. *Molecular and General Genetics MGG*, 221(2), 155–163.

Rousset, M., Magro, V., Forget, N., Guigliarelli, B., Belaich, J. P., and Hatchikian, E. C. 1998. Heterologous expression of the *Desulfovibrio gigas* [NiFe] hydrogenase in *Desulfovibrio fructosovorans* MR400. *Journal of Bacteriology*, 180(18), 4982–4986.

Rühle, T., Hemschemeier, A., Melis, A., and Happe, T. 2008. A novel screening protocol for the isolation of hydrogen producing *Chlamydomonas reinhardtii* strains. *BMC Plant Biology*, 8(1), 107.

Ruiyan, Z., Di, W., Yaoping, Z., and Jilun, Li. 2006. Hydrogen production by draTGB hupL double mutant of Rhodospirillum rubrum under different light conditions. *Chinese Science Bulletin*, 51, 2611–2618.

Sakurai, H., Masukawa, H., Zhang, Xh., Ikeda, H., and Inoue, K. 2008. Improvement of nitrogenase-based photobiological hydrogen production by cyanobacteria by gene engineering hydrogenases and homocitrate synthase. In *Photosynthesis. Energy from the Sun*, eds. J. F. Allen, E. Gantt, and J. H. Golbeck, pp. 1277–1280. The Netherlands: Springer.

Sato, N., Ohmori, M., Ikeuchi, M. et al. 2004. Use of segment-based microarray in the analysis of global gene expression in response to various environmental stresses in the cyanobacterium *Anabaena* sp. PCC 7120. *The Journal of General and Applied Microbiology*, 50(1), 1–8.

Shestakov, S. V. and Mikheeva, L. E. 2006. Genetic control of hydrogen metabolism in cyanobacteria. *Russian Journal of Genetics*, 42(11), 1272–1284.

Skizim, N. J. 2012. Enhancingthe metabolic capacity of cyanobacteria for biological hydrogen production: Biofuel application of *Cyanothece* and *Arthrospira* spp. *Ph. D. dissertation*, Princeton University.

Srirangan, K., Pyne, M. E., and Perry Chou, C. 2011. Biochemical and genetic engineering strategies to enhance hydrogen production in photosynthetic algae and cyanobacteria. *Bioresource Technology*, 102(18), 8589–8604.

Stripp, S. T., Goldet, G., Brandmayr, C. et al. 2009. How oxygen attacks [FeFe] hydrogenases from photosynthetic organisms. *Proceedings of the National Academy of Sciences*, 106(41), 17331–17336.

Tamagnini, P., Leitão, E., Oliveira, P. et al. 2007. Cyanobacterial hydrogenases: Diversity, regulation and applications. *FEMS Microbiology Reviews*, 31(6), 692–720.

Tanisho, S., Kamiya, N., and Wakao, N. 1989. Hydrogen evolution of *Enterobacter aerogenes* depending on culture pH: Mechanism of hydrogen evolution from NADH by means of membrane-bound hydrogenase. *Biochimica et Biophysica Acta (BBA)-Bioenergetics*, 973(1), 1–6.

Thauer, R. K., Jungermann, K., and Decker, K. 1977. Energy conservation in chemotrophic anaerobic bacteria. *Bacteriological Reviews*, 41(1), 100–180

Torzillo, G., Scoma, A., Faraloni, C., Ena, A., and Johanningmeier, U. 2009. Increased hydrogen photoproduction by means of a sulfur-deprived *Chlamydomonas reinhardtii* D1 protein mutant. *International Journal of Hydrogen Energy*, 34(10), 4529–4536.

Umen, J. G. and Goodenough, U. W. 2001. Control of cell division by a retinoblastoma protein homolog in *Chlamydomonas*. *Genes and Development*, 15(13), 1652–1661.

Uzumaki, T., Fujita, M., Nakatsu, T. et al. 2004. Crystal structure of the C-terminal clock-oscillator domain of the cyanobacterial KaiA protein. *Nature Structural and Molecular Biology*, 11(7), 623–631.

Vasilyeva, L., Miyake, M., Khatipov, E. et al. 1999. Enhanced hydrogen production by a mutant of *Rhodobacter sphaeroides* having an altered light-harvesting system. *Journal of Bioscience and Bioengineering*, 87(5), 619–624.

Vignais, P. M., Magnin, J. P., and Willison, J. C. 2006. Increasing biohydrogen production by metabolic engineering. *International Journal of Hydrogen Energy*, 31(11), 1478–1483.

Wang, D., Zhang, Y., Welch, E., Li, J., and Roberts, G. P. 2010. Elimination of Rubisco alters the regulation of nitrogenase activity and increases hydrogen production in *Rhodospirillum rubrum*. *International Journal of Hydrogen Energy*, 35(14), 7377–7385.

Xu, Q., Yooseph, S., Smith, H.O., and Venter, C. J. 2005. Development of a novel recombinant cyanobacterial system for hydrogen production from water. Paper presented at *Genomics: gtl program projects*, Rockville.

Yacoby, I., Pochekailov, S., Toporik, H., Ghirardi, M. L., King, P. W., and Zhang, S. 2011. Photosynthetic electron partitioning between [FeFe]-hydrogenase and ferredoxin: NADP þ-oxidoreductase (FNR) enzymes in vitro. *Proceedings of the National Academy of Sciences*, 108(23), 9396–9401.

Yang, D.H., Andersson, B., Aro, E.M., and Ohad, I. 2001. The redox state of the plastoquinone pool controls the level of the light-harvesting chlorophyll a/b binding protein complex II (LHCII) during photoacclimation: Cytochrome b_6f deficient *Lemna perpusilla* plants are locked in a state of high-light acclimation. *Photosynthesis Research*, 68, 163–174.

Yoshida, A., Nishimura, T., Kawaguchi, H., Inui, M., and Yukawa, H. 2005. Enhanced hydrogen production from formic acid by formate hydrogen lyase-overexpressing *Escherichia coli* strains. *Applied and Environmental Microbiology*, 71(11), 6762–6768.

7

Process and Culture Parameters

7.1 Introduction

Fermentative H_2 production is a complex process under the influence of several physicochemical factors. Type of reactor has a major effect on the final yield of the product, especially in the case of photofermentative/biophotolytic process as it determines the amount of light and agitation received by the light-dependent organisms growing inside the reactor. Therefore, an additional chapter (Chapter 8) has been completely dedicated to discuss all the aspects of photobioreactor design and operation and their effect on hydrogen production.

Voluminous literature is available on the optimization of various factors on biological hydrogen production. In this chapter the effect of each of these factors on hydrogen production by dark fermentative bacteria, photofermentative bacteria, and photosynthetic bacteria will be discussed in detail. Further, the chapter will throw some light on statistical optimization of factors affecting biohydrogen production.

7.2 Factors Affecting Dark Fermentation Process

The performance of a bioreactor in terms of hydrogen production is governed by various physicochemical parameters influencing the various biochemical pathways. Factors associated with environmental conditions, process-operating conditions, and chemical conditions have been critically reviewed in this section. The major physicochemical parameters affecting hydrogen production pathways include

- Inoculum
- Temperature
- pH
- Alkalinity

- Hydraulic retention time
- Hydrogen and CO_2 partial pressure
- Effect of micronutrients
- Light

Suitable physicochemical parameters required for hydrogen production as reported by various researchers are summarized in Table 7.1.

7.2.1 Effect of Inoculum on Fermentative Hydrogen Production

As already discussed in Chapter 2, microorganisms that are capable of hydrogen production occur in natural habitat such as soil, wastewater, sludge, compost, and so on. Thus, these materials can be used as potential source of inoculums for fermentative hydrogen production. The inoculum may either be a pure microbe or a mixed/constructed consortia. Members of *Clostridium* and *Enterobacter* are most widely used as inoculums for mesophilic fermentative hydrogen production. However, for thermophilic hydrogen production, species of *Thermotogales* is the preferred choice.

Inoculum may be defined as the preparation of a population of microorganisms from a dormant stock culture to an active state of growth that is suitable for inoculation in the final production stage. This is called inoculum development. Initially, the seed inoculum is obtained from a working culture to initiate growth in a suitable liquid medium. Inoculum development is generally done using culture flasks ranging in capacity from 0.05 to 12×10^{-3} m^3. For pilot and bench-scale operations, small reactors are used for the purpose of inoculum development. It is to be noted that inoculum development is carried out stepwise to increase the volume to the desired level (Figure 7.1). Usually, an inoculum size of 0.5–10% is used; this allows a 20–200-fold increase in inoculum volume at each step. Typically, the inoculum used for production stage is about 5–10% v/v. Working seed inoculum may be obtained from any microorganism or consortia obtained from nature or isolated in a pure state. Inoculums are incubated in a wide range of temperature depending on the type of organism used (mesophiles or thermophiles).

The physical state of the inoculum (lag, log, or stationary) and inoculum percentage are two critical factors for any fermentation industry. Since in dark fermentation hydrogen production is a growth-associated product, higher yield is expected when the seed inoculum is in a rapidly dividing state (mid-log phase). Optimization of inoculum volume is another aspect of maximization of hydrogen production. The optimal volume varies from 5% to 10% (v/v), depending on the species and medium used. Inoculum volumes, other than the optimal, drive the reductants away from hydrogen production. Kotay and Das (2007) studied the inoculum age, for *Bacillus subtilis* using sewage sludge as substrate. They found that up to 10% (v/v) stepwise increase in inoculum volume, there was a significant increase in H_2

TABLE 7.1

Suitable Physicochemical Parameters Required for Hydrogen Production as Reported by Various Researchers

Microorganism	Process	Substrate	Temperature (°C)	pH	Hydraulic Retention Time (10^3 S)	References
Anaerobic-digested sludge	CSTR[a]	Food waste	—	5.3	151–86	Wu and Chang (2003)
Sludge	UASBr[a-b]	Sucrose	312	6.3–3.4	47	Kotsopoulos et al. (2006)
Sludge	ASBR[c]	Starch	308	5.3	65	Venetsaneas et al. (2009)
Sludge from waste treatment plant	CSTR[a]	Glucose	310	5.5	43–2	Lee et al. (2007)
Waste from treatment plant	AnMBR[d]	Glucose	—	5.5	12–36	Guo et al. (2008)
Heat-treated soil	AFBR[e]	Sucrose	308	—	33–4	Yu and Mu (2006)
Sludge from waste treatment plant	UASBr[a-b]	Rice winery wastewater	328	6–4.5	86–7	Chang and Lin (2004)
Indigenous microflora	CSTR[a]	Cheese whey wastewater	308	5.2	86	Shen et al. (2009)
Acid-treated sludge	AnMBR[d]	Glucose	308	6.8–6.2	14–4	Ueno et al. (1996)
Mixed culture from treated molasses wastewater	EGSBr[f]	Molasses waste water	—	—	22–4	Barros et al. (2010)
Seed sludge	UASBR[b]	Sucrose	311	4.4	108–7	Peintner et al. (2010)
Seed sludge	UASBR[b]	Sucrose	308	6.7	86–14	Infantes et al. (2011)
Anaerobic digestor sludge	AnMBR[d]	Glucose	296	5.5	29	Li et al. (2011)
Anaerobic microflora	CSTR[a]	Sugary wastewater	333	6.8	11–2	Chin et al. (2003)
Heat-treated sludge	AFBR[e]	Glucose	303	—	29–4	Rajeshwari et al. (2000)
C. owensensis OL[T]	AFBR[e]	Glucose	343	—	72–36	Lee et al. (2003)

continued

TABLE 7.1 (continued)

Suitable Physicochemical Parameters Required for Hydrogen Production as Reported by Various Researchers

Microorganism	Process	Substrate	Suitable Physicochemical Parameters			References
			Temperature (°C)	pH	Hydraulic Retention Time (10^3 S)	
Manure from methanogenic reactor	CSTR[a]	Household solid waste	343	7	3.5×10^2 day	Liu et al. (2008)
Clostridium tyrobutyricum JM1	CSTR[a]	Glucose	310	—	7	Jo et al. (2008)
Ca. taiwanensis On1	CSTR[a]	Starch	310	7	7	Chen et al. (2006)

[a] CSTR: continuous stirred-tank reactor.
[b] UASBr: upflow anaerobic sludge blanket reactor.
[c] ASBR: anaerobic sludge bed reactor.
[d] AnMBR: anaerobic membrane bioreactor.
[e] AFBR: anaerobic-fluidized bed reactor.
[f] EGSBr: expanded granular sludge bed.

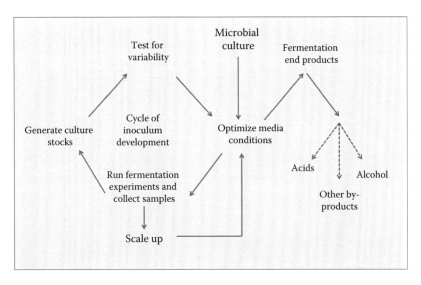

FIGURE 7.1
Model for inoculum development in fermentation industries.

production rate, however, any further increase decreased the hydrogen production rate by 15% and 20% (Figure 7.2). However, cell mass increased exponentially with increase in inoculum volume. Apparently, at higher inoculum volume, more carbon source is devoted for biomass production rather than hydrogen production. Further, they also studied the rate of inoculum age on hydrogen production and reported that the effect of inoculum age on rate of H_2 production was not very significant. All the same, 14 h inoculum age to some extent recorded higher rate of H_2 production and H_2 content; 14 h was mid-log phase of the growth of the bacterium; H_2 production was recorded to initiate in mid-log phase and reached maximum production rate in the stationary phase. For this reason, using 14-h old inoculum may have minimized the lag phase in the H_2 production (Kotay and Das, 2007).

Inoculum used for hydrogen production may be sporulating bacteria like *Bacillus* or *Clostridium* or nonsporulating bacteria for example *Enterobacter*, *Escherichia*, and *Klebsiella*. While working with spore enrichment for inoculum preparation, germination may be induced by amino acids, sugars, ribosides, enzymes, and hydrostatic pressure, but often a species-specific combination of germinants may be required; for example alanine and lactate for *Clostridium botulinum* spores. Besides, it is necessary that reactor conditions favoring sporulation are avoided as it may result in complete washout. It is well known that exposure to traces of oxygen initiates sporulation in continuous cultures of some species; for example *Clostridium beijerinckii* (Ross et al., 1990). It has been found that *Clostridium* strain C7 was more tolerant to trace concentrations of oxygen and to nutrient stress (Ross et al., 1990) while carbon starvation gave sporulation with *C. cellulolyticum*

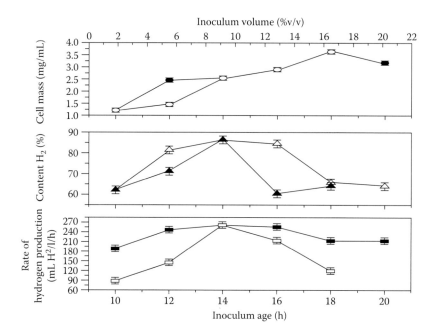

FIGURE 7.2
Effect of inoculum volume and inoculum age on rate of H_2 production, content H_2 (%), and cell mass. (Reprinted from Kotay, S. M. and Das, D. 2007. *Bioresource Technology*, 98, 1183–1190. With permission.)

on cellobiose (Gehin et al., 1995). Further, studies have shown that unfavorable environmental conditions such as accumulation of acetic and butyric acid fermentation end products due to substrate fermentation in the reactor may cause clostridia to switch to solvent production and sporulate (Gehin et al., 1995).

Studies thus far have been conducted using pure, constructed, or mixed consortia. The choice of the inoculum primarily depends on the substrate to be degraded. While using an organic substrate like glucose, it may be suffice to use a single organism, however, while using a complex substrate like organic waste, a mixed consortia may be preferred. Mixed consortia may obtained naturally such as by the use of sludge. While mixed or constructed consortia are that in which scientists use an array of organism whose properties are well known and which are known to co-inhabit to breakdown complex substrates. Consortia provide a diverse array of microbial enzymes to breakdown the complex substrate (Box 7.1). Thus consortia can provide greater substrate degradation over pure strain. Kotay and Das (2010) formulated a microbial consortium from three facultative H_2-producing anaerobic bacteria, *Enterobacter cloacae* IIT-BT 08, *Citrobacter freundii* IIT-BT L139, and *Bacillus coagulans* IIT-BT S1 to degrade sewage sludge. This consortium was tested as the seed culture for H_2 production

BOX 7.1 DEVELOPMENT OF CONSTRUCTED MICROBIAL CONSORTIUM USING H²-PRODUCING STRAINS (KOTAY, 2008)

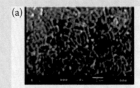 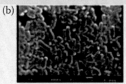 

(a) *Enterobacter cloacae* IIT-BT 08 (b) *Citrobacter freundii* IIT-BT L139
(c) *Bacillus coagulans* IIT-BT-S1

Enterobacter cloacae IIT-BT 08—isolated from leaf surface
Citrobacter freundii IIT-BT L139—isolated from oil-rich soil
Bacillus coagulans IIT-BT-S1—isolated from sludge

Enzyme Production Assays Conduction on the Three Strains

Enzyme	*E. cloacae* **IIT-BT 08**	*C. freundii* **IIT-BT L139**	*B. coagulans* **IIT-BT S1**
Amylase	+	+	+
Cellulase	+	−	+
Xylanase	+	−	+
Lipase	+	+	+
Protease	−	−	+

from sewage sludge. Initial studies showed that the constructed consortia in a ratio of 1:1:1 produced higher yield than the pure culture of *E. cloacae*. This is due to the different advantages that the organisms in the community provide to degrade the complex substrate. Use of mixed flora for fermentative H_2 production is more practical than pure culture as this mode of H_2 production is simpler to operate and easier to control and have broader source of feedstock.

At present, various works using mixed cultures of microorganisms as inoculums from wastewater, sludge, and soil are in progress for fermentative H_2 production. However, the major disadvantage of using mixed flora as inoculum for fermentative H_2 production is that in nature H_2-producing and H_2-consuming bacteria coexist. Thus, hydrogen produced gets consumed by H_2-consuming bacteria. However, appropriate use of pretreatment methods can thrive H_2-producing bacteria. The various pretreatment methods reported for enriching H_2-producing bacteria from mixed cultures mainly include heat-shock, acid-base aeration, freezing, and thawing, chloroform, sodium-2-bromoethane sulfonate or 2-bromoethane sulfonic acid, and iodopropane (Cheong and Hansen, 2006). Different pretreatment processes have different property and comparison of different pretreatment methods to get a better one for a given fermentative hydrogen production process has been conducted by many researches (Table 7.2). It has been observed that though heat-shock was the most widely used pretreatment method for enriching hydrogen-producing bacteria from inoculum (Li and Fang, 2007), was not always effective for enriching hydrogen-producing bacteria from mixed culture inoculum compared with other pretreatment methods, for it may inhibit the activity of some hydrogen-producing bacteria (Wang and Wan, 2008a). As observed from the table most of the studies are conducted in the batch mode, however, conducting these experiments in continuous mode is highly recommended. Table 7.2 shows that several authors have worked with different kinds of pretreatment processes depending on the substrate used. Generally, heat treatment is recommended for pretreatment of sewage sludge to eliminate the methanogens. Heat treatment kills nonspore-forming methanogens so that hydrogen losses due to methanogenesis can be prevented in batch tests (Brock et al., 1994). Besides, heat treatment activates clostridial spores to commence germination. Clostridia are known to be high hydrogen producers. Heat is known to activate the germination receptor. Thus, heat treatment is highly recommended for spore-former enrichment, although there is variation in the temperature attained and the time for which it is applied. In contrast to this, Hussy et al. (2005) used nonheat treated anaerobically digested sludge for H_2 production from sucrose and sugar beet. The experiments remained stable for 45 and 32 days before termination because of time constraints. Kotay (2008) used a host of various pretreatment methods to maximize hydrogen production from sewage sludge. The study concluded that COD, carbohydrate, protein, and lipids were significantly more solubilized in case of thermal and microwave pretreatment. This was confirmed by the decreasing trend in VSS/TSS ratio of sewage sludge achieved after heat treatment (Figure 7.3). However, on the downside, heat treatment may prevent development of nonspore-forming propionate producers, which convert hexose without forming H_2. However, Yusoff et al. (2009) successfully demonstrated heat treatment at pilot scale to degrade pome. They operated a 0.05×10^{-3} m^3 CSTR for 26 days. Further work is thus needed to establish whether the additional technical complexity of heat-treating the inoculum at industrial scale is cost-effective. Practically thinking, heat pretreatment of the inoculum is technically more difficult during scale-up as compared to acid/base pretreatment process. Analysis of inoculum conditions of mixed consortia can be conducted in various ways. Two methods are popular: 16S rRNA analysis by PCR-DGGE and Fluorescence-activated cell sorter (FACS) based on molecular markers. Besides, PCR-DGGE can be used to detect the changes in the community structure of mixed cultures after certain pretreatment. For example, using PCR-DGGE technique, Kim and Shin (2008) reported that base pretreatment of mixed cultures would prevent the microbial population shift to non-H_2-producing acidogens, thus was beneficial for fermentative hydrogen production.

TABLE 7.2

Effect of Various Pretreatment Processes on Biohydrogen Production

Pretreatment	Waste	Mode of Operation	Microorganism	Hydrogen Yield (mol/mol Hexose)	Reference
Mechanical	Sweet sorghum residues	Batch	*Rumicoccus albus*	2.59	Ntaikou et al. (2008)
Mechanical	Wheat straw	Batch	Mixed mesophilic culture	3.8	Ivanova et al. (2009)
Mechanical	Maize leaves	Batch	*Caldicellulosiruptor saccarolyticus*	3.6	Ivanova et al. (2009)
Mild acid treatment (1.8% H_2SO_4)	Barley straw	Batch	*Caldicellulosiruptor saccarolyticus*	—	Panagiotopoulos et al. (2009)
Akali-thermal hydrolysis (0.2–4 g/L NaOH, 100°C, 2 h)	Bagasse	Batch	Mixed thermophilic cultures	13.39 mmol H_2/g TVS	Chairattanamanokorn et al. (2009)
Acid-thermal hydrolysis 0.25–4 (v/v), 121°C, 30–180 min	Corn stover	Batch	*Thermoanaerobacterium thermosaccarolyticum*	2.24	Cao et al. (2009)
Steam explosion 90–220°C, 2–5 min	Corn stover	Continuous	Mixed mesophilic culture	3.0	Datar et al. (2007)
Chloroform	Glucose	Batch	Methanogenic granules	1.2	Hu and Chen (2007)
Sodium-2-bromoethan sulfonate	Dairy wastewater	Batch	Anaerobic sludge	0.0317 mmol/g COD	Mohan et al. (2008)
Heat conditioned	Glucose	Batch	Anaerobic sludge	1.75	Zheng and Yu (2005)
Heat shock	Glucose	Batch	Digested sludge	1.8	Wang et al. (2007)

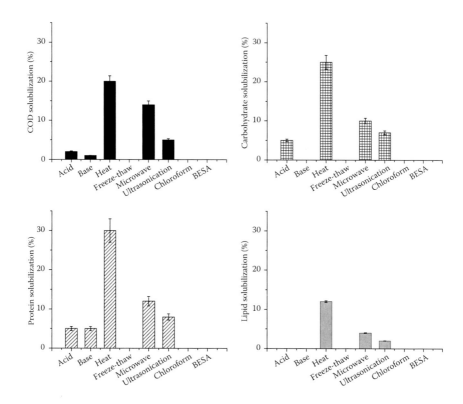

FIGURE 7.3
Effect of various pretreatments on sludge solubilization (Adapted from Kotay, S. M. 2008. Microbial hydrogen production from sewage sludge. Ph.D. thesis, IIT, Kharagpur.).

FLUORESCENCE-ACTIVATED CELL SORTER

The fluorescence-activated cell sorter is a machine that can rapidly separate the cells in a suspension on the basis of size and the color of their fluorescence.

The principle and working of FACS is illustrated in Figure 7.4.

- A cell suspension containing cells labeled with a fluorescent dye is directed into a thin stream so that all the cells pass in single file. The dye is coupled to a monoclonal antibody and binds to those cells coated with the antigen for which the antibody is specific.
- This stream emerges from a nozzle vibrating at some 40,000 cycles per second, which breaks the stream into 40,000 discrete droplets each second.
- Some of these droplets may contain a cell.

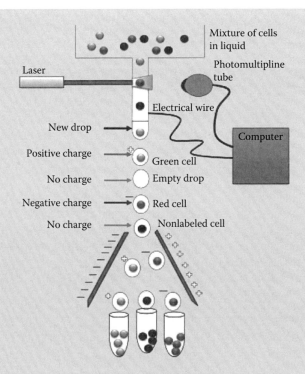

FIGURE 7.4
Principle and working of fluorescence-activated cell sorter.

- A laser beam is directed at the stream just before it breaks up into droplets.
- As each labeled cell passes through the beam, its resulting fluorescence is detected by a photocell.
- If the signals from the two detectors meet either of the criteria set for fluorescence and size, an electrical charge (+ or −) is given to the stream.
- The droplets retain this charge as they pass between a pair of charged metal plates.
  - Positively charged drops are attracted to the negatively charged plate and vice versa.
  - Uncharged droplets (those that contain no cell or a cell that fails to meet the desired criteria of fluorescence and size) pass straight into a third container and are later discarded.

This apparatus can sort as many as 300,000 cells per minute.

The cells are not damaged by the process. In fact, because the machine can be set to ignore droplets containing dead cells, the percent viability of the sorted cells can be higher than that in the original suspension.

DENATURING GRADIENT GEL ELECTROPHORESIS (DGGE)

Denaturing gradient gel electrophoresis (DGGE) is a molecular finger-printing method that separates polymerase chain reaction (PCR)-generated DNA products (Figure 7.5). The polymerase chain reaction of an environmental sample can produce a range of PCR products of identical size, but differing in a few nucleotides. In a traditional agarose gel, all the samples will run identical resulting in a single band. However, a DGGE can overcome the limitation by resolving bands that differ in a single nucleotide. During DGGE, PCR products encounter increasingly higher concentrations of chemical denaturant as they migrate through a polyacrylamide gel. Upon reaching a threshold denaturant concentration, the weaker melting domains of the double-stranded PCR product will begin to denature at which time migration slows dramatically. Ideally, differing sequences of DNA (from different bacteria) will denature at different denaturant concentrations resulting in a pattern of bands. Theoretically, each band represents a different bacterial population present in the community. Once generated, fingerprints can be uploaded into databases in which fingerprint similarity can be assessed to determine microbial structural differences between environments or among inoculum seed treatment.

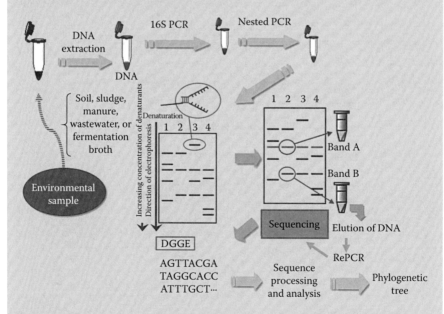

FIGURE 7.5
Schematic representation of various steps in DGGE analysis. (Adapted from Kotay, S. M. 2008. Microbial hydrogen production from sewage sludge. PhD thesis, IIT, Kharagpur.)

7.2.2 Temperature

Incubation temperature dramatically affects the growth rate of bacteria, because it affects the rates of all cellular reactions. Temperature may also affect the metabolic pattern, nutritional requirements, and composition of bacterial cells. The number of generations per hour can be plotted against temperature for any strain to determine the optimum temperature for growth. Although all bacteria have an optimal temperature at which they function best, their growth range is often quite varied. Operational temperature of the fermenter generally depends on the choice of microorganism used such as psychrophilic (273–293 K), mesophilic (293–315 K), or thermophilic (315–348 K). Studies have shown that suitable temperature range for effective H2 fermentation with pure mesophilic bacterial isolates (such as *Clostridium* and *Enterobacter* strains) is 310–318 K (Vindis, 2009). Temperatures above or below optimal severely affects growth rate of the organism. To this end, Nath et al. (2006) studied the effect of temperature on hydrogen production from *Enterobacter cloacae* IIT-BT 08. They studied the temperature in the range of 298–313 K. As the temperature trended toward the optimal temperature 310 K, hydrogen production increased (Figure 7.6). However, any further increase in temperature was found to significantly decrease the

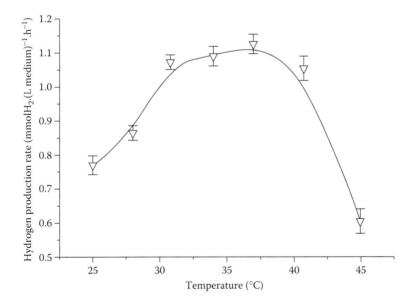

FIGURE 7.6

Effect of temperature on the rate of production of hydrogen in anaerobic fermentation by *Enterobacter cloacae* DM11. (Initial glucose concentration, 1% (m/v); initial pH, 6.0; and batch volume, 0.3×10^{-3} m³.) (Reprinted from Nath, K., Kumar, A., and Das, D. 2006. *Canadian Journal of Microbiology, 52,* 525–532. With permission.)

production. It has been suggested that the effect of temperature may be due to changes in the membrane architecture. This is suggested to be associated with a group of proteins known as hopanoids, which act as sterols. Besides, the increase in hydrogen production with increase in temperature, for the complex substrate, may be explained by the positive effect on the rate of hydrolysis of the complex particles (Veeken and Hamelers, 1999). Reactor operation at temperatures above 343 K increases the solubility of many polymeric substrates, decreases viscosity, increases bioavailability, and decreases the risk for contamination by methanogenic bacteria (Egorova and Antranikian, 2005).

Increase in hydrogen temperature under optimal temperature may further be attributed to the fact that increase in temperature doubles the enzymatic activity for every 10°C until the optimal temperature is reached, above which the enzymatic activity rapidly decreases. Hydrogenase enzymes are widely distributed in bacteria-enabling microorganism either to use H_2 as source of energy or to use H^+ as terminal electron acceptor. Optimal temperature is therefore required for their maximal activity. However, on further increasing the temperature beyond the optimal hydrogen production is found to decrease. Lee et al. (2006a) monitored the effects of temperature on hydrogen production using sucrose-based synthetic wastewater as feed in a carrier-induced granular sludge bed (CIGSB) reactor. He found that hydrogen production rate and yield trended to increase with increase in temperature from 303 to 313 K, while they trended to decrease with further increase in temperature from 313 to 318 K. The decrease in the rate of hydrogen production at higher temperature can be attributed to an increase in denaturation rate of the enzymes (Fabiano and Perego, 2002). In fermentation or enzymatic processes, it is known that the positive kinetic effect of an increase of temperature prevails over the negative effect on the biocatalyst activity, up to the threshold temperature beyond which thermal deactivation of biocatalyst takes place (Slininger et al., 1990). At higher temperatures, thermal inactivation of the enzymes controlling the metabolic pathway takes place.

Deactivation mechanism can be understood by calculating the activation enthalpy of fermentative hydrogen production and that of thermal deactivation using a modified Arrhenius approach (Fabiano and Perego, 2002). Nath et al. (2006) calculated the activation enthalpy of biohydrogenation process as 47.34 kJ/mol K. This was found to be lower than those reported for *E. aerogenes* NCIMB 10102 (67.3 kJ/mol/K) (Fabiano and Perego, 2002), *E. cloacae* IIT-BT-08 (53.79 kJ/mol K) (Kumar and Das, 2000), and *Citrobacter intermedius* (87.99 kJ/mol/K) (Brosseau and Zajic, 1982), but was almost in close proximity to *Citrobacter* sp. Y19 (48.0174 kJ/mol K) (Oh et al., 2003). Conversely, the activation enthalpy for thermal deactivation was 118.67 kJ/mol K, which was almost same as that of *E. aerogenes* NCIMB 10102 (118.1 kJ/mol K) (Fabiano and Perego, 2002). Large variations in the value of Arrhenius constant suggest the occurrence of a similar deactivation mechanism in connection with the progressive denaturation of the enzyme, which kinetically determines fundamental pathways in some other hydrogen-producing enteric bacteria such as *E. aerogenes* (Fabiano and Perego, 2002). This was well observed from the experimental results of Wang and Wan (2008b) who reported that the concentration of ethanol in batch tests increased with increasing temperature from 293 to 308 K, but it decreased with further increasing temperature from 308 to 328 K. Their results also showed that the concentration of acetic acid in batch tests increased with increasing temperatures from 293 to 308 K, but it trended to decrease with further increasing temperature from 308 to 328 K. The change in concentration of ethanol and acetic acid shows that change in temperature alters the preferred metabolic pathway.

Further, thermophiles have a great potential for producing hydrogen. High hydrogen yields have already been reported with thermophilic microorganisms obtained, for example, from hot spring (Koskinen et al., 2008), anaerobic digester sludge (Ueno et al., 1995), wastewater treatment plant (Thong et al., 2007), and cow dung (Yokoyama et al., 2007). Cow rumen contains microorganisms that can degrade cellulose at high rate. This can be explained by the fact that the conversion of 1 mol glucose to 2 mol of acetate and 4 mol of H_2 is thermodynamically more favorable at higher temperatures. Thermophiles are known to have several advantages over mesophiles. This is because (i) at higher temperature gases have lower solubility (Henry's law); (ii) the hydrogen synthesis pathways are less affected by the partial pressure of hydrogen (pH_2) (Levin et al., 2004) (iii) the rates of chemical and enzymatic reactions are higher (Hallenbeck, 2005); and (iv) high temperature wastewaters could be directly used without cooling. Further, thermophiles have enhanced capacity to consume complex substrate to produce high yields as well as show resistance to contamination by mesophiles. In accordance to this, Zieden and van Neil (2010) have reported 100% of the theoretical maximum yield, that is 4 mol per mol hexose by *Caldicellulosiruptor owensensis* strain OLT (DSM 13100) using synthetic medium. However, considering the economic perspectives, thermophiles may not be the most suitable choice as they require increased power and may greatly add to the cost of the process. Therefore, in future, it needs to be seen how the costs of the process can be reduced in order to make the thermophiles a suitable candidate for economic hydrogen production. As is shown in Table 7.3, glucose and sucrose were the most widely used substrate during the investigation of the effect of temperature on fermentative hydrogen production. Thus, investigating the effect of temperature on fermentative hydrogen production using organic wastes as substrate is recommended. In addition, most of the reviewed studies investigating the effect of temperature on fermentative hydrogen production were conducted in batch mode, and more studies conducted in continuous mode are required.

TABLE 7.3

Effect of Temperature on Fermentative Biohydrogen Production

Substrate	Microorganism	Temperature Range (K)	Optimal Temp. (K)	Hydrogen Yield (mol/mol)	Reference
Starch	Municipal sewage sludge	310–328	328	1.44 mol/kg starch	Lee et al. (2008a)
Sucrose	*Thermoan-aerobacterium thermosaccharo-lyticum* PSU-2	313–353	333	2.53 mol/mol hexose	Thong et al. (2007)
Sucrose	Municipal sewage sludge	303–328	213	3.88 mol/mol sucrose	Lee et al. (2006b)
Cow dung	Cow dung	310–348	333	0.7×10^{-3} m^3/kg cow dung	Yokoyama et al. (2007)
Organic waste	Anaerobic digester sludge	310–328	328	0.36 m^3/kg VS	Valdez-Vazquez et al. (2005)
Glucose	*Citrobacter* CDN1	300–313	303	2.1 mol/mol glucose	Pandey et al. (2009)
Glucose	*Ethanoligenes harbinense* YUAN-3	293–317	310	1.34 mol/mol glucose	Xing et al. (2008)
Glucose	Anaerobic sludge	310–328	328	0.27 m^3/kg glucose	Wang and Wan (2008a)

7.2.3 Effect of pH on Biohydrogen Production

pH is a dominant factor influencing the stability of the acid-producing fermentative bacteria. It not only regulates the metabolic pathway, but also determines the dominant microbe in a reactor having a mixed community. This occurs due to the fact that when pH of reactors is changed to some extent, the microorganism's proliferation rate and their metabolic pathways change. Khanal et al. (2004) stated that in a typical dark anaerobic process, hydrogen is produced only during the exponential growth phase. Thereafter, when the stationary phase is acquired, the reactions shift from a hydrogen/acid production phase to a solvent production phase. In dark fermentation, most researchers have reported that the phase shift occurs when the pH falls below 4.5 (Khanal et al., 2004). Grupe and Gottschalk (1992) conducted a detailed study on the shift from acidogenesis to solventogensis. The study showed that an intracellular acid concentration up to 440 mM induced solventogenesis. The study further showed that the switch in the metabolic pathway could be prevented by constantly maintaining the pH of the medium. In view of this, Khanna et al. (2011) studied the effect of maintaining constant pH in the extracellular medium during fermentation of glucose to produce hydrogen using *Enterobacter cloacae* IIT-BT 08. The studies showed that hydrogen production significantly increased during such trials as compared to the ones in which only the initial pH was maintained in the reactor. Further, the study showed that increased substrate consumption occurs under constant buffering of the extracellular medium, and the reaction is driven to a stop essentially by depletion of the substrate from the medium instead of inhibition to VFAs as occurs during studies with maintenance of initial pH alone.

Volatile fatty acids (VFAs) such as acetate, butyrate, propionate, and ethanol are produced concomitantly with hydrogen. Accumulation of the VFAs causes drop in pH of the extracellular medium as compared to the initial pH of the medium. Butyrate and acetate are two main products that are produced during the fermentation reaction. Studies have shown that, at lower pH, butyrate production is preferred over acetate. The concentration of the various end products produced is pH-dependent. At higher pH ethanol and propionate are found to accumulate in the extracellular medium. To further study the effect of the end products on hydrogen production, researchers have added undissociated acids (butyrate, propionate, acetate, or ethanol) into the feed (Zheng and Yu, 2004; Ginkel and Logan, 2005). Analysis of such studies shows that hydrogen production is inhibited to a large extent by such undissociated acids. Such studies suggest that at high concentration of undissociated acids, the ionic strength of the solution increases which may induce the shift from hydrogen production to solvent production. At high extracellular pH, the undissociated nonpolar acids can penetrate the cell (which is at lower pH) and release the protons. This causes an intercellular pH disturbance. The cell requires increased energy in the form of ATP for driving the protons out against proton gradient. Besides, increased intracellular pH may lead to denaturation of intracellular enzymes including hydrogenase leading to decreased hydrogen production. It may be interesting to correlate quantitative RT-PCR of hydrogenase transcripts with the change in pH conditions of the medium. Besides, the accumulation of acids and especially ethanol causes membrane instability affecting the solute uptake.

A wide range of initial optimal pH has been found suitable for different kinds of bacteria as shown in Table 7.4. For example, the optimal pH by Khanal et al. (2004) was 4.5 while that reported by Lee et al. (2002) was 9.0. It may be possible that the variation arises due to the difference in the inoculum, reactor type, and substrate. However, as a more general rule, the optimal pH for hydrogen production in dark fermentation is within 5.0–7.0 (Table 7.4), which well corresponds with Koku et al. (2002) finding that the optimum pH for hydrogenase activity is around 6.5–7.5. Van Ginkel and Logan (2005) reviewed the effects of undissociated acid concentration on the H_2/solvent shift and concluded the threshold value cannot be predicted because of the wide range reported. Thus, it is important to establish the optimal substrate concentration and minimum operating pH, a balance between lowest alkali cost and ion concentration in the effluent and highest

TABLE 7.4

Effect of pH on Hydrogen Production by Various Microorganisms Using an Array of Different Substrates

Organism	Substrate	Initial pH Range	Regulated pH Range	pH Controllers	Optimal pH	Reactor Size (10^{-3} m³)	Yield (mol H₂/ mol Glucose)	Reference
Natural inoculum	Sucrome	—	4.5–7.5	1 M HCl/1 M KOH	5.5	0.250	2.45	Ginkel and Sung (2001)
Sludge	Food waste	5–8	—	—	7.0	—	593[a]	Nazlina et al. (2009)
C. beijerinckii	Glucose	5.7–6.5	—	5 N NaOH/5 N HCl	6.1	0.100	—	Skonieczny and Yargeau (2009)
Compost	Starch sucrose	4.5–6.5	—	HCl/NaOH	5.0	0.25	125[b] 214[b]	Khanal et al. (2004)
Seed sludge	Rice	4.5–6.5	—	1 M HCl/1 M NaOH	6.5	0.125	113[a]	Ma et al. (2008)
E. cloacae IIT-BT 08	Glucose	5.5–7.5	—	5 N NaOH/5 N H₃PO₄	6.5	2.5	2.2	Khanna et al. (2011)
E. cloacae IIT-BT 08	Glucose		5.5–7.5	5 N NaOH/5 N H₃PO₄	6.5	2.5	3.1	Khanna et al. (2011)
Seed sludge	Glucose	4.0–7.0	—		5.5	3.0	2.1	Fang et al. (2002)
Sludge	Swine manure with glucose	4.7–5.9		0.5 N NaOH/0.5 N HCl	5.0	4.0	1.48	Li et al. (2010)

[a] 10^{-3} m³ H₂/kg COD.
[b] 10^{-3} m³ H₂/kg TV.

H_2 production. To this end, Zoetemeyer et al. (1982) found NaOH requirements for an acidogenic CSTR fed on $10\ kg/m^3$ glucose increased by 50% when operating pH increased from 4.5 to 5.0 and by more than 100% when the pH was increased from 4.5 to 5.7. However, the proportion of undissociated acetic and butyric acids increases as the pH falls toward the pKa's of 4.78 and 4.81, respectively. The undissociated forms can pass across the cell membrane, collapsing the membrane pH gradient, and the cell will convert into solvent formation, sporulate, or die.

Further, using a mixed consortia, a change in pH also causes changes in the microbial community structure. Fang et al. (2002) did a detailed study of the effect of pH on the microbial community of sewage sludge. They found that with varying pH, both the appearance and dominance of community changed. At lower pH of around 4.5, the samples looked creamy white while as the pH trended toward the neutral, the samples appeared dark. This was because of the action of sulfogenic bacteria that converted sulfur to sulfide at higher pH. Moreover, the DGGE band pattern varied from pH 4.0 to pH 7.0. The number of bands changed from 6 at lower pH to 14 at higher pH. Only four bands were found to be persistent in all the reactions.

From Table 7.4, it may also be observed that most of the studies on optimization of initial pH are based with sucrose or glucose as the fermentative substrate. Thus, investigating the effect of initial pH on fermentative hydrogen production using organic wastes as substrate is recommended.

7.2.4 Effect of Alkalinity on Biohydrogen Production

Alkalinity (buffering capacity) is considered as one of the most important parameters, which is governed by the volatile fatty acid accumulation (VFA) in the system. As VFA accumulates, the buffering capacity of the system decreases due to the drop in pH after neutralization of the alkalinity by VFA. In an interesting study, Tenca et al. (2011) maintained the process stability through endogenous alkalinity by mixing alkali-rich feedstock with swine manure to the maximize biohydrogen production without controlling the pH of the process. In another study, Shi et al. (2009) showed the importance of alkalinity for maximizing the hydrogen production in continuous stirred tank reactor (CSTR) and an integrative biological reactor (IBR). They found that alkalinity and the hydrogen production rate were directly correlated. In CSTR, the gas production rate was increased from 0 to 0.4 L/d when the alkalinity was increased from $1.8 \times 10^{-3}\ kg/m^3$ to $2.4 \times 10^{-3}\ kg/m^3$. The gas production rate was increased from $3 \times 10^{-3}\ m^3/d$ to $5 \times 10^{-3}\ m^3/d$, the alkalinity was found to be in the range of $0.5–1.0\ kg/m^3$. Thus, alkalinity of the process reflected the hydrogen production rate of the system. In order to improve the stability of the system, it is very essential to regulate the alkalinity of the inflow to the bioreactor.

7.2.5 Effect of Hydraulic Retention Time on Biohydrogen Production

Hydraulic retention time (HRT) is a measure of the average length of time that a soluble compound remains in a constructed bioreactor. The HRT may directly restrict the metabolic process. This is helpful to enrich a culture of mixed population with desired organism when the specific growth rates are known. For example, when sewage sludge is used as seed for biohydrogen production, it is co-dominated by methanogens and hydrogen producers. However, in a continuous process, it is possible to separate the two communities by controlling the HRT of the process. This may be a more cost-effective way of separating the two communities in a large-scale process as compared to heat or acid-base pretreatment of the inoculum. It is well known that the growth rate of methanogens is slower at about $0.046–0.056 \times 10^{-4}$ s^{-1} as compared to hydrogen producers (0.48×10^{-4} s^{-1}) (Zhang et al., 2006). Thus, a mechanical dilution may be used to separate the methane producers from the hydrogen producers to maximize the hydrogen production from the mixed consortium. By operating the reactor at lower dilution rates, the slower growing methanogens can be washed away, leaving an enriched hydrogen-producing community. Thus, by regulating the HRT, the hydrogen producers can be retained while washing out the methanogens. To this effect, Kim et al. (2006b) conducted experiments and showed that using a short HRT of three days terminated the methanogenesis and enriched an undefined consortia into mostly hydrogen producers. In another interesting study, HRT was coupled to low pH. Fang et al. (2002) showed that low pH may not suffice to completely eliminate methanogenic activity. Their findings showed that at low HRT the methanogens may stick to the walls of the reactor. However, operation of the reactor at low pH coupled with low HRT may help in completely eradicating the methanogens from the culture. This is in contrast to the results presented by Fan et al. (2008) who did not detect methane even at conditions of pH 6.5 and 6.5×10^3 s HRT. Besides, where a range of pH has been tested in continuous conditions, investigations point toward an optimum for hydrogen production in the pH range 5.2–5.8 over a variety of HRT ($22–115 \times 10^3$ s) and substrate types (sucrose, starch, and beer industry wastes). Similarly, Ueno et al. (1995) reported continuous hydrogen production from a mixed culture at the long HRT of 2.6×10^5 s (pH 6.4) without encountering problems with methanogenesis.

Besides, HRT also influences the end products formed. This is primarily due to change of the bacterial community at various HRTs. To this effect, Liu et al. (2008), using manure obtained from methanogenic reactor operated at 55°C as inocula, showed that a short HRT of 3 days eliminated the methanogens and accumulated VFA accompanied by a relatively low pH. Similarly, Zhang et al. (2006) effectively increased the hydrogen production by lowering the HRT from 29 to 22×10^3 s. This decreased the production of propionate resulting in higher hydrogen yields. In another study, Mariakakis et al. (2012) used inoculum from anaerobic sewage digester of a sewage treatment plant and showed the effect of a short optimal HRT of 8.6×10^4 s on biogas production and accumulation of volatile fatty acids (Figure 7.7).

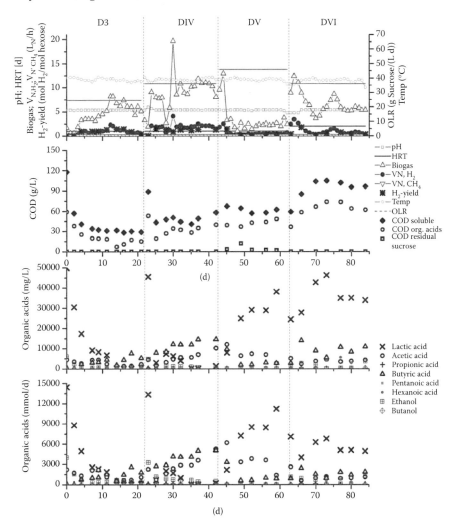

FIGURE 7.7
Operation parameters, gas production, hydrogen yield performance, butyric acid: acetic acid ratio, and production rate of relevant metabolites during different phases. Arrows indicate the sampling dates for microbial population analyses. (Reprinted from Mariakakis I., Meyer C., and Steinmetz, H. 2012. *Hydrogen Energy—Challenges and Perspectives*, InTech, DOI: 10.5772/47750. With permission.)

7.2.6 Hydrogen and CO_2 Partial Pressure

Hydrogen production pathways are very sensitive to hydrogen partial pressure. Accumulation of hydrogen in the reactor headspace may increase the partial pressure of hydrogen in the reactor system. According to Le Chatlier's principle, due to hydrogen buildup, the forward reaction will be inhibited. Thus, higher partial pressure of hydrogen in the reactor system will decrease the production of hydrogen. Concomitantly, metabolic pathways may also shift toward production of more reduced end products such as lactate, ethanol, acetone, and butanol. Studies have shown that partial pressure of hydrogen is an important factor in case of continuous hydrogen production (Hawkes et al., 2007). Different studies have shown a correlation between the optimal partial pressure of hydrogen and operational temperature of the reactor. Lee and Zinder (1988) obtained 50 kPa as the optimal operational pressure at 333 K. Similarly, different studies found 20 kPa at 343 K (van Niel et al., 2002), and 2 kPa at 98°C (Adams, 1990; Levin et al., 2004) as the optimal pressure.

There are several ways in which the hydrogen partial pressure can be decreased. Different methods can be applied depending on the type of reactor employed and the scale of the study (lab, bench, or pilot scale). Among the simplest way to decrease the hydrogen partial pressure is by providing optimal stirring in the culture. Lamed et al. (1988), showed the effect of hydrogen saturation in the gas phase of the reactor. Accumulation of hydrogen inhibited the formation of volatile fatty acids (VFAs), in particular acetate. It is well known that higher hydrogen production is associated with higher acetate production. To this respect, Lamed et al. found that acetate production decreases in unstirred cultures, as compared to stirred cultures due to accumulation of hydrogen in the broth of the unstirred reactors. Stirring the culture broth facilitates H_2 transfer to the gas phase, which relieves the inhibition of acetate formation caused by the high H_2 concentration in the medium. Further, the group showed that to reduce accumulation of hydrogen in the medium, a minimum critical speed is required which they showed to be higher than 2.5 rpm. This indicates that the degree of H_2 supersaturation is dependent on the rate of stirring. However, the limitation of the technique is that very high speeds cannot be used because it may cause mechanical sheer on the cells. Considering the upscale of the process, it may increase energy requirements and hence may not be cost-effective.

Second, hydrogen pressure in the liquid and gas phase can be removed by sparging with an inert gas. Generally, hydrogen-producing cultures are sparged with inert gases to create an anaerobic environment. Once hydrogen production has initiated, the anaerobic environment is maintained. However, researchers still provide intermittent sparging to their reactors. This is because hydrogen accumulates in the headspace of the reactor and causes product inhibition. However, sparging dilutes the hydrogen gas from the headspace and the forward reaction is accelerated. However, the disadvantage of the system is that it causes dilution of the final product (hydrogen), which may affect the economic viability of the process at the larger scales. Several studies have shown that decreasing the hydrogen partial pressure can enhance the hydrogen production (Tanisho et al., 1998; Lay et al., 2000). Mizuno et al. (2000) increased hydrogen yield by 68% by sparging nitrogen gas. In an interesting study, Tanisho et al. (1998) showed the effect of accumulation of hydrogen gas on the amount of residual NADH, volatile end products and hydrogen yield. The group showed that when carbon monoxide (CO) and trace amount of hydrogen had started to accumulate in the experimental set-up, *E. aerogenes* E.82005 shifted toward production of succinate. Later, when the system was sparged with an inert gas, argon, the amount of residual NADH and intermediate for hydrogen production increased. Subsequently, hydrogen production increased. Liu et al. (2008) used methane gas to remove the hydrogen and carbon dioxide from the liquid. As illustrated in Figure 7.8, gas sparging resulted in significant increase of the hydrogen production (88%).

The above two limitations are overcome by using membranes to adsorb the hydrogen. The membranes are made of silane or polyvinyl trimethylsilane and can selectively adsorb hydrogen from a mixture of gases. Teplyakov

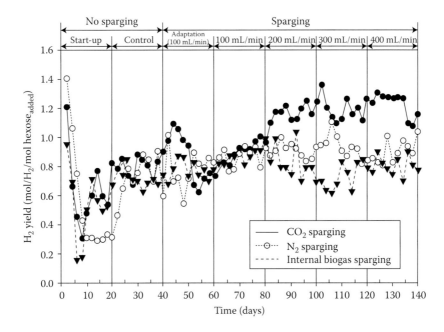

FIGURE 7.8
Effect of sparging the reactor on fermentative hydrogen production. (Reprinted from Kim, S. H., Han, S. K., and Shin, H. S. 2006b. *Process Biochemistry*, 41, 199–207. With permission.)

et al. (2002) have used a system of two polyvinyl trimethylsilane membranes to separate components of gas produced in a biohydrogen reactor, achieving gas purities of 90% for hydrogen and 99% for carbon dioxide. However, the main drawback of the technique is the development of biofilms on the membrane over the period of fermentation time. This may ultimately lead to clogging of the membrane. Besides, this kind of biofilm development especially in reactors where a mixed consortium is used, may cause the emergence of methanogenic bacteria.

Another indigenous way of reducing the partial pressure of the reactor especially in a continuous process was devised at Indian Institute of Technology, Kharagpur. The group devised a logic control system to reduce the partial pressure of the system to nearly half of the atmospheric pressure in a bench-scale reactor. The partial pressure of H_2 gas was reduced by decreasing the operating pressure of the bioreactor by adjusting the saline water level of the gas collector using a peristaltic pump, which was monitored by a U-tube mercury manometer connected with the system (Figure 7.9). The developed system comprises a U-tube manometer (containing Hg), a level sensor, and a logic controller (with a relay). One end of the U-tube manometer is connected to the gas collector; gas pressure inside collection

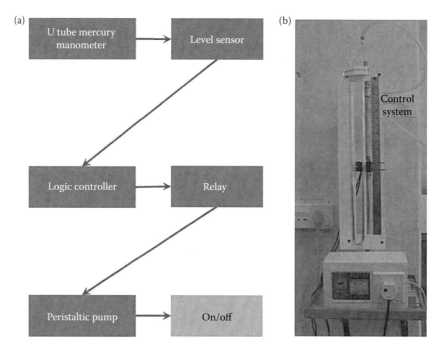

FIGURE 7.9
Automated logic control system. (a) Design of the system and (b) instrumental setup.

system is therefore directly reflected in the manometer. The level of the Hg is sensed by the level sensor comprising double "IR through-scan sensor" forking around the tube connected to the gas collection system. As the gas produced is collected during fermentation, the pressure inside increases thereby rendering the level of Hg in U-tube to fall. When the level of Hg falls below the sensor level, it transmits a signal to the logic controller to switch ON the relay. This consequently switches ON the peristaltic pump to drain out the saline water from gas collector, thereby balancing for the increase in gas pressure. The pump is switched OFF when the level of Hg rises back to the level of the IR sensor.

Besides hydrogen, partial pressure of CO_2 also affects the process yield and rate of hydrogen production. Cells synthesize succinate and formate from CO_2 both of which act as CO_2 sinks, stealing electrons away from hydrogen production. Thus, stripping the system off CO_2 increases hydrogen yields by reducing acetogenesis. Mizuno et al. (2000) reported that hydrogen production was increased 68% after sparging with N_2. This phenomenon could be directly explained by the decrease of hydrogen partial pressure and CO_2 concentration. In a contradiction to this, Kim et al. (2006a) compared sparging with CO_2 and with N_2 in a CSTR operating on 20 kg COD/m³ sucrose (pH 5.3, 4.3×10^4 s HRT). They showed that CO_2 sparging was more effective. Analysis of the microbial community from the experiments indicated acetogens and lactic acid bacteria were inhibited by the high CO_2 partial pressure and substrate conversion to microbial biomass was reduced. These results could prove useful in industrial situations if they can be extrapolated to other substrates.

7.2.7 Effect of Metal Ion on Fermentative Hydrogen Production

Biohydrogen production requires essential micronutrients for bacterial metabolism during fermentation. Sodium, magnesium, zinc, and iron are important trace metals that affect H_2 metabolism in microbes due to the fact that these elements are needed by bacterial enzyme's cofactor, transport processes, and dehydrogenases. Table 7.5 summarizes the effect of different metals on fermentative hydrogen production.

7.2.7.1 Effect of Iron and Nickel on Fermentative Hydrogen Production

Iron is an important nutrient element for the media preparation for dark hydrogen production. This is because hydrogen production primarily is catalyzed by hydrogenase, which contains either a dimetallic iron–iron center in [FeFe] hydrogenase or a nickel–iron center in [NiFe] hydrogenase. Moreover, the enzyme cannot receive the electrons directly, and must be channeled via an intermediate donor like ferredoxin. Ferredoxin is involved in pyruvate oxidation to acetyl-CoA, CO_2, and H_2. It is an iron–sulfur cluster-rich

TABLE 7.5

Effect of Traces of Different Metal Ions on Hydrogen Production

Inoculum	Starch	Reactor Type	Metal Ion	Concentration ($\times 10^{-3}$ kg/m³)		Hydrogen Yield	References
				Range Studied	Optimal		
Clostridium acetobutylicum	Glucose	Batch	Fe^{2+}	0–1000	25	0.41 m³/kg glucose	Alshiyab et al. (2008)
Anaerobic sludge	Starch	Batch	Fe^{2+}	0–4000	150	0.28 m³/kg glucose	Yang and Shen (2006)
Digested sludge	Sucrose	Continuous	Ca^{2+}	0–300	150	3.6 mo/mol	Chang and Lin (2006)
Anaerobic sludge	Glucose	Batch	Zn^{2+}	0–500	250	1.73 mol/mol glucose	Wang and Wan (2008c)
Municipal sewage sludge	Sucrose	Continuous	Ca^{2+}	0–27.2	27.2	2.19 mol/mol sucrose	Lee et al. (2004)
Clostridium saccharoperbutylacetonicum N1-4 (ATCC13564)	Glucose	Batch	Fe^{2+}	0–100	25	0.21×10^{-8} m³/s	Alalayah et al. (2009)
Anaerobic sludge	Palm oil	Batch	Fe^{2+}	2–400	257	6.33 m³/m³	O-Thong et al. (2008)
Anaerobic sludge	Glucose	Batch	Fe^{2+}	0–1500	350	0.31 m³/kg glucose	Wang et al. (2007)
Hydrogen-producing bacterial B49	Glucose	Batch	Mg^{2+}	1.2–23.6	23.6	2.36 m³/m³ culture	Wang et al. (2007)

protein. Vanacova et al. (2001) demonstrated that iron could induce metabolic change and be involved in the expression of both Fe–S and non-Fe–S proteins operating in hydrogenase. In view of this, most researchers have found that supplementation of iron in the media enhances hydrogen production, particularly when the enzyme is overexpressed in the cell. Kim et al. (2010) reported improved hydrogen production from the homologous overexpression of [NiFe] hydrogenase by supplementation of equal molar ratio of nickel and iron salts. Similarly, Ferchichi et al. (2005) demonstrated the influence of iron concentration on enhanced biological hydrogen production by *Clostridium saccharoperbutylacetonicum* ATCC 27021. Lee et al. (2008b) also studied the effect of iron concentration on the microbial metabolism. They reported that the addition of $FeSO_4$ between 2.7×10^{-3} kg/m^3 and 21.9×10^{-3} kg/m^3 was very favorable to H_2 production, because metabolic pathway shifted from lactic acid fermentation toward butyric acid fermentation. To this end, Khanna (2013) observed that with iron supplementation in the media, the metabolic pathway of *E. cloacae* IIT-BT 08/hydA$^+$ tended toward increased acetic acid production enhancing hydrogen production (Khanna, 2013) (Figure 7.10). Similarly, several studies are reported in literature, which show the effect of nickel–ion supplementation to enhance hydrogen production from [NiFe]-containing microorganisms. However, increase beyond the optimal conditions showed decrease in hydrogen production. Wang and Wan (2008d) in their study found that increasing the concentration of Ni–ion from 0 to 0.1×10^{-3} kg/m^3, H_2 production rate increases but it decreased with any further increase of Ni–ion concentration from 0.2 to 50×10^{-3} kg/m^3. The study of Li and Fang (2007) reported that hydrogen production potential tended to decrease with increasing Ni^{2+} concentration from 0.0 to 50×10^{-3} kg/m^3.

7.2.7.2 Effect of Magnesium on Fermentative Biohydrogen Production

Magnesium is one of the most abundant elements within any microorganism. It is one of the constituents of cell walls and membranes. Besides, magnesium is also principally required by the ribosomes. It functions as cofactor of several enzymes and also plays a very crucial role of activator of many kinases and synthetases. Most of the glycolytic enzymes require magnesium ion as cofactor such as hexokinase, phosphofructokinases, glyceraldehyde-3-phosphate dehydrogenases, and enolases (Wang et al., 2007). It is also well studied that Mg-chelated species of adenine nucleotides are the true substrates for cellular phosphate transfer reactions.

Lin and Lay (2005) carried out studies on the effect of nutrient supplementation on hydrogen production using the Taguchi orthogonal array. They found that magnesium, iron, sodium, and zinc were critical elements for high rate hydrogen production. Their findings indicated that among all these metals, the effect of magnesium was most significant. They found that

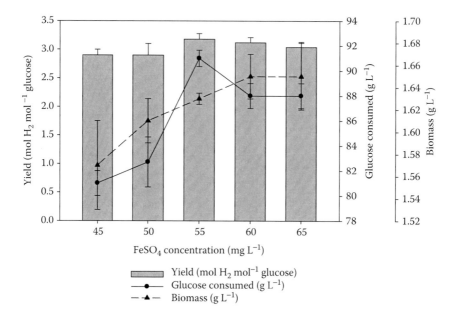

FIGURE 7.10
Effect of Fe^{2+} ion supplementation on biohydrogen production substrate consumption efficiency and biomass production from the developed strain of *Enterobacter cloacae* IIT-BT 08. (Batch operation, medium used: MYG {1% malt extract [M], 1% glucose [G], 0.4% yeast extract [Y]}, working volume 300 mL, temperature 310 K, agitation speed 3.33 rps, initial pH 6.0. The data are given as mean ± SD, $n = 3$.) (Adapted from Khanna, N. 2013. Strain development and determination of suitable process parameters maximization of hydrogen production using *Enterobacter cloacae* IIT-BT 08. Ph.D. thesis, IIT, Kharagpur.)

hydrogen production reached maximum at 3.52 mol H_2/mol sucrose when $MgCl_2$ and NaCl were supplemented at a concentration of 120 and 1000×10^{-3} kg/m³, respectively, into the medium.

7.2.7.3 Effect of Other Heavy Metals on Fermentative Biohydrogen Production

Heavy metals are present in significant concentrations in some industrial wastewaters and municipal sludge, and are often the leading cause for the upset of the wastewater treatment process (Lester et al., 1983; Stronach et al., 1986; Fang and Chan, 1997). Heavy metals can be stimulatory, inhibitory, or even toxic in biochemical reactions depending on their concentrations. It is well known that acidogenesis and methanogenesis of anaerobic processes are readily affected by chromium, copper, and zinc ions (Hickey et al., 1989; Lin, 1993; Yenigun et al., 1996). Though hydrogenesis is similar to acidogenesis in biochemical characteristics, its efficiency under the influence of heavy metals has not been studied extensively. Li and Fang (2007) investigated the toxicity of six electroplating metals on the H_2-producing activity of a granular sludge sampled from a H_2-producing upflow reactor treating sucrose-containing wastewater. The relative toxicity to H_2 production was found to be in the following order: Cu (most toxic) > Ni > Zn > Cr > Cd > Pb (least toxic).

7.2.7.4 Effect of Nitrogen and Phosphate on Fermentative Biohydrogen Production

The carbon to nitrogen (C/N) ratio is important in a biological process. It is necessary to maintain proper composition of the feedstock for efficient plant operation so that the C:N ratio in feed remains within desired range. Studies have shown that during anaerobic digestion, the microorganisms require 25–30 times faster carbon as compared to nitrogen. Thus to meet this requirement, microbes need a 20–30:1 ratio of C to N with the largest percentage of the carbon being readily degradable. Waste material that has a low carbon content can be made suitable by combining it with high nitrogen-based substances; this will balance the C/N ratio and make the substrate optimal for hydrogen production.

In a study conducted by Bisaillon et al. (2006) to investigate some limiting factors in microbial hydrogen fermentation by different strains of *E. coli* they found that limitation of phosphate or sulfate did not have any significant effect on hydrogen production. However, strains showed the highest yield of hydrogen per glucose when cultured at limiting concentrations of either ammonia or glucose. They reasoned the enhancement of production to C/N ratio on culture medium. However, most of the studies to evaluate the carbon and nitrogen ration on hydrogen production have been conducted at the batch mode, therefore studies on continuous hydrogen-producing systems is recommended.

Table 7.6 summarizes studies investigating the effect of different nitrogen sources on fermentative hydrogen production by *C. acetobutylicum* NCIMB13357 (Kalil et al., 2008). Mostly, ammonium salts as a source of nitrogen are the most widely investigated for fermentative hydrogen production. However, there exists certain disagreement on the optimal ammonia concentration for fermentative hydrogen production. For example, the optimal ammonia concentration for fermentative hydrogen production reported by Bisaillon et al. was 0.01 kg/m³, while that reported by Salerno et al. (2006) was 7.0 kg/m³. The possible reason for this disagreement was the difference in terms of inoculum used and nitrogen source concentration range that had been studied.

Phosphate is needed for hydrogen production as a micronutrient as well as for its buffering capacity. It has been demonstrated that in an appropriate range, increasing phosphate concentration enhances the hydrogen

TABLE 7.6

Effect of Organic and Inorganic Nitrogen on Hydrogen Production by
C. acetobutylicum NCIMB13357

Nitrogen Source	$Y^1_{p/s}$	$Y^2_{p/s}$	Biomass	$Y_{p/x}$	$Y_{x/s}$	$Y_{H_2/S}$
Organic						
Peptone	201	258	1.24	208	0.25	0.024
Tryptone	182	228	1.11	205	0.22	0.020
Yeast extract	240	308	1.10	280	0.22	0.028
Inorganic						
Ammonium sulfate	130	186	1.03	181	0.2	0.016
Ammonium nitrate	143	220	1.00	220	0.2	0.020
Ammonium chloride	120	200	1.01	198	0.2	0.0018

Source: Adapted from Kalil, M. S. et al. 2008. *American Journal of Biochemistry and Biotechnology*,
4, 393–401.

Note: $Y^1_{p/s}$: 10^{-3} H_2 m^3/kg glucose supplied; $Y^2_{p/s}$: 10^{-3} H_2 m^3/kg glucose utilized; biomass:
kg/m^3; $Y_{p/x}$: H_2 mL/g biomass L; $Y_{x/s}$: g biomass production/g glucose supplied; $Y_{H_2/s}$:
kg H_2/kg glucose utilized; glucose: 0.5 kg/m^3; nitrogen source concentration: 13 kg/m^3;
inoculum size: 10% (v/v); pH: 7.0, Temp. 303 K.

production potential, however, higher levels trend to decrease the hydrogen
production. It had been shown that an appropriate C/N and C/P ratios are
fundamental for fermentative hydrogen production.

7.3 Environmental Factors Affecting Hydrogen Production in Photosynthetic Organisms

7.3.1 Effect of Light Intensity

Though dark fermentation process is unaffected by the effect of light intensity, biophotolysis and photofermentation process are light-dependent processes. The photosynthetic organisms utilize the light to generate energy
(ATP), which is required by the photosynthetic organism to produce hydrogen. Therefore, optimal light intensity is required to generate a sufficient
pool of ATP. It has been suggested that light at the wavelength of 522, 805,
and 850 nm is critical for efficient photohydrogen production (Akkerman
et al., 2002; Chen et al., 2006). The light requirement like other physiological
parameters is an organism-dependent process. The light requirement of a
process therefore depends on the light-harvesting antenna pigments and
solar energy conversion efficiencies.

Based on the available studies, Miyake and Kawamura (1987) showed
that direct sunlight intensity measuring around 1 kW/m^2 may be too
strong for efficient hydrogen production. It has been observed that the

conversion efficiency from light to hydrogen decreases severely under the peak intensity of sunlight. Therefore several researchers have suggested light and dark cycle arrangements to enhance the light conversion efficiency. Wakayama et al. (2000) studied the effect of light and dark cycle from hours to seconds. They concluded that for *Rhodobacter sphaeroides* RV strain, cultured in aSy medium, a 1.8×10^3 s light and dark cycle was optimal and resulted in higher conversion efficiency as compared to continuous light supply for 4.3×10^4 s. This may be related to a stabler expression of nitrogenase which translated into higher hydrogen production. To further enhance the solar conversion efficiency, they proposed to combine the photobioreactor with solar cells that could utilize the light that was blocked for intermittent periods. However, Koku et al. (2003) suggested diurnal cycle for enhanced hydrogen production. To date, the highest light conversion efficiency reported is 8%, while the theoretical maximum is 10% (Akkerman et al., 2002; Kapdan and Kargi, 2006). The conversion was achieved using *Rhodobacter* sp. under laboratory growth conditions and light generated from a solar imulator (Miyake and Kawamura, 1987). However, this value was obtained in a laboratory-scale experiment with a low light intensity, and achieving high light conversion efficiency in large scale with photobioreactors is likely to be more difficult.

Further, to make photosynthetic hydrogen production, competitive with other resources, it is necessary to overcome the photosynthetic productivity and light utilization limitations of the system (Polle et al., 2003). For high biofuel production, it is essential that the biomass be harvested in high concentrations. However, for microalgae and photosynthetic bacteria, this imposes a technical barrier caused due to the "cell-shading" phenomenon of these pigmented cells. For microalgae, light limitation is a major bottleneck, because it cannot be stored and ideally should be distributed uniformly in the photobioreactor to provide, low-light intensity, ubiquitously in the reactor. However, practically this distribution occurs partially. The incident light energy decreases logarithmically with distance from light source and the concentration of a microorganism in the culture medium in accordance with the Beer-Lambert's law. Thus, shading effects of the biomass in the culture medium weaken the light energy. Photosynthetic yields are known to drop significantly along the reactor walls due to heat dissipation from high-light intensity. While further inside the reactor, light cannot penetrate because of cell-shading phenomenon. Since in dense microalgal mass culture light is rapidly attenuated, the cells acclimatize to a low-light environment (Polle et al., 2003). Such cells have large chlorophyll antenna size, driven by the evolutionary need to absorb large number of photons. Under bright light, the cells at the surface absorb most of the incoming light, however, heat dissipates most of it, as they absorb more photons than can be utilized by photosynthesis. This over-excitation causes dissipation of excess (~80%) photons as fluorescence or heat, lowering light conversion efficiency and cellular productivity. In view of this, Myers (1957) and later Radmer and Kok (1977) suggested the need of developing microalgae with truncated antenna size to increase the low-light conversion efficiency.

7.3.2 Effect of Temperature

The optimum reaction temperature plays an important role in shifting the metabolic pathways toward hydrogen production. The optimum temperature range for the hydrogen generation for photosynthetic bacteria *Rhodobacter* sp. lies between 304 and 309 K. Similarly, optimum temperature for hydrogen production for most cyanobacterial species is between 303 and 313 K. However, it is found to vary from species to species of cyanobacteria. For example, *Nostoc muscurom* showed maximum hydrogen production at 313 K (Datta et al., 2000). So a bioreactor with temperature controller system is always preferred to maintain the optimum temperature during the hydrogen generation process using photosynthetic organisms. The temperature controller may be a water jacket surrounding the photobioreactor, where the constant temperature is maintained by a circulating water bath kept at specified desired temperature. The economic viability of the temperature controller system is also important for reducing the cost of the entire process.

7.3.3 Effect of Nitrogen

Molecular nitrogen and several inorganic nitrogen compounds such as nitrite, nitrate, and ammonia have been reported to inhibit nitrogenase-dependent hydrogen production by reduction of hydrogen production rates and eventual time-dependent loss of hydrogen production activity. Studies have shown that in the presence of molecular nitrogen only 25% of the electrons are available to reduce protons to molecular hydrogen, the rest get transferred to the active site of nitrogenase to reduce nitrogen to ammonium (Rees and Howard, 2000; Igarashi and Seefeldt, 2003). Thus, nitrogenase effectively inhibits nitrogenase-dependent hydrogen production by competing for the same substrate. In accordance, several studies have shown inhibition of nitrogenase-mediated hydrogen production in *A. variabilis* SPU003 and *A. cylindrical* by addition of nitrite, nitrate, and ammonia (Datta et al., 2000; Lambert et al., 1979).

7.3.4 Effect of Sulfur

Sulfur is an essential constituent of proteins (methionine, cysteine), lipids, coenzymes (e.g., coenzyme A or lipoamide), molecules involved in photoprotection (glutathione), and electron carriers (e.g., in Fe-S clusters). In most cells, there is no specific sulfur storage, thus the growth of an organism mostly depends on sulfur that is available in the extracellular medium. *C. reinhardtii* senses the sulfur levels by the *sacI* gene (sulfur acclimation), a putative regulatory ion transporter (Davies et al., 1994). To increase hydrogen production, Melis and coworkers depleted sulfur from the medium (Melis et al., 2000). Sulfur depletion prevents the repair of the methionine containing PSII reaction center after photodamage. PSII is involved in biophotolysis of water. Therefore, absence of functional PSII induces anaerobic environment, optimal for hydrogen production by algae. Oxygen also competes with hydrogenase as an electron acceptor, making hydrogen production even more oxygen sensitive (Lee and Greenbaum, 2003). Consequently, Melis and coworkers included acetate in the medium to maintain a high level of respiration during the early stage of the sulfur depletion phase. This assisted in the consumption of residual O_2. Melis and coworkers concluded that "acetate is consumed by respiration for as long as there is O_2 in the culture medium ($0-1.08 \times 10^5$ s)" for wild-type cells, but that "it does not contribute significantly to the source of electrons in the H_2-production process ($1.08-4.3 \ 10^5$ s)" (Melis et al., 2000).

7.4 Statistical Optimization of Factors Effecting Biohydrogen Production

Maximization of hydrogen yields requires optimization of the environmental conditions. Either one or different variables can be optimized using different softwares. In conventional "one-variable-at-a-time" approach, the nutritional/cultural factors are optimized by changing one factor at a time, and keeping other variables constant. This approach is simplest to implement, and primarily helps in selection of significant parameters affecting the enzyme yield. However, this method is not only time restrictive, but it also ignores the combined interaction(s) among various physical and nutritional parameters (Vishwanatha et al., 2010). Therefore, researchers today opt for response surface methodology, a useful model by which certain selected factors are deliberately varied in a controlled manner to get their effects on a response of interest, often followed by the analysis of the experimental results. This also reduces the number of experiments required in growth medium optimization. Use of factorial designs and regression analyses for generating empirical models makes RSM a good statistical tool (Doddapaneni et al., 2007). Several softwares are available on the same. These include Taguchi design (fractional factorial design using orthogonal array, allows the effects of many factors with two or more levels on a response, to be studied in a relatively small number of runs) and Plackette Burman design (two level fractional factorial design, used to screen important factors for investigation) (Mohan et al., 2008). Software packages such as Design-Expert (Stat-Ease, Inc., Minneapolis, MN), Minitab (Minitab, Inc., State College, PA). are well known to conduct the factorial design such as

Taguchi design, Plackette Burman design. Besides, the concept of neural networks is also popular among the scientists. This has been applied successfully in multivariate nonlinear bioprocess as a useful tool to construct model. Neural network can be viewed as a powerful search and optimization technique to solve problems with objective function for optimization using genetic algorithm. More literature on the same is available in Wang and Wan (2009).

7.5 Comparison of Suspended Cell versus Immobilized Systems

Most dark fermentation has been reported via suspended-cell systems, but the high substrate feeding rate required for high-rate H_2 production may cause system instability due to severe cell washout. Besides, in order to produce H_2 continuously from biological resources for industrial use, the H_2 evolution rate is as important as its overall yield. Therefore, for the practical operation of a bioreactor, a high-cell density is required. To achieve the same, a variety of reactor systems with immobilized cells using several microbial support carriers have been reported. An immobilized system has several advantages over the suspended system as highlighted below.

Immobilization provides

- High-cell concentration in the reactor
- Eliminates the chance of cell washout at higher dilution rates
- Eliminates the costly processes of cell recovery and cell recycle
- Combination of high-cell concentrations and high-flow rates allows higher volumetric productivities
- Improves genetic stability
- Protects against shear damage caused by agitation

Recently, Ma et al. (2010) investigated the immobilization of mixed bacteria by microcapsulation for hydrogen production. Sodium cellulose sulfate (NaCS)/ploy-dimethyl-dially-ammonium-chloride (PDMDAAC) microcapsules were used as a novel pseudo "Cell Factory" to immobilize mixed bacteria for hydrogen production under anaerobic conditions. Compared to freely suspended cells, the hydrogen production from immobilized cells increased more than 30%. The biomass was increased from 1.5 kg/m^3 in free cell culture to 3.2 kg/m^3 in the pseudo "Cell Factory." The hydrogen yield maintained 1.73–1.81 mol H_2/mol glucose. The fermentation cycle was shortened from 1.73 h to 0.864×10^5 s, resulting in an increase of 198.6% in the hydrogen production rate.

The major problems associated with immobilized, whole cells system include mass transfer resistance in a fixed bed bioreactor, which affects the reactor performance and product yield. Further, for hydrogen production, the solid matrices reported for immobilization studies include mostly synthetic polymers or inorganic materials. However, these matrices cause disposal issues and add to the cost of the process. To overcome these limitations, Kumar and Das (2001) selected naturally occurring lignocellulose-rich fibers from ripened coconuts, that are regarded as waste, to serve as immobilization matrix. Experimental set-ups of immobilized and suspended cell cultures operated at IIT, Kharagpur are shown in Figure 7.11a and b, respectively. Further, to overcome mass transfer efficiency limitation, they worked with rhomboidal reactors. The particular configuration of the reactor as compared to tubular and tapered reactors helped in releasing the trapped gases in the void spaces of the matrix, enhancing the process performance (Kumar and Das, 2001) (Figure 7.12). From their studies using *E. cloacae* IIT-BT 08 immobilized over coconut coir and glucose as substrate, they obtained higher rate of hydrogen production as compared to the suspended system.

(a)

Rector with immobilized cells

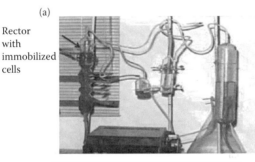

(b)

Rector with suspended cells

FIGURE 7.11
Experimental set-ups for biohydrogen production by dark fermentation using (a) immobilized cell and (b) suspended cell. (Reprinted from Khanna, N. and Das, D. 2013. *Wiley Interdisciplinary Reviews: Energy and Environment*, 2, 401–421. With permission.)

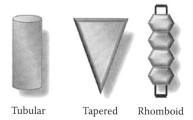

Tubular   Tapered   Rhomboid

FIGURE 7.12
Different configurations of bioreactors tested for hydrogen production.

7.6 Comparison of Batch Process versus Continuous Process

Chemical processes may be run in continuous or batch operation. In batch operation, production occurs in time-sequential steps in batches. A batch of feedstock(s) is fed into a process or unit for the chemical process to take place, followed by the recovery of the product(s) and any other outputs which may be removed. Such batch production may be repeated again and again with new batches of feedstock. Batch operation is commonly used in smaller-scale plants. However, if the process is to reach the commercial scale production, efficiency and product stability are desired. From an economic point of view, productivity gains can be achieved by continuous culture. Schematic diagrams for batch and continuous studies are shown in Figure 7.13a and b, respectively. Most studies on continuous biological hydrogen production have considered the effects of various parameters such as pH, HRT, biomass retention, reduction of hydrogen partial pressure, redox potential, and reactor type (Cohen et al., 1984; Fang et al., 2002).

7.7 Conclusion

The performance of a hydrogen-producing reactor is governed by various factors, namely, inoculum, pH, temperature, alkalinity, hydraulic retention time, hydrogen partial pressure, and light. Choice of inoculum determines the final maximum yield of the product. Both inoculum volume and age have been shown to have a significant influence on the final yield of the process. pH was considered a dominant factor that influenced the end metabolite and also governed the dominant microbe in a reactor with a mixed population. Temperature has been found to significantly affect the growth rate of the

(a)

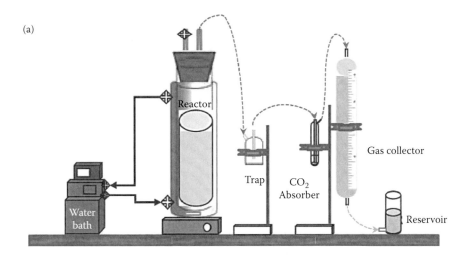

(b)

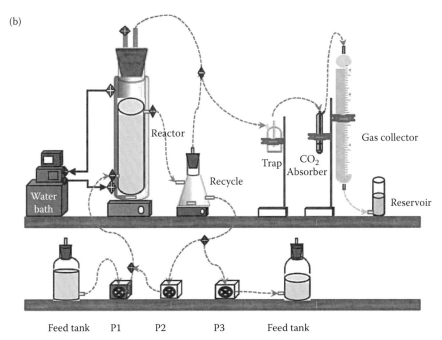

FIGURE 7.13
Schematic representation of (a) batch process and (b) continuous hydrogen production process. P1, P2, P3 represent the pumps. (Reprinted from Khanna, N. and Das, D. 2013. *Wiley Interdisciplinary Reviews: Energy and Environment*, 2, 401–421. With permission.)

microorganism because it determines the rates of all the cellular reactions. Alkalinity determines the fall of pH in the extracellular medium while HRT was found to enrich a culture of mixed population provided that the specific growth rates are known. Accumulation of hydrogen gas in the headspace of the reactor was found to inhibit the forward reaction in keeping with the Le Chatlier's principle. Photosynthetic and photofermentative bacteria have light-dependent metabolisms and have a critical requirement of optimal light intensity for maximal hydrogen production.

Glossary

AFBR	Anaerobic-fluidized bed reactor
AnMBR	Anaerobic membrane bioreactor
ASBR	Anaerobic sludge bed reactor
CIGSB	Carrier-induced granular sludge bed
CSTR	continuous stirred-tank reactor
DGGE	Denaturing gradient gel electrophoresis
EGSBr	Expanded granular sludge bed
FACS	Fluorescence-activated cell sorting
HRT	Hydraulic retention time
IBR	Integrative biological reactor
IR sensor	Infra-red sensor
PCR	Polymerase chain reactor
UASBr	Upflow anaerobic sludge blanket reactor
VFA	Volatile fatty acid

References

Adams, M. W. W. 1990. The structure and mechanism of iron-hydrogenases. *Biochemistry and Biophysics Acta, 1020,* 115.

Akkerman I., Janssen, M., Rocha, J., and Wijffels, R. H. 2002. Photobiological hydrogen production: Photochemical efficiency and bioreactor design. *International Journal of Hydrogen Energy, 27,* 1195–1208.

Alalayah, W. M., Kalil, M. S., Kadhum, A. A. H., Jahim, M., and Alauj, J. M. 2009. Effect of environmental parameters on hydrogen production using *Clostridium saccharoperbutylacetonicum* N1-4 (ATCC 13564). *American Journal of Environmental Science, 5,* 80–86.

Alshiyab, H., Kalil, M. S., Hamid, A. A., and Yusoff, W. M. W. 2008. Trace metal effect on hydrogen production using *C. acetobutylicum. Journal of Biological Sciences, 8,* 1–9.

Barros, A. R., Amorim, E. L. C., Reis, C. M., Shida, G. M., and Silva. E. L. 2010. Biohydrogen production in anaerobic fluidized bed reactors: Effect of support material and hydraulic retention time. *International Journal of Hydrogen Energy*, 35, 3379–3388.

Bisaillon, A., Turcot, J., and Hallenbeck, P. C. 2006. The effect of nutrient limitation on hydrogen production by batch cultures of *Escherichia coli*. *International Journal of Hydrogen Energy*, 31, 1504–1508.

Brock, T. D., Madigan, M. T., Martinko, J. M., and Parker, J. 1994. Archaeal cell walls. In *Biology of Microorganisms*, 7th edn, eds. T. D. Brock and M. T. Madigan. Upper Saddle River, NJ: Prentice Hall, pp. 824.

Brosseau, J. D. and Zajik, J. E. 1982. Hydrogen gas production with *Citrobacter intermedius* and *Clostrium pasteurianum*. *Journal of Chemical technology and Biotechnology*, 32, 496–502.

Cao, G., Ren, N., Wang, A. et al. 2009. Acid hydrolysis of corn stover for biohydrogen production using *Thermoanaerobacterium thermosaccharolyticum* W16. *International Journal of Hydrogen Energy*, 34, 7182–7188.

Chairattanamanokorn, P., Penthamkeerati, P., Reungsang, A., Lo, Y. C., Lu, W. B., and Chang, J. S. 2009. Production of biohydrogen from hydrolyzed bagasse with thermally preheated sludge. *International Journal of Hydrogen Energy*, 34, 7612–7617.

Chang, F. Y. and Lin, C. Y. 2004. Biohydrogen production using an up-flow anaerobic sludge blanket reactor. *International Journal of Hydrogen Energy*, 29, 33–39.

Chang, F. Y. and Lin, C. Y. 2006. Calcium effect on fermentative hydrogen production in an anaerobic up-flow sludge blanket system. *Water Science Technology*, 54, 105–112.

Chen, X., Sun, Y., Xiu, Z., Li, X., and Zhang, D. 2006. Stoichiometric analysis of biological hydrogen production by fermentative bacteria. *International Journal of Hydrogen Energy*, 31, 539–549.

Cheong, D. Y. and C. L. Hansen. 2006. Bacterial stress enrichment enhances anaerobic hydrogen production in cattle manure sludge. *Applied Microbiology and Biotechnology*, 72, 635–643.

Chin, H. L., Chen, Z. S., and Chou, C. P. 2003. Fedbatch operation using *Clostridium acetobutylicum* suspension culture as biocatalyst for enhancing hydrogen production. *Biotechnology Program*, 19, 383–388.

Cohen, A., Germert, J. M., Zoetemeyer, R. J., and Breure, A. M. 1984. Main characteristics and stoichiometric aspects of acidogenesis of soluble carbohydrate containing wastewater. *Process Biochemistry*, 19, 282–286.

Datar, R., Huang, J., Maness, P. C., Mohagheghi, A., Czernik, S., and Chornet, E. 2007. Hydrogen production from the fermentation of corn stover biomass pretreated with a steam-explosion process. *International Journal of Hydrogen Energy*, 32, 932–939.

Datta, M., Nikki, G., and Shah, V. 2000. Cyanobacterial hydrogen production. *World Journal of Microbiology and Biotechnology*, 16, 8–9.

Davies, J. P., Yildiz, F., and Grossman, A. R. 1994. Mutants of *Chlamydomonas* with aberrant responses to sulfur deprivation. *Plant Cell*, 6, 53–63.

Doddapaneni, K. K., Tatineni, R., Potumarthi, R., and Mangamoori, L. N. 2007. Optimization of media constituents through response surface methodology for improved production of alkaline proteases by Serratia rubidaea. *Journal of Chemical Technology and Biotechnology*, 82, 721–729.

Egorova, K. and Antranikian, G. 2005. Industrial relevance of thermophilic Archaea. *Current Opinion in Microbiology*, *8*, 649–655.

Fabiano, B. and Perego, P. 2002. Thermodynamic study and optimization of hydrogen production by *Enterobacter aerogenes*. *International Journal of Hydrogen Energy*, *27*, 149–156.

Fan, Y. T., Xing, Y., Ma, H. C., Pan, C. M., and Hou, H. W. 2008. Enhanced cellulose hydrogen production from corn stalk by lesser panda manure. *International Journal of Hydrogen Energy*, *33*, 6058–6065.

Fang, H. H. P. and Chan, O. C. 1997. Toxicity of electroplating metals on benzoate-degrading granules. *Environmental Technology*, *18*, 93–99

Fang, H. H. P., Liu, H., and Zhang, T. 2002. Characterization of a hydrogen-producing granular sludge. *Biotechnology and Bioengineering*, *78*, 44–52.

Ferchichi, M., Crabbe, E., Gil, G. H., Hintz, W., and Almadidy, A. 2005. Influence of initial pH on hydrogen production from cheese whey. *Journal of Biotechnology*, *120*, 402–409.

Gehin, A., Gelhaye, E., Raval, G., and Petitdemange, H. 1995. *Clostridium cellulolyticum* viability and sporulation under cellobiose starvation conditions. *Applied and Environmental Microbiology*, *61*, 868–871.

Ginkel, S. V. and Logan, B. E. 2005. Inhibition of biohydrogen production by undissociated acetic and butyric acids. *Environmental Science and Technology*, *39*, 9351–9356.

Ginkel, S. V. and Sung, S. 2001. Biohydrogen as a function of pH and substrate concentration. *Environmental Science and Technology*, *35*, 4726–4730.

Grupe, H. and Gottschalk, G. 1992. Physiological events in *Clostridium acetobutylicum* during the shift from acidogenesis to solventogenesis in continuous culture and presentation of a model for shift induction. *Applied and Environmental Microbiology*, *58*, 3896–3902.

Guo, W. Q., Ren, N. Q., Chen, Z. B. et al. 2008. Simultaneous biohydrogen production and starch wastewater treatment in an acidogenic expanded granular sludge bed reactor by mixed culture for long-term operation. *International Journal of Hydrogen Energy*, *33*, 7397–7404.

Hallenbeck, P. C. 2005. Fundamentals of the fermentative production of hydrogen technology. *Water Science Technology*, *52*, 21–29.

Hawkes, F. R., Hussy, I., Kyazze, G., Dinsdale, R., and Hawkes, D. L. 2007. Continuous dark fermentative hydrogen production by mesophilic microflora: Principles and progress. *International Journal of Hydrogen Energy*, *32*, 172–184.

Hickey, R. F., Vanderwielon, J., and Switzenbaum, M. S. 1989. The effect of heavy metals on methane production and hydrogen and carbon monoxide level during batch anaerobic sludge digestion. *Water Resources*, *23*, 207–18.

Hu, B. and Chen, S. L. 2007. Pretreatment of methanogenic granules for immobilized hydrogen fermentation. *International Journal of Hydrogen Energy*, *32*, 3266–3273.

Hussy, I., Hawkes, F. R., Dinsdale, R., and Hawkes, D. L. 2005. Continuous fermentative *Hydrogen Energy*, *33*, 6058–6065.

Igarashi, R. Y. and Seefeldt, L. C. 2003. Nitrogen fixation: The mechanism of the Mo-dependent nitrogenase. *Critical Reviews in Biochemistry and Molecular Biology*, *38*, 351–84.

Infantes, D., Campo, A. G., Villaseñor J., and Fernández, F. J. 2011. Continuous biohydrogen production in immobilized biofilm system versus suspended cell culture. *International Journal of Hydrogen Energy*, *36*, 15595–15601.

Ivanova, G., Rakhely, G., and Kovacs, K. L. 2009. Thermophilic biohydrogen production from energy plants by *Caldicellulosiruptor saccharolyticus* and comparison with related studies. *International Journal of Hydrogen Energy, 34,* 3659–3670.

Jo, H, J., Sung, D., Park, D., and Moon, J. 2008. Biological hydrogen production by immobilized cells of *Clostridium tyrobutyricum* JM1 isolated from a food waste treatment process. *Database, 99,* 6666–6672.

Kalil, M. S., Alshiyab, H. S., Mohtar, W., Yusoff, W., and Selangor, B. 2008. Effect of nitrogen source and carbon to nitrogen ratio on hydrogen production using *C. acetobutylicum*. *American Journal of Biochemistry and Biotechnology, 4,* 393–401.

Kapdan, I. K. and Kargı, F. 2006. Bio-hydrogen production from waste materials. *Enzyme and Microbial Technology, 38,* 569–582.

Khanal, S. K., Chen, W. H., and Sung, L. L. S. 2004. Biological hydrogen production: Effects of pH and intermediate products. *International Journal of Hydrogen Energy, 29,* 1123–1131.

Khanna, N. 2013. Strain development and determination of suitable process parameters maximization of hydrogen production using *Enterobacter cloacae* IIT-BT 08. Ph.D. thesis, IIT, Kharagpur.

Khanna, N. and Das, D. 2013. Biohydrogen production by dark fermentation. *Wiley Interdisciplinary Reviews: Energy and Environment, 2,* 401–421.

Khanna, N., Kotay, S. M., Gilbert, J. J., and Das, D. 2011. Improvement of biohydrogen production by *Enterobacter cloacae* IIT-BT 08 under regulated pH. *Journal of Biotechnology,152,* 15–30.

Kim, D. H, Han, S. K., Kim, S. H., and Shin, H. S. 2006a. Effect of gas sparging on continuous fermentative hydrogen production. *International Journal of Hydrogen Energy, 31,* 2158–2169.

Kim, J. Y. H., Jo, B. H. and Cha, H. J. 2010. Production of biohydrogen by recombinant expression of [NiFe]-hydrogenase 1 in *Escherichia coli*. *Microbial Cell Factories, 9,* 1–10.

Kim, S. H., Han, S. K., and Shin, H. S. 2006b. Effect of substrate concentration on hydrogen production and 16S rRNA based analysis of the microbial community in a fermenter. *Process Biochemistry, 41,* 199–207.

Kim, S. H. and Shin, H. S. 2008. Effects of base-pretreatment on continuous enriched culture for hydrogen production from food waste. *International Journal of Hydrogen Energy, 33,* 5266–5274.

Koku, H., Eroglu, I., Gunduz, U., Yucel, M., and Turker, L. 2002. Aspects of the metabolism of hydrogen production by *Rhodobacter sphaeroides*. *International Journal of Hydrogen Energy, 27,* 1315–1329.

Koku, H., Eroglu, I., Gunduz, U., Yucel, M., and Turker, L. 2003. Kinetics of biological hydrogen production by the photosynthetic bacterium *Rhodobacter sphaeroides* O.U.001. *International Journal of Hydrogen Energy, 28,* 381–388.

Koskinen, P. E. P., Lay, C. H., Puhakka, J. A., Lin, P. J. Wu, S. Y., and Örlygsson, J. 2008. High efficiency hydrogen production by an anaerobic, thermophilic enrichment culture from Icelandic hot spring. *Biotechnology and Bioengineering, 101,* 665–678.

Kotay, S. M. 2008. Microbial hydrogen production from sewage sludge. Ph.D. thesis, IIT, Kharagpur.

Kotay, S. M. and Das, D. 2007. Microbial hydrogen production with Bacillus coagulans IIT-BT S1 isolated from anaerobic sewage sludge. *Bioresource Technology, 98,* 1183–1190.

Kotay, S. M. and Das, D. 2010. Microbial hydrogen production from sewage sludge bioaugmented with a constructed microbial consortium. *International Journal of Hydrogen Energy*, 35, 10653–10659.

Kotsopoulos, T. A., Zeng, R. J., and Angelidaki, I. 2006. Biohydrogen production in granular up-flow anaerobic sludge blanket (UASB) reactors with mixed cultures under hyper-thermophilic temperature (70°C). *Biotechnology and Bioengineering*, 94, 296–301.

Kumar, N. and Das, D. 2000. Production and purification of alpha-amylase from hydrogen producing *Enterobacter cloacae* IIT-BT 08. *Bioprocess Engineering*, 23, 205–208.

Kumar, N. and Das, D. 2001. Continuous hydrogen production by immobilized *Enterobacter cloacae* IIT-BT 08 using lignocellulosic materials as solid matrices. *Enzyme and Microbiol Technology*, 29, 280–287.

Lambert, G. R., Daday, A., and Smith, G. D. 1979. Effects of ammonium ions, oxygen, carbon monoxide, and acetylene on anaerobic and aerobic hydrogen formation by *Anabaena cylindrica* B629. *Applied and Environmental Microbiology*, 38, 521–529.

Lamed, J., Lobos, J. H., and Su, T. M. 1988. Effects of stirring and hydrogen on fermentation products of *Clostridium thermocellum*. *American Society for Microbiology*, 54, 1216–1221.

Lay, J. J. 2000. Modelling and optimization of anaerobic digested sludge converting starch to hydrogen. *Biotechnology and Bioengineering*, 68, 269–278.

Lee, H. S., Salerno, M. B., and Rittmann, B. E. 2008a. Thermodynamic evolution on H_2 production in glucose fermentation. *Environmental and Science Technology*, 42, 2401–2407.

Lee, J. W. and Greenbaum, E. 2003. A new oxygen sensitivity and its potential application in photosynthetic H_2 production. *Applied Biochemistry and Biotechnology*, 105, 303–313.

Lee, K. S, Hsu, Y. F., Lo, Y. C., Lin, P. J., Lin, C. Y., and Chang, J. S. 2008b. Exploring optimal environmental factors for fermentative hydrogen production from starch using mixed anaerobic microflora. *International Journal of Hydrogen Energy*, 33, 1565–1572.

Lee, K. S., Lin, P. J. and Chang, J. S. 2006a. Temperature effects on biohydrogen production in a granular sludge bed induced by activated carbon carriers. *International Journal of Hydrogen Energy*, 31, 465–472

Lee, K. S., Lin, P. J., Fangchiang, K., and Chang, J. S. 2007. Continuous hydrogen production by anaerobic mixed microflora using a hollow-fiber microfiltration membrane bioreactor. *International Journal of Hydrogen Energy*, 32, 950–957.

Lee, K. S., Lo, Y. C., Lin, P. J., and Chang, J. S. 2006b. Improving biohydrogen production in a carrier-induced granular sludge bed by altering physical configuration and agitation pattern of the bioreactor. *International Journal of Hydrogen Energy*, 31, 1648–1657.

Lee, K. S., Lo, Y. S., Lo, Y. C., Lin, P. J., and Chang, J. S. 2003. H_2 production with anaerobic sludge using activated-carbon supported packed-bed bioreactors. *Biotechnology Letters* 25, 133–138.

Lee, K. S., Lo, Y. S., Lo, Y. C., Lin, P. J., and Chang, J. S. 2004. Operation strategies for biohydrogen production with a high rate anaerobic granular sludge bed bioreactor. *Enzyme and Microbial Technology*, 35, 605–612.

Lee, M. J. and Zinder, S. H. 1988. Hydrogen partial pressures in a thermophilic acetate-oxidizing methanogenic co-culture. *Applied and Environmental Microbiology*, 54, 1457–1461.

Lee, Y. J., Miyahara, T., and Noike, T. 2002. Effect of pH on microbial hydrogen-fermentation. *Journal of Chemical Technology and Biotechnology*, 77, 694–698.

Lester, J. N., Sterritt, R. M., and Kirk, P. W. 1983. Significance and behavior of heavy metals in wastewater treatment process. *Science and Total Environment*, 30, 45–83.

Levin, D. B., Pitt, L., and Love, M. 2004. Biohydrogen production: Prospects and limitations to practical application. *International Journal of Hydrogen Energy*, 29, 173–185

Li, C. L. and Fang, H. H. P. 2007. Fermentative hydrogen production from wastewater and solid wastes by mixed cultures. *Critical Reviews in Environmental and Science Technology*, 37, 1–39.

Li, Y., Zhu, J., Wu, X., Miller, C., and Wang, L. 2010. The effect of pH on continuous biohydrogen production from swine wastewater supplemented with glucose. *Applied Biochemistry and Biotechnology*, 162, 1286–1296.

Li, Y. F., Chen, H., Han, W., Liu, F. J., and Wang, Z. Q. 2011. Influence of initial pH and temperature on fermentative biohydrogen production of *Biohydrogenbacterium* R3 sp. nov. from glucose. *Applied Mechanical Mater*, 71, 2929–2932.

Lin, C. Y. 1993. Effect of heavy metals on acidogenesis in anaerobic digestion. *Water Resource*, 127, 147–52.

Lin, C. Y. and Lay, C. H. 2005. A nutrient formation for fermentative hydrogen production using anaerobic sewage sludge microflora. *International Journal of Hydrogen Energy*, 30, 285–92.

Liu, D., Zeng, R. J., and Angelidaki, I. 2008. Effects of pH and hydraulic retention time on hydrogen production versus methanogenesis during anaerobic fermentation of organic household solid waste under extreme-thermophilic temperature (70 degrees C). *Biotechnology and Bioengineering*, 100, 1108–14.

Ma, J., Ke, S., and Chen, Y. 2008. Mesophilic biohydrogen production from food waste. *Bioinfomatics and Biomedical Engineering*, 2841–2844.

Ma, Q., Lin, D., and Yao, S. 2010. Immobilization of mixed bacteria by microcapsulation for hydrogen production—A trial of pseudo "Cell Factory". *Chinese Journal of Biotechnology*, 26, 1444–1450.

Mariakakis I., Meyer C., and Steinmetz, H. 2012. Fermentative hydrogen production by molasses; effect of hydraulic retention time, organic loading rate and microbial dynamics. In *Hydrogen Energy—Challenges and Perspectives*, ed. D Minic, InTech, DOI: 10.5772/47750.

Melis, A., Zhang, L., Forestier, M., Ghirardi, M. L., and Seibert, M. 2000. Sustained photobiological hydrogen gas production upon reversible inactivation of oxygen evolution in the green alga *Chlamydomonas reinhardtii*. *Plant Physiology*, 122, 127–136.

Miyake, J. and Kawamura, S. 1987. Efficiency of light energy conversion to hydrogen by the photosynthetic bacterium *Rhodobacter sphaeroides*. *International Journal of Hydrogen Energy*, 12, 147–149.

Mizuno, O., Dinsdale, R., Hawkes, F. R., Hawkes, D. L., and Noike, T. 2000. Enhancement of hydrogen production from glucose by nitrogen gas sparging. *Bioresource Technology*, 73, 59–65.

Mohan, S. V., Babu, V. L., and Sarma, P. N. 2008. Effect of various pretreatment methods on anaerobic mixed microflora to enhance biohydrogen production utilizing dairy wastewater as substrate. *Bioresource Technology*, 99, 59–67.

Myers, J. 1957. Algal culture. In *Encyclopedia of chemical technology*. New York: Interscience.

Nath, K., Kumar, A., and Das, D. 2006. Effect of some environmental parameters on fermentative hydrogen production by *Enterobacter cloacae* DM11. *Canadian Journal of Microbiology*, *52*, 525–532.

Nazlina, H. M. Y., NorAini, A. R., Ismail, F., Yusof, M. M., and Hasan, M. A. 2009. Effect of different temperature, initial pH and substrate composition on biohydrogen production from food waste in batch farm. *Asian Journal of Biotechnology*, *1*, 1–9.

Ntaikou, I., Gavala, H. N., Kornaros, M., and Lyberatos, G. 2008. Hydrogen production from sugars and sweet sorghum biomass using *Ruminococcus albus*. *International Journal of Hydrogen Energy*, *33*, 1153–1163.

Oh, Y. K., Seol, E. H., Kim, J. R., and Park, S. 2003. Fermentative biohydrogen production by a new chemoheterotrophic bacterium *Citrobacter* sp. Y19. *International Journal of Hydrogen Energy*, *28*, 1353–1359

O-Thong, S., Prasertsan, P., Intrasungkha, N., Dhamwlchukorn, S., and Birkeland, N. K. 2008. Optimization of simultaneous thermophilic fermentative hydrogen production and COD reduction from palm oil mill effluent by *Thermoanaerobacterium*-rich sludge. *International Journal of Hydrogen Energy*, *33*, 1221–1231.

Panagiotopoulos, I. A., Bakker, R. R., Budde, M. A. W., de Vrije, T., Claassen, P. A. M., and Koukios, E. G. 2009. Fermentative hydrogen production from pretreated biomass: A comparative study. *Bioresource Technology*, *100*, 6331–6338.

Pandey, A., Sinha, P., Kotay, S. M., and Das, D. 2009. Isolation and evaluation of a high H_2-producing lab isolate from cow dung. *International Journal of Hydrogen Energy*, *34*, 7483–7488.

Peintner, C., Zeidan, A. A., and Schnitzhofer, W. 2010. Bioreactor systems for thermophilic fermentative hydrogen production: Evaluation and comparison of appropriate systems. *Journal of Cleaner Production*, *18*, 15–22.

Polle, J. E. W., Kanakagiri, S. D., and Melis, A. 2003. tla1, a DNA insertional transformant of the green alga *Chlamydomonas reinhardtii* with a truncated light-harvesting chlorophyll antenna size. *Planta*, *217*, 49–59.

Radmer, R. and Kok, B. 1977. Photosynthesis: Limited yields, unlimited dreams. *Bioscience*, *29*, 599–605.

Rajeshwari, K. V., Balakrishnan, M., Kansal, A., Lata, K., and Kishore, V. V. N. 2000. State of the art of anaerobic digestion technology for industrial wastewater treatment. *Renewable and Sustainable Energy Reviews*, *2*, 135–156.

Rees, D. C. and Howard, J. B. 2000. Nitrogenase: Standing at the crossroads. *Current Opinion in Chemical Biology*, *4*, 559–66.

Ross, R., D'Elia, J., Mooney, R., and Chesbro, W. 1990. Nutrient limitation of two saccharolytic clostridia; secretion, sporulation, and solventogenesis. *FEMS Microbiology Letters*, *74*, 153–163

Salerno, M. B., Park, W., Zuo, Y., and Logan, B. E. 2006. Inhibition of biohydrogen production by ammonia. *Water Research*, *40*, 1167–1172

Shen D. M., Bagley, M. D., and Liss, S. N. 2009. Effect of organic loading rate on fermentative hydrogen production from continuous stirred tank and membrane bioreactors. *International Journal of Hydrogen Energy*, *34*, 3689–3696.

Shi, Y., Gai, G., Zhao, X., and Hu, Y. 2009. Influence and simulation model of operational parameters on hydrogen bioproduction through anaerobic microorganism fermentation using two kinds of wastes. In *Proceedings of the World Congress on Engineering and Computer Science, WCECS 2009*, Vol. 2. San Francisco, 2009, pp. 1–5.

Skonieczny, M. T., and Yargeau, V. 2009. Biohydrogen production from wastewater by *Clostridium beijerinckii*: Effect of pH and substrate concentration. *International Journal of Hydrogen Energy*, 34, 3288–3294.

Slininger, P. J., Bothast, R. J., Ladisch, M. R., and Okos, M. R. 1990. Optimum pH and temperature conditions for xylose fermentation by *Pichia stipitis*. *Biotechnology and Bioengineering*, 35, 727–731.

Stronach, S. M., Rudd, T., and Lester, J. N. 1986. Anaerobic digestion process. In *Industrial Wastewater Treatment*. Berlin: Springer Verlag.

Tanisho, S., Kuromoto, M., and Kadokura, N. 1998. Effect of CO_2 removal on hydrogen production by fermentation. *International Journal of Hydrogen Energy*, 23, 559–563.

Tenca, A., Schievano, A., Perazzolo, F., Adani, F., and Oberti, R. 2011. Biohydrogen from thermophilic co-fermentation of swine manure with fruit and vegetable waste maximizing stable production without pH control. *Bioresource Technology*, 102, 8582–8588.

Teplyakov, V. V, Gassanova, L. G., Sostina, E. G., Slepova, E. V., Modigell, M., and Netrusov, A. I. 2002. Lab scale bioreactor integration with active membrane system for hydrogen production: Experience and prospects. *International Journal of Hydrogen Energy*, 27, 1149–1155.

Thong, S. O., Prasertsan, P., Intrasungkha, N., Dhamwichukorn, S., and Birkeland, N. K. 2007. Improvement of biohydrogen production and treatment efficiency on palm oil mill effluent with nutrient supplementation at thermophilic condition using an anaerobic sequencing batch reactor. *Enzyme and Microbial Technology*, 41, 583–590.

Ueno, Y., Kawai, T., Sato, S., Otsuka, S., and Morimoto, M. 1995. Biological production of hydrogen from cellulose by natural anaerobic microflora. *Journal of Fermentation Bioengineering*, 79, 395–397.

Ueno, Y., Otsuka, S., and Morimoto, M. 1996. Hydrogen production from industrial wastewater by anaerobic microflora in chemostat culture. *Journal of Fermentation Bioengineering*, 82, 194–197.

Valdez-Vazquez, I., Rios-Leal, E., Esparza-Garcia, F., Cecchi, F., and Poggi-Varaldo H. A. 2005. Semi-continuous solid substrate anaerobic reactors for H_2 production from organic waste: Mesophilic versus thermophilic regime. *International Journal of Hydrogen Energy*, 30, 1383–1391.

Vanacova, S., Rasoloson, D., Razga, J., Hrdy, I., Kulda, J., and Tachezy, J. 2001. Iron-induced changes in pyruvate metabolism of *Tritrichomonas foetus* and involvement of iron in expression of hydrogenosomal proteins. *Microbiology*, 147, 53–62.

van Niel, E. W. J., Budde, M. A. W., de Haas, G. G., van der Wal, F. J., Claasen, P. A. M., and Stams, A. J. M. 2002. Distinctive properties of high hydrogen producing extreme thermophiles, *Caldicellulosiruptor saccharolyticus* and *Thermotoga elfii*. *International Journal of Hydrogen Energy*, 27, 1391–1398.

Veeken, A. and Hamelers, B. 1999. Effect of temperature on hydrolysis rates of selected biowaste components. *Bioresource Technology*, 69, 249–254.

Venetsaneas, N., Antonopoulou, G., Stamatelatou, K., Kornaros, M., and Lyberatos, G. 2009. Using cheese whey for hydrogen and methane generation in a two-stage continuous process with alternative pH controlling approaches. *Bioresource Technology*, 100, 3713–3717.

Vindis, P. 2009. The impact of mesophilic and thermophilic anaerobic digestion on biogas production. *Journal of Achievements in Materials and Manufacturing Engineering*, 36, 192–198.

Vishwanatha, K. S., Rao, A. G. A., and Singh, S. A. 2010. Acid protease production by solid-state fermentation using *Aspergillus oryzae* MTCC 5341: Optimization of process parameters. *Journal of Industrial Microbiology and Biotechnology, 37,* 129–138.

Wakayama, T., Nakada, E., Asada, Y., and Miyake, J. 2000. Effect of light/dark cycle on bacterial hydrogen production by *Rhodobacter sphaeroides* RV. From hour to second range. *Applied Biochemistry and Biotechnology, 84,* 431–440.

Wang, J. and Wan, W. 2008c. Effect of Fe^{2+} concentration on fermentative hydrogen production by mixed cultures. *International Journal of Hydrogen Energy, 33,* 1215–1220.

Wang, J. and Wan W. 2008d. Influence of Ni^{2+} concentration on biohydrogen production. *Bioresource Technology, 99,* 8864–8868.

Wang, X. J., Ren, N. Q., Xiang, W. S., and Guo, W. Q. 2007. Influence of gaseous end-products inhibition and nutrient limitations on the growth and hydrogen production by hydrogen-producing fermentative bacterial B49. *International Journal of Hydrogen Energy, 32,* 748–754.

Wang, J. L. and Wan, W. 2008b. Effect of temperature on fermentative hydrogen production by mixed cultures. *International Journal of Hydrogen Energy, 33,* 5392–5397.

Wang, J. L. and Wan, W. 2008a. Comparison of different pretreatment methods for enriching hydrogen-producing cultures from digested sludge. *International Journal of Hydrogen Energy, 33,* 2934–2941.

Wang, J. and Wan, W. 2009. Factors influencing fermentative hydrogen production: A review. *International Journal of Hydrogen Energy, 2,* 799–811.

Wu, C. N. and Chang, J. S. 2003. Hydrogen production with immobilized sewage sludge in three-phase fluidized-bed bioreactors. *Biotechnology Program, 19,* 828–832.

Xing, D. F., Ren, N. Q., Wang, A. J, Li, Q. B, Feng, Y. J., and Ma, F. 2008. Continuous hydrogen production of auto-aggregative *Ethanoligenens harbinense* YUAN-3 under non-sterile condition. *International Journal of Hydrogen Energy, 33,* 1489–1495.

Yang, H. and Shen, J. 2006. Effect of ferrous ion concentration on anaerobic bio-hydrogen production from soluble starch. *International Journal of Hydrogen Energy, 31,* 2137–2146.

Yenigun, O., Kizilgun, F., and Yilamazer, G. 1996. Inhibition effect of zinc and copper on volatile fatty acid production during anaerobic digestion. *Environmental Technology, 17,* 1269–1274.

Yokoyama, H., Waki, M., Ogino, A., Ohmori, H., and Tanaka, Y. 2007. Hydrogen fermentation properties of undiluted cow dung. *Journal of Biosciences and Bioengineering, 104,* 82–85.

Yu, H. Q. and Mu, Y. 2006. Biological hydrogen production in a UASB reactor with granules II: Reactor performance in 3-year operation. *Biotechnology and Bioengineering, 94,* 988–995.

Yusoff, M. Z. M., Hassan, M. A., Abd-Aziz, S., and Rahman, N. A. A. 2009. Start-up of biohydrogen production from palm oil mill effluent under non-sterile condition in 50 L continuous stirred tank reactor. *International Journal of Agricultural Research, 4,* 163–168.

Zeidan, A. A. and van Niel, W. J. 2010. A quantitative analysis of hydrogen production efficiency of the extreme thermophile *Caldicellulosiruptor owensensis* OLT. *International Journal of Hydrogen Energy, 35,* 1128–1137.

Zhang, H., Bruns, M. A., and Logan, B. E. 2006. Biological hydrogen production by *Clostridium acetobutylicum* in an unsaturated flow reactor. *Water Resources*, 40, 728–734.

Zheng, X. J. and Yu, H. Q. 2005. Inhibitory effects of butyrate on biological hydrogen production with mixed anaerobic cultures. *Journal of Environmental Management*, 74, 65–70.

Zheng, X. J. and Yu, H. Q. J. 2004. Biological hydrogen production by enriched anaerobic cultures in the presence of copper and zinc. *Environmental Science and Health*, 39, 89–101.

Zoetemeyer, R. J., Vandenheuvel, J. C., and Cohen, A. 1982. pH influence on acidogenic dissimilation of glucose in an anaerobic digester. *Water Resources*, 16, 303–11.

8

Photobioreactors

8.1 Introduction

Hydrogen production through biophotolysis or photofermentation is usually a two-stage process. In photofermentation, small-chain organic acids are used by photosynthetic bacteria as electron donors for the production of hydrogen at the expense of light energy. Phototrophic bacteria have an advantage over their fermentative counterparts, both in terms of a high theoretical conversion and the ability to use light energy in a wide range of absorption spectra.

Several types of photobioreactors (PBRs) have been designed to improve the biomass production and with a little modification for hydrogen production. Among them the tubular, flat panel, and vertical-column type of PBRs are widely used for hydrogen production. However, the reactors come with their merits as well as demerits.

In this chapter, several geometrical variants of PBRs and their performances have been compared. The design criteria associated with the development of these PBRs have been explained. The effects of physical and physiochemical parameters on the performance of a PBR have been discussed and energy analysis of the photofermentation process in a PBR has been performed.

8.2 Types of PBRs

PBR design is a subject of great relevance for the attainment of a sustained development in modern technology, and has also considerable interest from the basic scientific and technological point of view. From the 1940s, PBRs have developed from an open pond to various closed systems. The main parameter that affects reactor design is provision for light penetration, which implies a high surface-to-volume (A/V) ratio; such penetration is crucial if one wants to improve the photosynthetic efficiency, which is essential to reach high product and biomass productivities. In order to achieve said high surface-to-volume ratio, several shapes have been developed that met with success. There are various types of PBRs designed and tested for biomass production and few have been successful in large-scale operations.

The PBRs can be broadly classified into two major types:

1. Open system: raceway ponds, lakes, etc.
2. Closed system: tubular, flat panel, conical, pyramidal, fermenters, etc.

PBRs of open system are considered for biomass production but not for the hydrogen production. So, the present chapter deals with mostly the close system.

8.2.1 Closed System PBRs

8.2.1.1 Tubular Reactors

The tubular reactors are of different configurations falling under the following major categories (Carvalho et al., 2006):

1. Airlift and bubble column (vertical type) agitated by bubbling CO_2
2. A horizontal tubular reactor (HTR) in which the light-harvesting and gas exchange units are separated
3. A helical tubular reactor in which the material of construction is transparent and flexible and is coiled in a desired fashion
4. α-Shaped reactors
5. Fermenter type

8.2.1.1.1 Vertical Tubular Reactors (VTR)

The airlift and bubble column reactors fall under this category. It consists of vertical transparent tubes (polyethylene or glass tubes) in which the agitation is achieved with the help of bubbling at the bottom. Airlift bioreactor possesses good mixing properties while bubble column configuration has efficient aeration without any internal constructions (Figure 8.1). In few bioreactors, the bubbling is done by sparging from the sides. Supply of CO_2 and removal of O_2 are very efficient in this type of reactors (Kumar et al., 2013).

The advantages include low-material cost, high transparency and area to volume ratio, biomass productivity, and low contamination risk. Usage of such systems with various capacities and modifications to overcome practical difficulties in biomass production has been reported (Tredici et al., 1993; Martnez-Jeronimo and Espinosa-Chavez, 1994). Miron et al. (2000) studied the hydrodynamics and mass transfer in bubble column, split cylinder airlift, and concentric draft tube sparged airlift reactors for *Phaeodactylum*

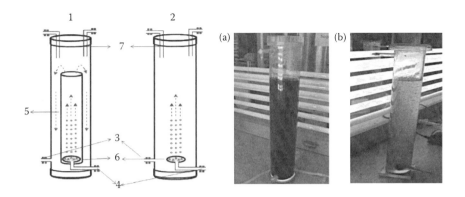

FIGURE 8.1
(See color insert.) Schematic diagram and photographs of (a) airlift and (b) bubble column PBRs for algal biomass production using *C. sorokiniana* (1: airlift PBR, 2: bubble column PBR, 3: sampling port, 4: gas inlet, 5: inert draft tube, 6: sparger, 7: gas outlet).

tricornutum. In all the three bioreactors, a biomass concentration of 4 kg/m^3 was achieved with a specific growth rate of $6.1 \times 10^{-6} \text{ s}^{-1}$.

The major drawbacks in scaling up are fragility, gas transfer at the top regions of the reactors temperature control, and gas holdup. Cells are carried along with the bubbles and face high shear when the bubble bursts (Chalmers, 1994). Use of VTR for the hydrogen production stage is challenged by the type of agitation system, in which the bubbling of an inert gas will dilute the stream of hydrogen (Figures 8.2 and 8.3).

8.2.1.1.2 Horizontal Tubular Reactors

HTRs are widely used for their orientation toward sunlight that results in high-light conversion efficiency. The gas is introduced into the tube connection or via a dedicated gas exchange unit. The various configurations of HTR are long tubular with parallel sets of tubes, looped tubes, and near HTRs (NHTR) of which NHTR has been reported for hydrogen production. The major drawback of this type of reactor is temperature control. Oxygen buildup due to photosynthetic activity results in photobleaching and thus reduced photosynthetic efficiency (Miron et al., 1999)

8.2.1.1.3 Near-Horizontal Tubular Reactor

The NHTR designed by Tredici consists of parallel tubes made of plexiglass connected at the top and bottom ends by tubular plexiglass manifolds and tilted at 5° from the surface (Tredici, 1999). The elevation helps in reducing gas holdup and improves oxygen removal. With *Arthrospira platensis*, a volumetric productivity of $1.46 \times 10^{-5} \text{ kg m}^{-3} \text{ s}^{-1}$ was achieved. With *R. capsulatus*, the hydrogen production rate has been found to be around $9.16 \times 10^{-7} \text{ s}^{-1}$ (volume/volume) (Gebicki et al., 2009).

FIGURE 8.2
Photograph of the vertical tubular PBR using *Rhodobacter sphaeroides* O.U.001.

FIGURE 8.3
Customized double jacket vertical tubular PBR for hydrogen production using *Anabaena* sp. PCC 7120.

8.2.1.1.4 Helical Tubular Reactors

Helical tubular reactors are constructed by coiling straight tubes made of flexible plastic into three-dimensional helical frameworks with a desired inclination (Figure 8.4). It is externally coupled with a gas exchanger and a heat exchanger. A centrifugal pump is used to drive the feed through the tubes in an ascending fashion. Due to its high A/V ratio, it is possible to achieve photosynthetic efficiency (PE) of up to 6.6% with a volumetric productivity of 1.04×10^{-5} kg/m^3 s with *A. platensis* (Tredici et al., 1998). A helical photobioreactor has been used in two-staged hydrogen production using *Anabaena azollae* with rates of up to 3.61×10^{-6} m^3/m^3 s hydrogen. Polyurethane foam balls were used to prevent the deposition of the culture in the inner walls (which is known as biofouling) (Tsygankov et al., 1999). Many modifications of helical framework have been proposed to improve the design and light distribution. It was reported that for a given area 60° cone angle of the conical helical layout had the maximal photoreceiving area and photosynthetic efficiency of 6.84% (Morita et al., 2000).

8.2.1.1.5 α-Shaped Reactors

The α-shaped reactor is another type of tubular PBRs designed and constructed based on the algal physiology and the sunlight (Lee et al., 1995). In this reactor, the culture is lifted 5 m by air to a receiver tank and flows down

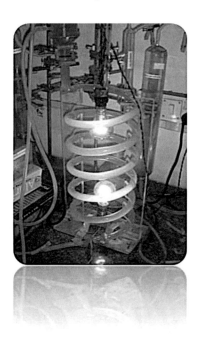

FIGURE 8.4
Helical tubular bioreactor using *Rhodobacter sphaeroides* O.U.001.

an inclined PVC tube (2.5×10^{-2} m ID $\times$ 25 m) to reach another set of air riser tubes and the process repeated for the next set of tubes. The unidirectional and high-liquid flow rate occurs at relatively low-air flow rates and resulted in biomass concentration of about 10 kg/m³. This reactor design comprises all the basic requirements for hydrogen production. If needed, hydrogen production in large scale can be achieved by replacing air with an inert gas in the system.

8.2.1.1.6 Fermenter Type of PBR

Commercial bioreactors for heterotrophic organisms have been modified and used as *PBRs*. In controlled bioreactors, every parameter can be monitored and controlled precisely with higher degree compared to other reactor types. It is mainly used for optimization studies, and their scaling up is limited due to low A/V ratio. For very large volumes, illumination is provided internally (Pohl et al., 1988). Agitation is provided with the help of an impeller (marine or ribbon type) (Figure 8.5) or a magnetic stirrer (in smaller units). It offers open gas exchange but has many restrictions toward scaling up.

8.2.1.2 Flat Panel PBRs

Flat panel PBRs are of special interest due to their high A/V ratio when the thickness is minimal (Caravalho et al., 2006). The orientation of the plates is either vertical (transparent sides facing east–west), tilted (north–south), or horizontal. In large-scale production, several plates are arranged parallel over an area. The flat panel PBRs are characterized by an open gas transfer area thus reducing the need for a dedicated degassing unit. In a comparison between the different types of outdoor reactors by Tredici and Zittelli, it was

FIGURE 8.5
Bioengineering AG control fermentor used as PBR using *C. sorokiniana*.

hypothesized that the better performance of the flat plate reactors [2.23×10^{-5} kg/m^3 s and 5.30% photosynthetic efficiency (PE)] was due to lack of susceptibility to the orthogonal rays during mid-day that cause light saturation effects on the culture. Vertical flat panel PBRs can be divided into three generations. The first generation comprises transparent sheets sandwiching a frame while the second generation was distinguished by alveolar panels. The flat plate air-lift reactor of Subitec GmbH consisting of two deep-drawing film half-shells welded together to form internal static mixers is considered to be the third generation. The mixing is usually achieved by bubbling air from the nozzles at the bottom or from the sides. The rate of mixing exerts little influence on the productivity and photosynthetic efficiency in low density cultures while it has great impact in high-density cultures for efficient utilization of light. It should be noted that when the mixing rate is too low, maximum utilization of light is affected while higher rates result in cell damage (Hu et al., 1996). A V-shaped flat panel reactor has a very high-mixing rate and very low-shear stress owing to its engineering features eliminating escape corners, providing low shear and lack of cell adhesion to the walls of the reactor (Iqbal et al., 1993). It was a design based on a model on fluidized bed reactors. A polysaccharide concentration of 25.96 kg/m^3 was achieved with *P. cruentum*.

In the case of these PBRs being used in hydrogen production, mixing is achieved by recirculation of the evolved gas (Lee, 1986; Hoekema et al., 2002). A 6.5×10^{-3} m^3 capacity flat plate solar bioreactor with temperature control, tilted 30° to the horizontal and facing south, was used for hydrogen production in outdoor (Eroglu et al., 2008). A maximum rate of 2.77×10^{-6} m^3/m^3 s H$_2$ was achieved in outdoor illumination. Another vertical flat plate PBR of Zhang et al. (2001) consists of baffles to improve the agitation.

8.2.1.2.1 Alveolar Panels

The modular flat panel PBR (vertical alveolar plates) with A/V ratio of 80 m^{-1} was reported for mass cultivation of *A. azollae* and *Spirulina platensis*. These bioreactors are constructed with transparent PVC or polycarbonate sheets that are internally portioned to form rectangular channels called alveoli. The major disadvantage of this system is the oxygen buildup due to high photosynthetic activity. The use of alveolar panels with internal or external gas exchange has also been mentioned for hydrogen production (Richmond, 1987). Unlike tubular reactors the alveolar plates do not achieve light dilution (unless placed at high inclination to the horizontal), thus cultures suffer from light saturation and from photoinhibition.

8.2.1.2.2 Flat Panel PBRs with Rocking Motion

The use of flat panel PBRs for hydrogen production is limited by the poor agitation achieved with recirculation systems. To overcome the problem of agitation, the flat plate reactor has been mounted on a stand and gave it a rocking motion with the help of motor with eccentric cam motion (Figure 8.6). A 3.3×10^{-2} rps can provide sufficient agitation (unpublished work of the

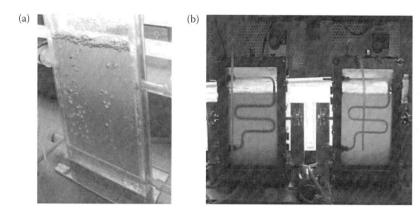

FIGURE 8.6
Photograph of the (a) flat panel reactor and (b) flat panel-rocking PBRs with temperature and pH-monitoring system using algae.

authors). The type of agitation in the system is found to a pulsation motion at greater tilting angles. The system is completely balanced and hence requires very low energy for the movement of the entire reactor (Gilbert et al, 2011).

A floating type of reactor which consists of a moving bed on a triangular roof and having see-saw type of motion constructed on sea or lakes and agitated by the motion of the waves was reported for hydrogen production using *Rhodopseudomonas palustris* strain R_1 (Otsuki et al., 1998). Although the light conversion efficiency is low (0.308%), this reactor is very innovative and the water surrounding provides the temperature control (Tables 8.1 and 8.2).

8.2.2 Other Reactors Geometries

8.2.2.1 Torus-Shaped Reactor

The torus-shaped PBRs build at Legrand's laboratory is an innovative tool for methodical optimization of H_2 production using photosynthetic microorganisms. It consists of a marine impeller that produces a three-dimensional swirling action. It is fully automated and can be operated in batch and continuous modes. The light is provided externally by tubes placed parallel to the illuminated surface (Pottier et al., 2005; Fouchard et al., 2008).

8.2.2.2 Annular Triple-Jacketed Reactor

The annular triple-jacketed PBR consists of three concentric chambers (Figure 8.7). Fluorescent bulbs are inserted in the innermost chamber (Basak and Das, 2009). Surrounding this is the culture chamber agitated with the help of a magnetic stirrer. Temperature is controlled by circulating water in the outermost

TABLE 8.1

Main Design Features of Closed PBRs

Reactor Type	Microbial Species	Light-Harvesting Efficiency	Degree of Control	Area of Land Required	Scalability	Cell Productivity (kg/m³ s)	References
Vertical tubular	P. cruentum	Medium	Medium	Medium	Medium	5.75×10^{-6}	Guan et al. (2004)
Horizontal tubular	S. platensis	High	Medium	High	Medium	2.89×10^{-5}	Tsygankov et al. (1999)
Helical	S. platensis	Medium	High	Very low	High	4.62×10^{-6}	Molina et al. (2001)
α-Shaped	—	Very high	High	High	Very low	—	Hu et al. (1996)
Flat-plate	S. platensis	Very high	Medium	Low	Medium	2.48×10^{-6}	Wykoff et al. (1998)
Fermenter type	Several	Low	Very high	Very low	Low	3.47×10^{-7} 5.75×10^{-7}	Melis (2002)

TABLE 8.2

Types of PBRs with Optimal Features for Hydrogen Production

Type of PBR	Agitation System	S/V Ratio	Temperature Control	Gas Exchange	References
Vertical tubular	Airlift, bubble column	Small	—	Open gas exchange	Tamagnini et al. (2002); Martnez-Jeronimo and Espinosa-Chavez (1994)
Horizontal tubular	Recirculation	Large	Shading, overlapping, water spraying	Injection (feed) and dedicated degassing units	Miron et al. (1999); Richmond (1987); Tredici et al. (1998)
Helical tubular	Centrifugal pumps	Large	Heat exchanger	Injection (feed) and dedicated degassing units	Tsygankov et al. (1999); Morita et al. (2000)
α-Shaped reactor	Airlift	Large	Heat exchanger	Injection (vertical units) and degassing (top)	Lee et al. (1995)
Flat panel reactor	Bubbling and recirculation	Medium	Heat exchange coils	Bubbling	Tredici et al. (1998); Hu and Richmond (1996)
V-shaped panel	Bubbling	Medium	Heat exchange coils	Bubbling	Iqbal et al. (1993)
Alveolar panel	Bubbling	Large	Water circulation	Bubbling	Tredici et al. (1991); Hu et al. (1996)
Flat panel (pivoted at center)	Pulsating motion	Medium	Heat exchange coils	Degasser	(Unpublished data of the authors)
Floating type bioreactor	See-saw motion	Medium	Not required	Degasser	Otsuki et al. (1998)
Fermenter-type with lighting	Impellers	Small	Heat exchange coils	Sparger	Pohl et al. (1988)
Torus-shaped reactor	Marine impeller	Medium	Cooling fans	CO_2 inlet (after impeller) and outlet (top)	Pottier et al. (2005); Fouchard et al. (2008)
Annular triple-jacketed (internal lighting)	Magnetic stirrer	Medium	Outer water jacket	Open gas exchange	Basak and Das (2007, (2009)
Induced-diffused PBR	Not required	Large	—	—	El-Shishtawy et al. (1997, 1998a)

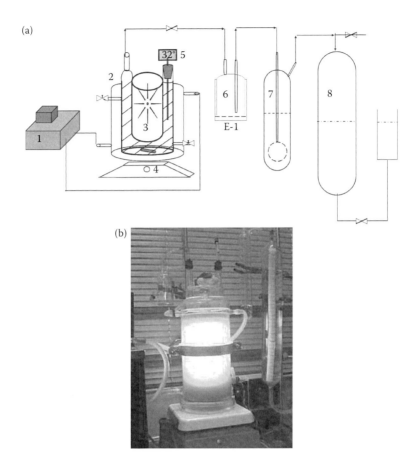

FIGURE 8.7
(See color insert.) (a) Schematic diagram of triple-jacketed reactor (1: Temperature control water reservoir, 2: Triple-jacketed PBR, 3: Light source, 4: Magnetic stirrer, 5: Temperature monitoring, 6: Gas trap, 7: CO_2 absorber, 8: Gas collector). (b) Photograph of the triple-jacketed PBR using *Rhodobacter sphaeroides* O.U.001.

chamber. The reactor has a high A/V ratio and a light conversion efficiency of 3.7% was achieved for hydrogen production using *R. sphaeroides* O.U.001. Similar annular reactor involving *Anabaena variabilis* PK84 resulted in a light conversion efficiency of 1% (Liu et al., 2006).

8.2.2.3 Induced–Diffused PBR

The main design criterion of the PBRs is to maximize the A/V ratio. Due to this, the thickness of the reactor is small and hence economically not feasible. El-Shishtawy et al. (1998a) introduced the concept of induced and diffused photobioreactor (IDPBR) for distributing the light homogenously inside the bioreactor. It consists of two parts, first a diffusion plate made of two transparent polyacrylate (PMMA) sheets. One of the sheets is treated with dot printing on one side for diffusion of light. A reflection sheet (PET) is placed on the printed surface and fastened tightly between the two plates to make up the diffusion plate. A maximum light energy conversion to H_2 of about 9.23% has been reported for this type of PBRs.

8.3 Physicochemical Parameters

8.3.1 Physical Parameters Affecting Performance of a PBR

The physical factors affecting the performance of a PBR are

- Light penetration
- High area to volume ratio
- Temperature control
- Transparency and durability of the material of construction
- Gas exchange
- Agitation system

8.3.2 Physicochemical Parameters Affecting the Performance of a PBR

The hydrogen production in a PBR is affected not only by the physical parameters (e.g., quantity of light penetrating into the reactor), but also several physicochemical parameters (Table 8.3). The physicochemical parameters are found to influence the biochemical pathways toward hydrogen production. The major physicochemical parameters affecting the higher production rate (HPR) of a PBR are

- pH: Optimum pH of the medium is usually around 7, except for chlorella and spirulina, which require a higher pH around 10 for biomass production
- Temperature: Temperature requirements are in the range of 298–308 K
- Light intensity: Light intensity in the range of 50–200 mE/m^2 s was generally found to be optimal and intensities >200 mE/m^2 s may cause inhibitory effects. It was shown that the light conversion efficiency to hydrogen decreased with increasing light intensity (Nath et al., 2006)
- Dissolved oxygen
- Dissolved CO_2
- Shear due to agitation

TABLE 8.3

Optimal Physicochemical Conditions for Achieving Maximum HPR for Different Hydrogen-Producing Microorganisms

Organism	Strains	Optimum Physiochemical Conditions				HPR	References
		pH	T (°C)	Light Intensity	Carbon Source		
Green algae	*Chlamydomonas* MGA 161	8	303	25 W/m²	5% CO_2	1.24×10^{-6} m³/ m³ s	Ohta et al. (1987)
Heterocystous cyanobacteria	*Anabaena variabilis*	—	303	150 µE/m² s	1% CO_2	6.94×10^{-5} mol/ kg_{dw} s	Berberoglu et al. (2008)
Nonheterocystous cyanobacteria	*Oscillotoria miami* BG7	7.75	308	100 µE/m² s	CO_2	6.94×10^{-5} mol/ $kg_{protein}$ s	Phlips and Mitsui (1983)
Purple sulfur photosynthetic bacteria	*Chromatium* sp. Miami PSB 1071	—	—	140 µE/m² s	Succinate, thiosulphate	1.67×10^{-3} mol/ $kg_{protein}$ s	Ohta et al. (1981)
Purple nonsulfur photosynthetic bacteria	*Rhodobacter capsulatus* ST410	—	—	66 W/m²	Malate	2.77×10^{-5} m³/ m³ s	Ooshima et al. (1998)

8.3.3 Other Factors Affecting Hydrogen Production and Biomass Production

Parameters for biomass and hydrogen production are different. In *R. sphaeroides*, a carbon-to-nitrogen (C:N) ratio of 7.5:10 induces biomass formation while 15:2 induces hydrogen production (Eroglu et al., 1999). Apart from C:N ratio, the source of nitrogen–ammonia, glutamate, or yeast extract can also affect the process as nitrogenases are reversibly inhibited by ammonia (Oh et al., 2004). Carbon fixation rate of 56.6% for green algae and 68.8% for cyanobacteria was observed when nitrate was added in fed-batch mode. This was due to an increase of the extension of the exponential growth phase by more than 3 days (Jin et al., 2006). In green algae, growth can be either autotrophic (using CO_2 from the atmosphere) or photoheterotrophic (using an organic substrate). However, hydrogen production by indirect biophotolysis usually requires that the entire cells are exposed to anaerobic conditions. By optimizing concentrations of key nutrients in the media of *Synechocystis* sp. PCC 6803, a 150-fold increase H_2 production was achieved mainly due to a 44-fold increase in glycogen concentration (Burrows et al., 2008). The mode of operation, batch, continuous or fed-batch, also influences the yield mainly due to substrate or product inhibition.

8.4 Design Criteria

8.4.1 Features of an Efficient PBR

The efficiency of PBRs is determined by the integration of light capturing, light transportation, light distribution, and light utilization by microalgae through photosynthesis (Zijffers et al., 2008). An efficient PBR design should achieve the following:

1. Harvest as much sunlight as possible and transport, channel, and distribute it into the cultivation vessel in such a way that maximum light energy is used for biomass formation.
2. Allow convenient and precise control of important operational parameters so that cells are entertained in an environment that encourages best utilization of light energy.
3. Minimize the capital costs and operational costs.
4. Minimize energy consumption during operation.

8.4.2 Light-Related Design Considerations

Light capturing, distribution, and utilization of light intensity play an important role in microalgal photosynthesis. While light limitation could occur at low-light intensity, it should be noted that light saturation and photoinhibition might occur when light intensity goes beyond a critical level. Photoinhibition can be reversible or irreversible, depending on the light stress and the length of time the microalgae are exposed to the stress. Light inhibition should be avoided as much as possible.

Light spectral quality is another important factor to consider in PBR design. While sunlight covers a wide spectral range, only light within the range of 400 and 700 nm is photosynthetically active radiation (PAR), which accounts for approximately 50% of sunlight (Suh and Lee, 2003). There is a natural barrier for enhancing the photosynthetic efficiency, and the actual photosynthetic efficiency is even lower due to losses such as light reflection and cellular respiration.

The most important design considerations that concern light capture are the transparency of the materials and the surface/volume ratio. Currently, common materials used for PBRs include glass, plexiglass, poly vinyl chloride (PVC), acrylic-PVC, and, the most common, polyethylene. These materials all satisfy the transparency requirement and are mechanically sufficient for PBR construction. However, they all have their pros and cons and need to be evaluated case by case for use in the construction of particular PBRs. For instance, glass is strong and transparent and very good material for the construction of small-scale PBRs. However, it requires many connection parts for the construction of large-scale PBRs, which could be costly. The ability of the material surface to prevent the formation of biofilm is another important feature to consider. Biofilms are not only difficult to clean but could also drastically reduce the light transmission through PBR.

Different designs have been developed to maximize the surface/volume ratio. For instance, flat panel PBR (FP-PBR) and tubular PBR (TPBR) have been established as two of the most promising types of PBRs for industrial production of microalgal biomass, primarily because they offer large A/V ratios. Some parameters that could affect the light distribution include light scattering by algal cells in solution and absorption by intracellular pigments. In high-density microalgal cultures, mutual shading between different cells becomes the main factor influencing light distribution among cells. In an under-agitated culture, an undesirable scenario could occur where some other cells (i.e., these on the light-receiving surface) are exposed to overdose of light whereas some cells (i.e., these at regions at distance from the surface) receive less or zero light. To this end, limiting the length of the light path (e.g., thin PF-PBR or small diameter TPBR) and improving mixing are the most commonly adopted strategies to improve light distribution. More complicated technologies include using internally illuminated PBRs.

8.4.3 Temperature as a Design Criterion

The space and light requirements of microalgal farming imply that a commercial cultivation system will most likely be located outdoors and be exposed to a large range of day/night and seasonal temperature changes. Devising cost-effective and reliable temperature control mechanisms is a significant challenge in PBR design. It was demonstrated that, without temperature control, the temperature in a closed PBR could reach a level 10–30 K higher than the ambient temperature. Thus, additional cooling mechanisms are often employed to maintain the culture within a favorable range. Some of these mechanisms are submersion of the entire culture in a water pool, spraying with water, shading, or incorporating a heat exchanger with PBR for cooling. Interested readers could refer to a recent review focusing on this topic (Mehlitz, 2009).

8.4.4 Sterility (Species Control) and Cleanability

It is well accepted that certain extent of impurity in microalgal cultures must be tolerated when the processes are designed for low-value objectives such as biofuel production and CO_2 sequestration. Nevertheless, cautions must be taken to avoid excessive contamination. Fortunately, for autotrophic microalgal farming facilities, contamination of heterotrophic microorganism is usually not of significant concern due to the lack of organic carbon sources in the system. However, the control of exotic and invasive algal species and predators is critical for stable and continuous operations and also for stable quality of products. Species control could be particularly difficult for cultivation of relatively slow-growing microalgal species.

Cleanability is of critical importance to a PBR for the following reasons:

1. Preventing formation of biofilm on the wall and therefore maintaining high-light transmission.
2. Minimizing the chance of contamination.

To increase the cleanability, the following principles should be observed:

1. The internal surface of a PBR should be smooth.
2. Minimizing the number of internals and bends.
3. The internal dimensions of a PBR should be large enough to allow convenient cleaning.

8.4.5 Surface Area to Volume (A/V) Ratio

The A/V ratio is the major concern which determines the amount of light (which is the limiting factor) entering the system per unit volume of the reactor. The higher the A/V ratio, the higher the cell concentration and greater will be the volumetric productivity. Reactors with same volume but different A/V ratios were evaluated for biomass productivity (Gebicki et al., 2009). It was seen that the photosynthetic photon flux density (PPFD) at the surface of the culture was significantly lower for reactors with higher A/V ratios as the impinging radiation was distributed over greater surface area. This resulted in 18% higher volumetric productivity in flat reactor than the curved chamber.

8.4.6 Oxygen Removal

Oxygen evolution during photosynthesis causes toxic effects like photobleaching and even inhibition of photosynthesis. Hence, it is necessary that an efficient degassing system should be present. Accumulation of O_2 is a serious problem in reactors with high A/V ratio. It is necessary to have a separate degassing unit in which the distance between the entrance and exit is such that even smallest bubbles can disengage. O_2 removal is not of major concern in flat plate, airlift, and bubble column reactors. A high performance gas–liquid transmission device described for enclosed flat plate *Spirulina* culture system claims to satisfy the demand for carbon supplement and excessive oxygen removal (Su et al., 2010). In reactors during hydrogen production, degassing is only necessary when the O_2 production (due to basal activity of PSII) is higher than respiration.

8.4.7 Mixing

Mixing ensures the homogenous distribution of nutrients and avoids light and temperature gradients across the bioreactor. The flashing light effect provided by movement between light to dark zones has been found to be essential for higher biomass productivity (Becker, 1994). Inadequate mixing will result in settling of biomass, stagnant or dead zones, cell aggregation, and formation of multiphase systems thus affecting the mass transfer rate. The different types of mixing systems are pumping, mechanical stirring, and airlift type. The pumps used for recirculation cause larger shear forces, which are lethal to the cells. It is learnt that the degree of stress in the centrifugal and rotary displacement pumps is proportional to the rotational speed (Lee, 1986).

In mechanical stirring, maximum shearing occurs in the area surrounding the impeller. Bubble column reactors offer the minimum stress. The cell damages associated with bubbling are

1. Cell interactions with bubble generation at the sparger
2. Cell interactions with bubbles coalescing and breaking up in the region of bubble rise
3. Cell interactions with bubbles at the air-medium interface (Chamlers, 1994; Barbosa et al., 2004)

Mixing can be improved by using horizontal and vertical baffles in flat plate airlift PBRs (Zhang et al., 2001). The baffles create a defined circulation path exposing the cells to intermittent light and thus creating a flash light effect. Good mixing is achieved by giving a simple rocking action to the flat panel reactor mounted on a stand.

Mixing time is the time required to achieve the terminal mixing that is when the concentration of the mixture remains uniform. The mixing time of the reactor was determined by injecting a color tracer (methylene blue) and monitoring dye concentration in periodically drawn samples. This was done for different agitation rates at 91% filling. The power input was determined by measuring the voltage applied and current drawn by the dc motor.

$$\text{Power input (W)} = \text{Voltage (V)} \times \text{Current (A)} \qquad (8.1)$$

The energy input to achieve mixing (J) = Power input (J/s) × Mixing time (s)
$$(8.2)$$

It has been observed that the mixing time reduces with the increase in rocking frequency from 4 to 0.25 cycles per s (cps). The lowest possible mixing time reported as 17 s at 0.25 cps. This value is the lowest as compared to 100–150 s for flat panel PBR, 60 s for bubble column, 37.7 s for torus-shaped reactor with marine propeller and comparable with 9–264 s as obtained in Bio Wave[R] bag bioreactor used in animal cell culture (Gilbert et al., 2011).

8.4.8 Material of Construction

The materials used for construction of PBRs play a major role in deciding the establishment and maintenance costs. The property of the materials should be such that it is stable and long-lasting. The materials used should have the following characteristics:

- High transparency
- Flexible and durable
- Nontoxic
- Resistance to chemicals and metabolites produced by the microorganisms
- Resistance to weathering
- Low cost

It was found that only PMMA and Teflon retained their transparency over months when exposed to natural climatic conditions (Tredici, 1999). By comparing the properties of the different types of materials, it was found that PMMA and PET have more advantages for bioreactor construction with still PVC remaining an alternate when chemical resistance is of more concern than transparency (Table 8.4). It is seen that materials like natural rubber, silicon rubber, and polypropylene are found to have toxic effects on S-deprived algal cells (Skjånes et al., 2008).

TABLE 8.4

Study of Materials Used in the Construction of PBRs

Material	Molecular Formula	Density (kg/m³)	Transparency (%)	Melting Point (K)	Advantages	Disadvantages
Glass	—	2.35–2.52	95	—	High transparency Chemically stable Durable	Fragile High-installation cost
Polymethyl methyl acrylate (PMAA)	$(C_5O_2H_8)_n$	1.19	92	403–413	Impact resistant High-optical transmittance Environmentally stable	Not autoclavable Scratch prone Low-chemical resistance
Polycarbonate (PC)	$(C_{16}H_{14}O_3)_n$	1.22	92	540	Impact resistant Autoclavable Absorbs UV rays	Susceptible to solvents, concentrated acids, and alkalis Diffusible to gases
Polyethylene (PE)	$(C_2H_4)_n$	0.91–0.97	80–85	388, 408	Chemically inert Resistant to solvents, acids, and alkalis	Weakens on exposure to light and oxygen
Polypropylene	$(C_3H_6)_n$	0.85–0.94	80	433	Durable Autoclavable Resistant to chemical attacks	Coloration and loss of transparency on exposure to environment
Polyvinyl Chloride	$(C_2H_3Cl)_n$	1.2–1.34	80	353	High-tensile strength Resistant to acids and alkalis Low permeability to gases	Low transmittance
Polystyrene	$(C_8H_8)_n$	1.05	—	513	Thermoplastic	—
Polyethylene terephthalate (PET)	$(C_{10}H_8O_4)_n$	1.37	—	533	Impact resistant Low permeability to gases	Hygroscopic
Polyurethane (transparent)	$C_{25}H_{42}N_2O_6$	1.34	—	353	High optical transmittance High UV stability No coloration	Expensive

Source: Adapted from Nag Dasgupta, C. et al., 2010. *International Journal of Hydrogen Energy*, 35, 10218–10238.

8.5 Comparison of the Performance of the PBRs

Different working principles of the PBRs and the operational conditions that prevail in them lead to varied performance levels. Airlift systems remain as to provide a promising method of agitation as it provides the minimum shear compared to the other types. CO_2-enriched stream is used in the biomass production stage and recirculation of the evolved gases in airlift mode is effective in a hydrogen production stage. With *P. tricornutum*, hydrogen and biomass productivities were 1.37×10^{-5} kg/m^3 s and 2.31×10^{-7} kg/m^2 s, respectively (Fernandez et al., 2001). It was seen that although the alveolar panels achieved high-volumetric productivity of 1.67×10^{-5} kg/m^3 s, the photosynthetic efficiency is low due to oxygen toxicity at high photosynthetic rate (Eroglu et al., 1999; Tredici et al., 1991). A helical tubular reactor offers the highest A/V ratio (200 m^{-1}) and is considered to be the most effective and also possesses some operational difficulties. Biomass productivity of 1.04×10^{-5} kg/m^3 s for *A. platensis* (Tredici and Zittelli, 1998) and a hydrogen production rate of 3.61×10^{-6} m^3/m^{3}s for *A. azollae* (Tsygankov et al., 1999) is reported for the helical tubular reactor (Table 8.5). Flat panel reactors are reported to give the greater light conversion efficiency, which reduces drastically on increase in thickness in the direction of light. With an A/V ratio of 40 m^{-1}, a biomass productivity of 1.26×10^{-5} kg/m^3 s was achieved using *A. platensis*. Induced–diffused reactors take advantage of the spatial distribution of the light in the deeper regions of the reactor and hence the greater reactor thicknesses are achievable. The efficiency of light conversion to hydrogen is found to be maximum (9.23%) for

TABLE 8.5

Comparison of the Performance of Reactors with Respect to Hydrogen Production

Reactor Type	Organism	S/V (m^{-1})	Volumetric Productivity (kg/m^3 s)	Photosynthetic Efficiency (%)	References
Airlift	*Phaeodactylum tricornutum* UTEX 640	—	1.39×10^{-5}	—	Fernandez et al. (2001)
Flat plate reactor	*Arthrospira platensi*	40	1.26×10^{-5}	4.84	Terdici and Zittelli (1998)
Alveolar	*Nannochloropsis* sp.	80	1.68×10^{-5}	0.48	Tredici et al. (1993)
Helical tubular	*A. rthrospira platensis*	53	1.04×10^{-5}	6.6	Terdici and Zittelli (1998)
Near horizontal tubular	*A. rthrospira platensis*	70	1.62×10^{-5}	5.6	Terdici and Zittelli (1998)
Conical tubular	*Chlorella*	—	1.17×10^{-5}	6.84	Morita et al. (2000)

Source: Adapted from Nag Dasgupta, C. et al., 2010. *International Journal of Hydrogen Energy*, 35, 10218–10238.

TABLE 8.6

Comparison of the Hydrogen Yield and Light Conversion Efficiency to Hydrogen in Different PBRs

Type of PBRs	Organism	H₂ Yield (mol/ m³ s)	Light Conversion Efficiency (%)	References
Vertical tubular reactor	*Rhodobacter sphaeroides*	2.47×10^{-4}	1.1	Eroglu et al. (1998); Akkerman et al. (2002)
Annular triple jacketed	*R. sphaeroides*	8.05×10^{-5}	3.7	Basak and Das (2009)
Helical tubular reactor	*Anabaena azollae*	1.61×10^{-4}	—	Tsygankov et al. (1999)
Near horizontal tubular reactor	*Rhodobacter capsulatus*	4.08×10^{-5}	—	Gebicki et al. (2009)
Flat plate tilted solar bioreactor	*R. sphaeroides*	1.24×10^{-4} (max. HPR)	—	Eroglu et al. (2008)
Flat panel with gas recirculation	*Rhodopseudomonas* sp.	3.1×10^{-4}	—	Hoekema et al. (2002)
Fermenter type	*Rhodospirillum rubrum*	—	8.67	Syahidah et al. (2008)
Induced-diffused PBR	*R. sphaeroides*	5.03 mol H₂/ mol lactate	9.23	El-Shishtawy et al. (1998a,b)
Floating type PBR	*Rhodopseudomonas palustris*	$(1.25–1.5) \times 10^{-4}$	0.308	Otsuki et al. (1998)

Source: Adapted from Nag Dasgupta, C. et al., 2010. *International Journal of Hydrogen Energy*, 35, 10218–10238.

this PBR (El-Shishtawy et al., 1998a,b) (Table 8.6). Fermenter type of reactors offers the advantage of precise control of physicochemical parameters. The hydrogen production rate was comparatively high with *Rhodospirulum ruburum* in bioreactors with internal lighting (Syahidah et al., 2008).

8.6 Energy Analysis

The net energy ratio (NER) of a system is defined as the ratio of the total energy produced (in terms of biomass and hydrogen) over the energy required for plant operations such as mixing, pumping, aeration, and cooling.

$$\text{NER} = \frac{\sum \text{Energy produced (biomass/hydrogen)}}{\sum \text{Energy input (mixing, aeration, pumping, cooling, etc.)}} \tag{8.3}$$

Open pond systems require the minimal energy input only during harvesting while raceway ponds consume energy for the paddle wheels for mixing. Tubular reactors involve pumps for recirculation of media and compressed gas for aeration and additional cooling system. Bubble column and airlift reactor systems are advantageous in mixing only by sparging-compressed air while temperature control is achieved by regulating the temperature of the inlet gas. Tubular systems are estimated to have NER > 1 by Burgess and Fernandez-Velasco (2007). Only raceways ponds are reported to have an NER > 1 by Huesemann and Beneman (2009). Rodolfi et al. (2009) reported the NER > 1 for flat panel reactor (alveolar panels). The NER for biomass production for second-generation flat-panel bioreactor (alveolar panels) was estimated to be 4.51, 8.34 for raceway ponds, and 0.20 for HTR (Jorquera et al., 2010). Burgess and Fernandez-Velasco (2007) estimated the NER of tubular PBR for production of H_2 by microalgae. For a hypothetical microalgal H_2 energy generation efficiency of 5%, the estimated NER $\approx$ 6 for low-density polyethylene film and glass. The floating type of reactor requires no external energy for agitation (Otsuki et al., 1998). Hence, it is expected to have an NER > 1 for hydrogen production if the light conversion efficiency to hydrogen can be improved.

8.7 Conclusion

PBR design is evolving rapidly to meet the need of industrial production. A large variety of different PBRs have been developed in the past three decades, and flat panel PBR and tubular PBR are the two most promising configurations for industrial process. Building on these two basic designs, developments aiming to improve light capturing and distribution (e.g., internal illumination and spectral shifting), mass transfer (e.g., MPBR), and other aspects of PBR operation are expected to elevate the overall PBR efficiency to a new level in the near future.

Glossary

FP-PBR	Fat panel photobioreactor
HTR	Horizontal tubular reactors
IDPBR	Induced and diffused photobioreactor
MPBR	Membrane photobioreactor
NER	Net energy ratio
NHTR	Near horizontal tubular reactors
PAR	Photosynthetically active radiation

PBR   Photobioreactor
PE   Photosynthetic efficiency
PET   Polyethylene terephthalate
PPFD   Photosynthetic photon flux density
PSI   Photosystem I
PSII   Photosystem II
PVC   Poly vinyl chloride
TPBR   Tubular photobioreactor
VTR   Vertical tubular reactors

References

Akkerman, I., Janssen, M., Rocha, J., and Wijffels R. H. 2002. Photobiological hydrogen production: photochemical efficiency & bioreactor design. *International Journal of Hydrogen Energy, 27*, 1195–11208.

Barbosa, M. J., Hadiyanto, S., and Wijffels R. H. 2004. Overcoming shear stress of microalgae cultures in sparged photobioreactors. *Biotechnology and Bioengineering, 85*, 78–85.

Basak, N. and Das, D. 2007. The prospect of purple non-sulfur (PNS) photosynthetic bacteria for hydrogen production: The present state of the art. *World Journal of Microbiology and Biotechnology, 23*, 31–42.

Basak, N. and Das, D. 2009. Photofermentative hydrogen production using purple non-sulfur bacteria *Rhodobacter sphaeroides* O.U.001 in an annular photobioreactor: A case study. *Biomass and Bioenergy, 33*, 911–919.

Becker E. W. 1994. Large-scale cultivation. In *Microalgae: Biotechnology & microbiology*, ed. E. W. Becker, pp. 63–171. New York: Cambridge University Press.

Berberoglu, H., Jenny, J., and Laurent, P. 2008. Effect of nutrient media on photobiological hydrogen production by *Anabaena variabilis* ATCC 29413. *International Journal of Hydrogen Energy, 33*, 1172–1184.

Burgess, G. and Fernandez-Velasco, J. G. 2007. Materials, operational energy inputs, & net energy ratio for photobiological hydrogen production. *International Journal of Hydrogen Energy, 32*, 1225–1234.

Burrows, E. H., Chaplen, F. W. R., and Ely, R. L. 2008. Optimization of media nutrient composition for increased photofermentative hydrogen production by *Synechocystis* sp. PCC 6803. *International Journal of Hydrogen Energy, 33*, 6092–6099.

Carvalho, A. P., Meireles, L. A., and Malcata, F. X. 2006. Microalgal reactors: A review of enclosed system designs & performances. *Biotechnology Progress, 22*, 1490–1506.

Chalmers, J. J. 1994. Cells & bubbles in sparged bioreactors. *Cytotechnology, 15*, 311–320.

El-Shishtawy, R. M. A., Kawasaki, S., and Morimoto, M. 1997. Biological H_2 production using a novel light-induced & diffused photoreactor. *Biotechnology Techniques, 11*, 403–407.

El-Shishtawy, R. M. A., Kawasaki, S., and Morimoto, M. 1998a. Cylindrical-type induced & diffused photobioreactor. In *Biohydrogen*, ed. Zaborsky O. R., pp. 353–358. London: Plenum Press.

El-Shishtawy, R. M. A., Kitajima, Y., Otsuka, S., Kawasaki, S., and Morimoto, M. 1998b. Study on the behavior of production & uptake of photobiohydrogen by photosynthetic bacterium *Rhodobacter sphaeroides* RV. In *Biohydrogen*, ed. Zaborsky O. R., pp. 117–138. London: Plenum Press.

Eroglu, I., Aslan, K., Gunduz, U., Yucel, M., and Turker, L. 1998. Continuous hydrogen production by *hodobacter sphaeroides* O.U. 001. In *Biohydrogen*, ed. Zaborsky O. R., pp. 143–151. London: Plenum Press.

Eroglu, I., Aslan, K., Gunduz, U., Yucel, M., and Turker, L. 1999. Substrate consumption rates for hydrogen production by *Rhodobacter sphaeroides* in a column photobioreactor. *Journal of Biotechnology, 70*, 103–113.

Eroglu. I., Tabanoglu, A., Gunduz, U., Eroglu, E., and Yucel, M. 2008. Hydrogen production by *Rhodobacter sphaeroides* O.U.001 in a flat plate solar bioreactor. *International Journal of Hydrogen Energy, 33*, 531–41.

Fernandez, F. G. A., Sevilla, J. M. F., Perez, J. A. S., Molina, G. E., and Christi, Y. 2001. Airlift-driven external-loop tubular photobioreactors for outdoor production of microalgae: Assessment of design & performance. *Chemical Engineering Science, 56*, 2721–2732.

Fouchard, S., Pruvost, J., and Legrand, J. 2008. Investigation of H_2 production by microalgae in a fully-controlled photobioreactor. *International Journal of Hydrogen Energy, 33*, 3302–3310.

Gebicki, J., Modigell, M., Schumacher, M., Burgb, J. V. D., and Roebroeck, E. 2009. Development of photobioreactors for anoxygenic production of hydrogen by purple bacteria. *Chemical Engineering Trans, 18*, 363–366.

Gilbert, J. J., Ray, S., and Das, D. 2011. Hydrogen production using *Rhodobacter sphaeroides* (O.U.001) in a flat panel rocking photobioreactor. *International Journal of Hydrogen Energy, 36*, 3434–3441.

Guan, Y. F., Deng, M. C., Yu, X. J., and Zhang, W. 2004. Two-stage photobiological production of hydrogen by marine green alga *Platymonas subcordiformis*. *Biochemical Engineering Journal. 19*, 69–73.

Hoekema, S., Bijmans, M., Janssen, M., Tramper, J., and Wijffels, R. H. 2002. A pneumatically agitated flat-panel photobioreactor with gas re-circulation: Anaerobic photoheterotrophic cultivation of a purple non-sulfur bacterium. *International Journal of Hydrogen Energy, 27*, 1331–1338.

Hu, Q., Gutermann, H., and Richmond, A. 1996. A flat inclined modular photobioreactor for the outdoor mass cultivation of photoautotrophs. *Biotechnology and Bioengineering, 51*, 51–60.

Hu, Q. and Richmond, A. 1996. Productivity & photosynthetic efficiency of *Spirulina platensis* as affected by light intensity, algal density & rate of mixing in a flat plate photobioreactor. *Journal of Applied Phycology, 8*, 139–145.

Huesemann, M. and Benemann, J. R. 2009. Biofuels from microalgae: Review of products, process & potential, with special focus on *Dunaliella* sp. In *The Alga Dunaliella: Biodiversity, Physiology, Genomics & Biotechnology*, eds. Ben-Amotz, A., Polle, J. E. W., and Subba Rao, V. D.. New Hampshire: Science Publishers.

Iqbal, M., Grey, D., Stepan-Sarkissian, F., and Fowler, M. W. 1993. A flat sided photobioreactor for continuous culturing microalgae. *Aquaculture Engineering, 12*, 183–190.

Jin, H. F., Lim, B. R., and Lee, K. 2006. Influence of nitrate feeding on carbon dioxide fixation by microalgae. *Journal of Environmental Science Health A, 41*, 2813–2824.

Jorquera, O., Kiperstok, A., Sales, E. A., Embirucu, M., and Ghirardi, M. L. 2010. Comparative energy life-cycle analyses of microalgal biomass production in open ponds & photobioreactors. *Bioresource Technology, 101,* 1406–1413.

Kumar, K., Roy, S., and Das, D. 2013. Continuous mode of carbon dioxide sequestration by *C. sorokiniana* & subsequent use of its biomass for hydrogen production by *E. cloacae* IIT-BT 08. *Bioresource Technology, 145,* 116–122.

Lee, Y. K. 1986. Enclosed bioreactors for the mass cultivation of photosynthetic microorganisms: The future trend. *Trends in Biotechnology. 14,* 186–189.

Lee, Y. K., Ding, S. Y., Low, C. S., and Chang, Y. C. 1995. Design & performance of an α-type tubular photobioreactor for mass cultivation of microalgae. *Journal of Applied Phycology, 7,* 47–51.

Liu, J., Bukatin, V. E., and Tsygankov, A. A. 2006. Light energy conversion into H_2 by *Anabaena variabilis* mutant PK84 dense cultures exposed to nitrogen limitations. *International Journal of Hydrogen Energy, 31,* 1591–1596.

Martnez-Jeronimo, F. and Espinosa-Chavez, F. 1994. A laboratory-scale system for mass culture of freshwater microalgae in polyethylene bags. *Journal of Applied Phycology, 6,* 423–425.

Mehlitz, T. H., 2009. Temperature influence & heat management requirements of microalgae cultivation in photobioreactors. Master Thesis. California Polytechnic State University, The USA.

Melis, A. 2002. Green alga hydrogen production: Progress, challenges & prospects. *International Journal of Hydrogen Energy, 27,* 1217–1228.

Miron, A. S., Gomez, A. C., Camacho, F. G., Molina, G. E., and Chisti, Y. 1999. Comparative evaluation of compact photobioreactors for large-scale monoculture of microalgae. *Journal of Biotechnology, 70,* 249–270.

Miron, A. S., Camacho, F. G., Gomez, A. C., Molina, G. E., and Chisti, Y. 2000. Bubble column & airlift photobioreactors for algal culture. *American Institute of Chemical Engineers, 46,* 872–1887.

Molina, G. E., Fernandez, J., Acien, F. G., and Chisti, Y. 2001. Tubular photobioreactor design for algal cultures. *Journal of Biotechnology, 92,* 113–131.

Morita, M., Watanable, Y., and Saiki, H. 2000. Investigation of photobioreactor design for enhancing the photosynthetic productivity of microalgae. *Biotechnology and Bioengineering, 69,* 693–698.

Nag Dasgupta, C., Gilbert, J. J., Lindblad, P., Heidorn, T., Borgvang, S. A., Skjanes, K., and Das, D. 2010. Recent trends on the development of photobiological processes & photobioreactors for the improvement of hydrogen production. *International Journal of Hydrogen Energy, 35,* 10218–10238.

Nath, K., Kumar, A., and Das, D. 2006. Hydrogen production by *Rhodobacter sphaeroides* strain O.U.001 using spent media of *Enterobacter cloacae* DM11. *Applied Microbiology and Biotechnology, 68,* 533–541.

Oh, Y. K., Seol, E. H., Kim, M. S., and Parka, S. 2004. Photoproduction of hydrogen from acetate by a chemoheterotrophic bacterium *Rhodopseudomonas palustris* P4. *International Journal of Hydrogen Energy, 29,* 1115–1121.

Ohta, S., Miyamoto, K., and Miura, Y. 1987. Hydrogen evolution as a consumption mode of reducing equivalents in green algal fermentation. *Plant Physiology, 83,* 1022–1026.

Ohta, Y., Frank, J., and Mitsui, A. 1981. Hydrogen production by marine photosynthetic bacteria: Effect of environmental factors & substrate specificity on the growth of a hydrogen-producing marine photosynthetic bacterium, *Chromatium* sp. Miami PBS 1071. *International Journal of Hydrogen Energy*, *6*, 451–460.

Ooshima, H., Takakuwa, S., Katsuda, T., Okuda, M., Shirasawa, T., and Azuma, M. 1998. Production of hydrogen by a hydrogenase deficient mutant of *Rhodobacter capsulatus*. *Journal of Fermentation and Bioengineering*, *85*, 470–474.

Otsuki, T., Uchiyama, S., Fujiki, K., and Fukunaga, S. 1998. Hydrogen production by a floating-type photobioreactor. In *Biohydrogen*, ed. Zaborsky O. R., pp. 369–374. London: Plenum Press.

Phlips, E. J. and Mitsui, A. 1983. Role of light intensity & temperature in the regulation of hydrogen photoproduction by the marine cyanobacterium *Oscillatoria* sp. strain Miami BG7. *Applied Environmental Microbiology*, *45*, 1212–1220.

Pohl, P., Kohlhase, M., and Martin, M. 1988. Photobioreactors for the axenic mass cultivation of icroalgae. In *Algal Biotechnology*, eds. Stadler, T., Mollion, J., Verdus, M. C., Karamanos, Y., Morvan, H., Christiaen, D., pp. 209–218. New York: Elsevier.

Pottier, L., Pruvost, J., Deremetz, J., Cornet, J. F., Legr, and J., and Dussap, C. G. 2005. A fully predictive model for one-dimensional light attenuation by *Chlamydomonas reinhardtii* in a torus photobioreactor. *Biotechnology and Bioengineering*, *91*, 569–582.

Richmond, A. 1987. The challenge confronting industrial microagriculture: High photosynthetic efficiency in large scale reactors. *Hydrobiology*, *151&152*, 117–121.

Rodolfi, L., Zittelli, G. C., Bassi, N., Padovani, G., Biondi, N., Bonini G and Tredici M. R. 2009. Microalgae for oil: Strain selection, induction of lipid synthesis & outdoor mass cultivation in a low-cost photobioreactor. *Biotechnology and Bioengineering*, *102*, 100–112.

Skjånes, K., Knutsen, G., Källqvist, T., and Lindblad, P. 2008. H_2 production from marine & freshwater species of green algae during sulfur starvation & considerations for bioreactor design. *International Journal of Hydrogen Energy*, *33*, 511–521.

Su, Z., Kang, R., Shi, S., Cong, W., and Cai, Z. 2010. An effective device for gas-liquid oxygen removal in enclosed microalgae culture. *Applied Biochemistry and Biotechnology*, *160*, 428–437.

Suh, I. S. and Lee, C.-G. 2003. Photobioreactor engineering: Design & performance. *Biotechnology and Bioprocess Engineering*, *8*, 313–321.

Syahidah, K. I. K., Najafpour, G., Younesic, H., Mohamed, A. R., and Kamaruddin, A. Z. 2008. Biological hydrogen production from CO: Bioreactor performance. *Biochemical Engineering*, *39*, 468–477.

Tamagnini, P., Axelsson, R., Lindberg, P., Oxelfelt, F., Wunschiers, R., and Lindblad, P. 2002. Hydrogenases & hydrogen metabolism of cyanobacteria. *Microbiology and Molecular Biology Review*, *66*, 1–20.

Tredici, M. R. 1999. Bioreactors, photo. In *Encyclopedia of Bioprocess Technology: Fermentation, Biocatalysis & Bioseparation*, eds. Flickinger, M. C., and Drew, S. W., vol. 1, pp. 395–419. New York: Wiley.

Tredici, M. R. and Materassi, R. 1992. From open ponds to vertical alveolar panels: The Italian experience in the development of reactors for the mass cultivation of phototrophic microorganisms. *Journal of Applied Phycology*, *4*, 221–231.

Tredici, M. R. and Rodolfi, L. 2004. Reactor for Industrial Culture of Photosynthetic Microorganisms. Italy: University of Florence. PCT WO 2004/074423 A2; 2004.

Tredici, M. R., Carlozzi, P., Zittelli, G. C., and Materassi, R. 1991. A vertical alveolar panel (VAP) for outdoor mass cultivation of microalgae & cyanobacteria. *Bioresource Technology*, *38*, 153–159.

Tredici, M. R., Zittelli, G. C., and Benemann, J. R. 1998. A tubular integral gas exchange photobioreactor for biological hydrogen production. In *Biohydrogen*, ed. Zaborsky O. R., pp. 391–401. London: Plenum Press.

Tredici, M. R. and Zittelli, G. C. 1998. Efficiency of sunlight utilization: Tubular versus flat photobioreactors. *Biotechnology and Bioengineering*, 57, 187–197.

Tredici, M. R., Zittelli, G. C., Biagiolini, S., and Materassi, R. 1993. Novel photobiore-actor for the mass cultivation of *Spirulina* spp. *Bull. L'Institut Océanographique, Monaco*, 12, 89–96.

Tsygankov, A. A., Hall, D. O., Liu, J., and Rao, K. K. 1999. An automated helical pho-tobioreactor incorporating cyanobacteria for continuous hydrogen production. In *Biohydrogen*, pp. 431–440. New York: Plenum Press.

Wykoff, D. D., Davies, J. P., Melis, A., and Grossman, A. R. 1998. The regulation of photosynthetic electron transport during nutrient deprivation in *Chlamydomonas reinhardtii*. *Plant Physiology*, 117, 129–139.

Zhang, K., Miyachi, S., and Kurano N. 2001. Evaluation of a vertical flat-plate photobi-oreactor for outdoor biomass production & carbon dioxide bio-fixation: Effects of reactor dimensions, irradiation & cell concentration on the biomass produc-tivity & irradiation utilization efficiency. *Applied Microbiology and Biotechnology*, 55, 428–433.

Zijffers, J. W. F., Janssen, M., Tramper, J., and Wijffels, R. H. 2008. Design process of an area-efficient photobioreactor. *Marine Biotechnology*, 10, 404–415.

9

Mathematical Modeling and Simulation of the Biohydrogen Production Processes

9.1 Introduction

It is widely accepted that hydrogen is capable of being a "universal" fuel with many social, economic, and environmental benefits to its credit as compared to fossil fuels. It is plentiful, clean, and compatible with existing energy conversion technologies. Dark fermentation has great promise for biohydrogen production process in terms of its higher rate of H_2 evolution and also the versatility of the substrates used.

Hydrogen may be produced by a number of processes, including electrolysis of water, thermocatalytic reformation of hydrogen-rich organic compounds, and biological processes (Das et al., 2008). Currently, hydrogen is produced, almost exclusively, by electrolysis of water or by steam reformation of methane. Biohydrogen technology provides a wide range of approaches to generate hydrogen, including dark fermentation, direct biophotolysis, indirect biophotolysis, and photofermentation. In this chapter, the mathematical modeling of two biohydrogen production technologies, dark fermentation and photofermentation, has been discussed (Figure 9.1).

A mathematical model may be used to explain a system and to study the effects of different parameters on the same. More importantly, a mathematical model finds its application in making predictions about the behavior of the system. It is imperative to have a predicted outcome if a process is going to be scaled up from the laboratory to an industry. The quality of a mathematical model depends on how well theoretical predictions agree with the observations obtained from repeated experiments. In no case, must the disagreement between the theoretical values and experimental values be greater than 5%. If the disagreement is greater than 5%, the model is unsuitable and needs to be modified.

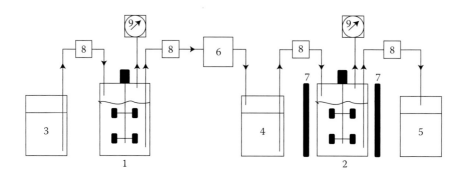

FIGURE 9.1
Schematic description of the two-stage process: dark fermentation followed by photo fermentation. [(1) Reactor for dark fermentation, (2) reactor for photofermentation, (3) feed tank, (4) spent media tank, (5) drain tank, (6) cell separator, (7) illumination system, (8) peristaltic pump, and (9) gas analyzer].

9.2 Development of Mathematical Models to Correlate Substrate and Biomass Concentration with Time

9.2.1 Monod's Model for Cell Growth Kinetics

The Monod equation is a mathematical model that deals with the growth kinetics of microorganisms. Monod (1949) proposed a relation between microbial growth rates in an aqueous environment and the concentration of a limiting substrate, given by the equation

$$\mu = \mu_{max} \frac{S}{K_s + S} \tag{9.1}$$

where

- μ is the specific growth rate of microorganisms.
- μ_{max} is the maximum specific growth rate of the microorganisms.
- S is the concentration of the limiting substrate for growth.
- K_S is the saturation constant.

The specific growth rate of microorganisms (μ) can be expressed mathematically as

$$\mu = \frac{1}{X} \frac{dX}{dt} \tag{9.2}$$

where X is the cell mass concentration at any point of time.

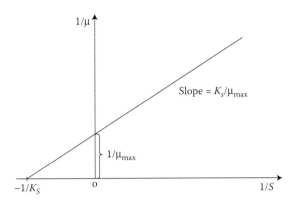

FIGURE 9.2
Lineweaver–Burk's plot of $1/\mu$ versus $1/S$.

The Monod equation may be written in the Lineweaver–Burk plot

$$\frac{1}{\mu} = \frac{K_s}{\mu_{max}S} \frac{1}{\mu_{max}} \tag{9.3}$$

Regression analysis is used to find the best fit for a straight line of the $1/\mu$ versus $1/S$. The values of K_s and μ_{max} are estimated from the intercept and the slope of the straight line, respectively (Figure 9.2).

9.2.2 Determination of Cell Mass Concentration and Substrate Concentration

Overall yield coefficient, $Y_{X/S}$ is defined as grams of cell mass generated per grams of substrate degraded and can be expressed as

$$Y_{X/S} = -\frac{dX}{dS} \tag{9.4}$$

For experimental data, $Y_{X/S}$ can be calculated by averaging the values obtained using the expression

$$Y_{X/S} = \frac{X - X_o}{S_o - S} \tag{9.5}$$

Again, Equation 9.1 may be written as

$$\frac{dX}{dT} = \mu X = \frac{\mu_{max}S}{K_s + S} \cdot X \tag{9.6}$$

$$\frac{dS}{dt} = \frac{-1}{Y_{X/S}} \cdot \frac{\mu_{max}S}{K_s + S} \cdot X \tag{9.7}$$

X in Equation 9.7 is replaced by an expression involving S, S_0, X_0, and $Y_{X/S}$ from Equation 9.5. It can then be integrated to give an expression for simulated values of S as a function of t.

$$\mu_{max}(X_0 + Y_{X/S}S_0)t = [X_0 + Y_{X/S}(S_0 + K_S)]\ln\left(\frac{X_0 + Y_{X/S}(S_0 - S)}{X_0}\right) - K_S Y_{X/S}\ln\frac{S}{S_0} \tag{9.8}$$

Now, X as a function of time can be obtained by substituting the value of S from Equation 9.5 in Equation 9.8

$$\mu_{max}(X_0 + Y_{X/S}S_0)t = [X_0 + Y_{X/S}(S_0 + K_S)]\ln\left(\frac{X}{X_0}\right) - K_S Y_{X/S}\ln\left(1 - \frac{X - X_0}{S_0 Y_{X/S}}\right) \tag{9.9}$$

9.2.3 Modeling and Simulation of the Fermentation Process

The experimental data of the batch biohydrogen production process reported by Nath et al. (2008) (Table 9.1) were used to simulate the substrate and cell mass concentration profiles. These are shown in Figures 9.3 and 9.4.

TABLE 9.1

Modeling of Biomass Growth Kinetics Based on Monod's Model for the Hydrogen Production Using *E. cloacae* DM11 in Stage 1. Experimental Conditions: Batch Operation, Volume of Medium: 2.50×10^{-4} m³, Temperature 310 K, and Initial Medium pH: 6.0 ± 0.2

Time (h)	Substrate Concentration (kg/m³)	Biomass Concentration (kg/m³)
0	10.0	0.05
2	8.7	0.14
4	7.9	0.26
6	7.6	0.41
8	6.7	0.49
10	5.1	0.51
12	3.9	0.53
14	3.5	0.57
16	3.3	0.60
18	2.9	0.62

Source: Adapted from Nath, K. et al., D. 2008. *International Journal of Hydrogen Energy*, 33, 1195–11203.

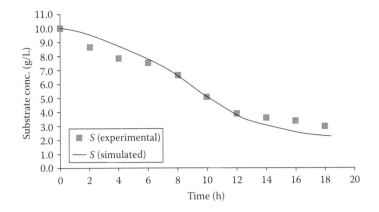

FIGURE 9.3
Plot of simulated (using Monod's model) and experimental values of substrate concentration versus time.

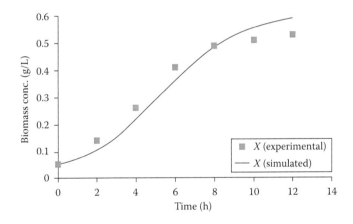

FIGURE 9.4
Plot of simulated (using Monod's model) and experimental values of biomass concentration versus time.

9.2.4 Regression Analysis of Simulated Values Obtained from Monod's Model and Experimentally Obtained Values

9.2.4.1 Coefficient of Determination (R^2)

The coefficient of determination is a statistic that indicates how well data points fit a line or curve. It provides a measure of how well observed outcomes are replicated by the model, as the proportion of total variation of outcomes explained by the model.

R^2 is mathematically expressed as

$$R^2 = \frac{\text{Total sum of squares due to regression (SSR)}}{\text{Total sum of squares (SST)}} \tag{9.10}$$

$$SSR = \sum (\hat{y}_i - \bar{y})^2 \tag{9.11}$$

$$SST = \sum (y_i - \bar{y})^2 \tag{9.12}$$

where $\hat{y}_i$ represents the simulated values, y_i represents the experimental value, and $\bar{y}$ represents the mean of the experimental data.

The closer the value of R^2 is to 1, the better is the fit of the model. A model having R^2 equal to 1 is a perfect-fit. However, a value of $R^2 > 1$ indicates the absence of regression.

From the values of SSR and SST as calculated in Table 9.2, we have

$$R^2 = 0.987$$

From the values of SSR and SST as calculated in Table 9.3, we have

$$R^2 = 0.979$$

The values of R^2 obtained indicate that the model is a good fit.

TABLE 9.2

Table for Calculation of SST and SSR from the Experimental Values of Substrate Concentration y_i and Simulated Values of Substrate Concentration $(\hat{y}_i)$

S. No.	y_i	$\hat{y}_i$	SST	SSR
1	10.0	10.0	16.32	16.32
2	8.7	8.9	7.51	8.64
3	7.9	8.1	3.76	4.58
4	7.6	7.7	2.69	3.03
5	6.7	6.3	0.55	0.12
6	5.1	5.5	0.74	0.21
7	3.9	4.2	4.24	3.10
8	3.5	3.7	6.05	5.11
9	3.3	3.3	7.08	7.08
10	2.9	2.9	9.36	9.36
$\bar{y}$	6.0	Σ	58.30	57.54

TABLE 9.3

Table for Calculation of *SST* and *SSR* from the Experimental Values of Biomass Concentration y_i and Simulated Values of Biomass Concentration $(\hat{y}_i)$

S. No.	y_i	$\hat{y}_i$	*SST*	*SSR*
1	0.050	0.050	0.135	0.135
2	0.140	0.120	0.077	0.089
3	0.260	0.250	0.025	0.028
4	0.410	0.400	0.000	0.000
5	0.490	0.500	0.005	0.007
6	0.510	0.520	0.008	0.010
7	0.530	0.550	0.013	0.017
8	0.570	0.560	0.023	0.020
9	0.600	0.570	0.033	0.023
10	0.620	0.570	0.041	0.023
$\bar{y}$	0.418	Σ	0.361	0.354

9.2.5 Other Monod's Type Models

9.2.5.1 Monod-Type Model Including pH Inhibition Term

The Monod-type kinetic expression, incorporating the empirical lower pH inhibition term I_{pH}, was put forward by a number of authors (Lin et al., 2008; Ntaikou et al., 2008, 2009).

$$r_{Glu} = \frac{q_{Glu}^{max} S_{Glu} X}{K_{Glu} + S_{Glu}} \cdot I_{pH} \tag{9.13}$$

where X denotes biomass concentration, r_{Glu} denotes rate of glucose consumption, S_{Glu} denotes residual glucose concentration q_{Glu}^{max} notes maximum specific glucose consumption rate, and K_{Glu} represents saturation constant.

9.2.5.1.1 Monod-Type Model with Biomass-Decay Constant

Together with pH inhibition coefficient (I_{pH}), Ntaikou et al. (2008) modified the Monod equation with one additional term, biomass decay constant k_d.

$$\frac{dx}{dt} = \frac{\mu_{max} \cdot s}{K_s + s} \times I_{pH} - k_d X \tag{9.14}$$

9.2.5.1.2 Monod-Type Model Including pH Inhibition Term and Substrate Inhibition Factors

The growth of fermentative hydrogen producing *R. albus* on glucose was modeled by a modified Monod's equation including both pH inhibition (I_{pH}) and substrate inhibition (I_S) factors

$$\frac{ds}{dt} = -k_m \frac{S}{K_s + S} \times I_{pH} \times I_s \tag{9.15}$$

9.3 Substrate Inhibition Model

9.3.1 Modified Andrew's Model

In the biohydrogen production process, substrate inhibition plays an important role because the product is gas (hydrogen). Monod's model is found unsuitable for the substrate inhibition process. Andrew proposed a mathematical model for the substrate inhibition in the microbial fermentation as given below (Andrews, 1968):

$$\mu = \frac{\mu_{max} S}{K_S + S + K_i S^2} \tag{9.16}$$

Kumar et al. (2000) proposed a modified Andrew's model for the biohydrogen production process with substrate inhibition:

$$\mu = \frac{\mu_{max} S}{K_S + S - K_i S^2} \tag{9.17}$$

A modification of Andrew's model suggests a nonlinear relationship between the specific growth rate μ, and substrate concentration, S.

9.3.1.1 Simulation of Cell Mass Concentration and Substrate Concentration Profiles

Now,

$$\frac{dX}{dT} = \mu X = \frac{\mu_{max} S}{K_S + S - K_i S^2} \cdot X \tag{9.18}$$

Thus,

$$\frac{dS}{dt} = \frac{-1}{Y_{X/S}} \cdot \frac{\mu_{max} S}{K_S + S - K_i S^2} \cdot X \tag{9.19}$$

X in Equation 9.19 is replaced by an expression involving S, S_0, X_0, and $Y_{X/S}$ (Equation 9.5).

It can then be integrated to give an expression for simulated values of S as a function of t.

$$
\begin{aligned}
\mu_{max}(X_0 + Y_{X/S}S_0)t \\
= [X_0 + Y_{X/S}(S_0 + K_S)]\ln\left(\frac{X_0 + Y_{X/S}(S_0 - S)}{X_0}\right) - K_S Y_{X/S}\ln\frac{S}{S_0} \\
-K_i(X_0 + Y_{X/S}S_0)\left[S - S_0 + \frac{(X_0 + Y_{X/S}S_0)}{Y_{X/S}}\ln\left(\frac{X_0 + Y_{X/S}(S_0 - S)}{X_0}\right)\right]
\end{aligned}
\tag{9.20}
$$

Now, X as a function of time can be obtained by substituting the value of S from Equation 9.5 in Equation 9.19.

$$
\begin{aligned}
\mu_{max}(X_0 + Y_{X/S})t \\
= [X_0 + Y_{X/S}(S_0 + K_S)]\ln\left(\frac{X}{X_0}\right) - K_S Y_{X/S}\ln\left(1 - \frac{X - X_0}{S_0 Y_{X/S}}\right) \\
-K_i(X_0 + Y_{X/S}S_0)\left[\frac{X_0 - X}{Y_{X/S}} + \frac{(X_0 + Y_{X/S}S_0)}{Y_{X/S}}\ln\left(\frac{X}{X_0}\right)\right]
\end{aligned}
\tag{9.21}
$$

9.3.2 Simulation of the Biohydrogen Production Process with Substrate Inhibition

The experimental values reported by Nath et al. (2008) and Nath and Das (2011) (Table 9.4) was simulated using Andrew's model and shown in Figures 9.5 and 9.6.

TABLE 9.4

Modeling of Biomass Growth Kinetics Based on the Substrate Inhibition Model for the Hydrogen Production Using *E. cloacae* DM11 in Stage 1 [Experimental Conditions: Batch Operation, Volume of Medium: 2.5×10^{-4} m³, Temperature 310 K, and Initial Medium pH: 6.0 ± 0.2]

Time (h)	Substrate Concentration (kg/m³)	Biomass Concentration (kg/m³)
0	10.2	0.04
1	9.9	0.08
2	9.0	0.14
3	8.6	0.19
4	8.0	0.31
5	7.7	0.39
6	7.3	0.40
7	6.4	0.46
8	5.7	0.52

Source: Adapted from Nath, K. et al., D. 2008. *International Journal of Hydrogen Energy*, 33, 1195–11203.

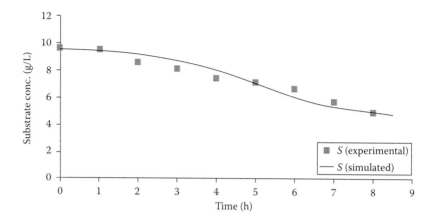

FIGURE 9.5
Plot of simulated (using substrate inhibition model) and experimental values of substrate concentration versus time.

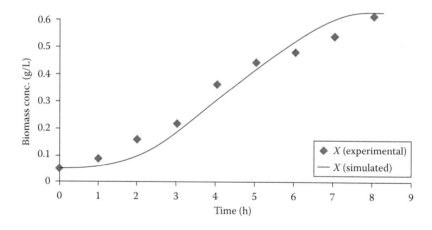

FIGURE 9.6
Plot of simulated (using substrate inhibition model) and experimental values of biomass concentration versus time.

9.3.3 Regression Analysis of Simulated Values Obtained from Substrate Inhibition Model and Experimentally Obtained Values

The values of R^2 for the substrate concentration profile and biomass concentration profile were calculated as discussed in Section 9.1 for the Monod model.

For substrate concentration profile,

$$R^2 = 0.99$$

For biomass concentration profile,

$$R^2 = 0.97$$

The values of R^2 obtained indicate that the model is a good fit.

9.4 Determination of Cell Growth Kinetic Parameters: K_S, μ_{max}, K_i

9.4.1 Kinetic Parameters and Their Estimation

Kinetic equations, which describe the activity of the enzymes or a microorganism on a particular substrate, are crucial in understanding many phenomena in biotechnological processes. Quantitative experimental data are required for the design and optimization of biological transformation processes. A variety of mathematical models have been proposed to describe the dynamics of metabolism of compounds exposed to pure cultures of microorganisms or microbial populations of the natural environment. The Monod equation has been widely used to describe growth-linked substrate utilization. Characterization of microbe–substrate interactions involves estimation of several parameters in the kinetic models from experimental data. In order to describe the true behavior of the system, it is important to obtain accurate estimates of the kinetic parameters in these models.

Different approaches have been proposed for estimating the kinetic parameters, but the progress curve analysis is the most popular because substrate depletion or product formation data from a single experiment are enough for parameter estimation. In this approach, substrate profile or product profile is used in the integrated form of the kinetic model for parameter estimation. It is important to note that most kinetic models and their integrated forms are nonlinear. This makes parameter estimation relatively difficult. However, some of these models can be linearized. Various linearized forms of the integrated expressions have been used for parameter estimation. However, the use of linearized expression is limited because it transforms the error associated with the dependent variable making it not to be normally distributed, thus inaccurate parameter estimates. Therefore, nonlinear least-squares regression is often used to estimate kinetic parameters from nonlinear expressions. However, the application of nonlinear least-squares regression to the integrated forms of the kinetic expressions is complicated. The parameter estimates obtained from the linearized kinetic expressions can be used as initial estimates in the iterative nonlinear least-squares regression. The kinetic parameters of Andrew's equation (μ_{max}, q_{max}, K_s, or K_i) can be estimated by the application of the reduced form of the generalized substrate inhibition model reduced to the form of Andrew's equation. The linearized expression of this model was used to obtain initial parameter estimates for use in nonlinear regression.

9.4.2 Calculation of Kinetic Parameters Using the Method of Least Squares

For calculating the kinetic parameter values from Equation 9.17, a numerical method approach viz. *method of least squares* is used, where the following substitutions are made

$$S \equiv p \tag{9.22}$$

$$\frac{1}{\mu} \equiv q \tag{9.23}$$

$$\frac{-K_i}{\mu_{max}} \equiv a \tag{9.24}$$

$$\frac{1}{\mu_{max}} \equiv b \tag{9.25}$$

$$\frac{K_S}{\mu_{max}} \equiv c \tag{9.26}$$

Using the above relations, Equation 9.2.1 can be rewritten in the form:

$$q = ap + b + \frac{c}{p} \tag{9.27}$$

Now, when a data set of p and q is available, the coefficients a, b, and c can be calculated by solving the following three equations simultaneously:

$$an + b\Sigma \frac{1}{p_i} + c\Sigma \frac{1}{p_i^2} = \Sigma \frac{q_i}{p_i} \tag{9.28}$$

$$a\Sigma p_i + bn + c\Sigma \frac{1}{p_i} = \Sigma_{qi} \tag{9.29}$$

$$a\Sigma p_i^2 + b\Sigma p_i + cn = \Sigma q_i p_i \tag{9.30}$$

Here, n is the number of data for p and q and i refers to the ith data of p or q.

TABLE 9.5

Summation of p_i, q_i, $1/p_i$, $1/p_i^2$, q_i/p_i, p_i^2, p_iq_i for Calculating the Values of a, b, and c

T	$S\,(\text{exp}) = p_i$	$1/\mu = q_i$	$1/p_i$	$1/p_i^2$	q_i/p_i	p_i^2	p_iq_i
1	9.9	5.33	0.10	0.01	0.54	98.01	52.80
2	9	4.67	0.11	0.01	0.52	81.00	42.00
3	8.6	7.60	0.12	0.01	0.88	73.96	65.36
4	8	5.17	0.13	0.02	0.65	64.00	41.33
5	7.7	10.86	0.13	0.02	1.41	59.29	83.60
6	7.3	27.33	0.14	0.02	3.74	53.29	199.53
7	6.4	18.40	0.16	0.02	2.88	40.96	117.76
8	5.7	17.33	0.18	0.03	3.04	32.49	98.80
Σ	62.6	96.69	1.05	0.14	13.66	503.00	701.19

Once values for a, b, and c are obtained, values of kinetic parameters K_S, μ_{max}, and K_i can be calculated by using Equations 9.24 through 9.26.

Using the method of least squares as given in Table 9.5, K_S, μ_{max}, and K_i are estimated from data given in Table 9.4 as shown below:

Substituting values in Equations 9.28, 9.29 and 9.30, we get

$$8a + 1.05b + .14c = 13.66$$

$$62.6a + 8b + 1.05c = 96.69$$

$$503a + 62.6b + 8c = 701.19$$

On solving for a, b, and c, we get

$$a = -2.78, \quad b = 23.18, \quad c = 81.16$$

Now, from Equations 9.24 through 9.26

$$K_S = 3.49 \text{ kg/m}^3, \quad \mu_{max} = 1.2 \text{ s}^{-1}, \quad K_i = 0.12 \text{ kg/m}^3$$

9.5 Cumulative Hydrogen Production by Modified Gompertz's Equation

The products from fermentative biohydrogen processes are broadly grouped into two major categories. The first category is the gaseous products (primarily H_2 and CO_2), while the second group includes volatile fatty acids and solvents. During hydrogen production by anaerobic fermentation, the distribution of the acidogenic products varies substantially such as acetic acid, butyric acid, ethanol, and so on.

9.5.1 Modified Gompertz's Equation

The modified Gompertz's equation has been extensively used to describe the progress of hydrogen production and some soluble metabolite production in a batch fermentative hydrogen production process. The equation can be expressed as

$$H(t) = P \cdot \exp\left\{-\exp\left[\frac{R_m e}{p}(\lambda - t) + 1\right]\right\} \tag{9.31}$$

where $H(t)$ represents cumulative volume of hydrogen production (m³), P the gas production potential (m³), R_m (m³/s) the maximum production rate, λ(s) the lag time, t incubation time (s), and e the exp(1) 2.718.

The R^2 value of the simulated data as shown in the Figure 9.7 was found equal to 0.996.

9.5.2 Modified Gompertz's Equation for Modeling Hydrogen, Butyrate, and Acetate Production

Mu et al. (2006) used the Modified Gompertz's equation to model the production of hydrogen as well as butyrate and acetate, during a batch hydrogen production process by mixed anaerobic culture.

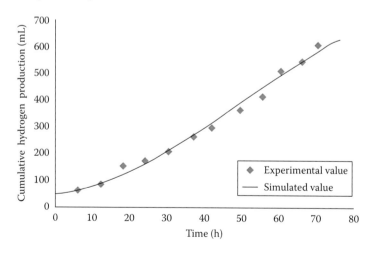

FIGURE 9.7
Cumulative hydrogen production curve during photofermentation of spent medium in stage 2, fitted by the Modified Gompertz's equation (volume of spent medium: 2.50×10^{-4} m³, light intensity = 5000 ± 500 lux, temperature = 310 K, initial pH: 6.8). (Adapted from Kumar, N. and Das, D. 2000. *Process Biochemistry*, 35, 589–593.)

TABLE 9.6

Kinetic Parameters of Cumulative Hydrogen Production for Different Initial Glucose Concentrations Calculated from Nonlinear Regression of Gompertz's Equation

Glucose Conc. (% w/v)	P (m^3 H$_2$)	R_m (m^3/s)
0.2	4.4×10^{-5}	2.69×10^{-9}
0.4	4.5×10^{-5}	2.36×10^{-9}
0.6	5.6×10^{-5}	2.77×10^{-9}
0.8	7.5×10^{-5}	0.64×10^{-9}
1.0	1.0×10^{-4}	3.86×10^{-9}
1.2	9.2×10^{-5}	4.88×10^{-9}
1.4	8.0×10^{-5}	3.94×10^{-9}

Source: Adapted from Kumar, N. and Das, D. 2000. *Process Biochemistry*, 35, 589–593; Nath, K. et al., 2008. *International Journal of Hydrogen Energy*, 33, 1195–11203.

$$P_i = P_{max,i} \exp\left\{-\exp\left[\frac{R_{max,i}e}{P_{max,i}}(\lambda - t) + 1\right]\right\} \qquad (9.32)$$

where i represents hydrogen, butyrate, and acetate, respectively; P_i is the product i formed per liter of reactor volume at fermentation time t; P_{max} is the potential maximum product i formed per liter of reactor volume; R_{max} is the maximum rate of product formed (Table 9.6).

A summary of recent studies where the modified Gompertz's equation has been used to model the cumulative production of biohydrogen is shown in Table 9.7.

TABLE 9.7

Comparative Studies on the Reported Values of R_m Using Modified Gompertz's Equation

Substrate	Mode of Operation	Microorganism	Substrate Conc.	Max. Rate of H$_2$ Production (R_m)	References
Glucose	Batch	*Enterobacter cloacae*	1.4 kg/m^3	3.93×10^{-9} m^3/s	Nath et al. (2008)
Sucrose	Batch	Mixed microflora	25 kg/m^3	2.94×10^{-9} m^3/s	Chen et al. (2006)
Lactose	Batch	Anaerobic Microflora	15 kg/m^3	1.89×10^{-9} m^3/s	Davila-Vazquez et al. (2008)
Starch	Batch	Mixed anaerobic culture	20 kg$_{COD}$/m^3	1.05×10^{-8} m^3/s	Lin et al. (2008)
Food waste	Batch	Mixed microflora	25 kg$_{COD}$/m^3	8.97×10^{-9} m^3/s	Chen et al. (2006)
Waste water	Continuous	Seed sludge	20 kg$_{COD}$/m^3	3.34×10^{-9} m^3/s	Jung et al. (2010)

9.5.3 Product Formation Kinetics by the Luedeking–Piret Model

The relationship between biomass and the products for the anaerobic hydrogen production by mixed anaerobic cultures can be modeled by the Luedeking–Piret model (Mu et al., 2006; Lo et al., 2008; Obeid et al., 2009):

$$\frac{dP_i}{dt} = \alpha_i \frac{dx}{dt} + \beta_i X \qquad (9.33)$$

where α_i is the growth-associated formation coefficient of the product i; β_i is nongrowth-associated formation coefficient of product i. Biohydrogen production is reported as growth associated production (Kumar et al., 2000) which may be represented as in Equation 9.34:

$$\frac{1}{x}\frac{dC_{H_2}}{dt} = \alpha\mu = \alpha\frac{1}{x}\frac{dX}{dt} \qquad (9.34)$$

where C_{H_2} represents H_2 concentration (mol), x represents cell concentration (kg VSS/m), and μ represents the specific growth rate (s⁻¹).

On dividing Equation 9.33, by x (cell concentration) for a specific product, we have

$$\frac{1}{x}\frac{dP}{dt} = \alpha\frac{1}{x}\frac{dx}{dt} + \beta \qquad (9.35)$$

Equation 9.35 can be written as

$$v = \alpha\mu + \beta \qquad (9.36)$$

where, $v = \dfrac{1}{x}\dfrac{dP}{dt}$, is the specific product formation rate, and μ is the specific growth rate of microorganism, as defined in Equation 9.2.

Depending on the values of α and β, three types of biohydrogen production patterns can be expected:

Growth-associated product formation ($\beta = 0$):
In this case, the specific product formation rate is directly proportional to the specific growth rate of microorganism (Figures 9.8, through 9.10).

Mixed growth-associated product formation ($\alpha \neq 0$, $\beta \neq 0$):
In this case, the specific product formation rate depends not only on the specific growth rate of microorganism but also a constant (β).

Nongrowth-associated product formation ($\alpha = 0$):
In this case, the specific product formation rate is independent of the specific growth rate of the microorganism.

Comparing the above plot with Figure 9.8, it can be established that biohydrogen production is a growth-associated process.

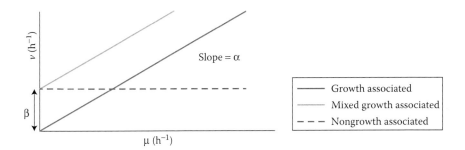

FIGURE 9.8
Plot of the specific product formation rate (v) versus specific growth rate of microorganism (μ).

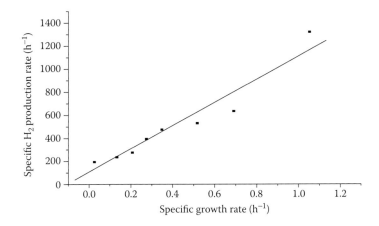

FIGURE 9.9
Plot of specific hydrogen production rate versus specific growth rate using the Leudeking–Piret model.

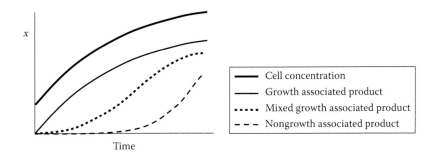

FIGURE 9.10
Plot of cell concentration and product formation versus time.

9.6 Development of Mathematical Models for Cell Growth Kinetics in Photofermentation Process

9.6.1 Logistic Equation

Logistic equation and its modified version have also been used to describe the microbial growth during light (as well as dark) fermentation process for biohydrogen production (Wang and Wan, 2008, 2009; Nath and Das, 2011). The logistic equation for cell concentration for a batch reactor can be represented as

$$X = \frac{X_0 \exp(k_c t)}{1 - \dfrac{X_0}{X_{max}}(1 - \exp(k_c t))} \tag{9.37}$$

where k_c is the apparent specific growth, X is the biomass concentration, and X_{max} is the maximum biomass concentration. The logistic model is a fair approximation of representing the entire growth curve including the lag phase (if present), the exponential growth, and the stationery phases even though a substrate term is not taken into consideration. The values of k_c are shown in Table 9.8.

9.6.2 Modified Logistic Model

Wang and Wan (2008) used the modified logistic model (Equation 9.37) to describe the progress of bacterial growth during fermentative hydrogen production by mixed cultures in batch tests using glucose as substrate.

TABLE 9.8

Kinetic Model Parameters of the Logistic Equation for Various Initial Acetic Acid Concentrations

Initial Acetic Acid Conc. (g/L)	X_{max} (kg/m³)	X_o (kg/m³)	k_c (s⁻¹)
2.50	0.439	0.057	2.75×10^{-5}
2.00	0.420	0.050	2.22×10^{-5}
0.95	0.396	0.045	1.47×10^{-5}
0.90	0.370	0.042	1.38×10^{-5}
0.55	0.311	0.061	1.25×10^{-5}
0.50	0.300	0.060	1.11×10^{-5}
0.35	0.256	0.038	1.50×10^{-5}
0.30	0.270	0.050	1.38×10^{-5}

Source: Adapted from Kumar, N. and Das, D. 2000. *Process Biochemistry, 35*, 589–593; Nath, K. et al., 2008. *International Journal of Hydrogen Energy, 33*, 1195–11203.

$$\int_0^t \frac{dX}{dt}\, dt = \frac{X_m}{1 + \exp\left[4\mu_m \dfrac{(\lambda - t)}{X_m} + 2\right]} \tag{9.38}$$

The coefficients of determination (R^2) of all the fittings were all over 0.99, which indicated that the modified logistic model could describe the progress of cumulative hydrogen production in the batch tests adequately.

9.7 Modeling of Hydrogen Production by Photofermentation

The cumulative H_2 production by photofermentation is simulated by modified Gompertz's equation, and the kinetic parameters for photo-H_2 production can be estimated by using computerized plotting applications.

9.7.1 Modified Gompertz's Equation

$$H(t) = P \cdot \exp\left\{-\exp\left[\frac{R_{m,e}}{p}(\lambda - t) + 1\right]\right\} \tag{9.39}$$

where $H(t)$ represents cumulative volume of hydrogen production (m^3), P is the gas production potential (m^3), R_m (m^3/s) is the maximum production rate, $\lambda(s)$ is the lag time, t is the incubation time (s), and e is the exp(1) 2.718.

9.7.2 Overall Biohydrogen Production Rate and Hydrogen Yield

The hydrogen production performance is assessed by maximum cumulative H_2 production, overall H_2 production rate, and H_2 yield. The definition of overall H_2 production rate and H_2 yield are given below:

$$\text{Overall } H_2 \text{ production rate} = \frac{\text{Maximum cumulative } H_2 \text{ production (m}^3)}{\text{Culture time for } H_2 \text{ evolution (s)}} \times \text{Culture volume (m}^3) \tag{9.40}$$

$$H_2 \text{ yield} = \frac{\text{Amount of } H_2 \text{ produced (mol)}}{\text{Amount of substrate consumed (mol)}} \tag{9.41}$$

9.7.3 Monod-Type Kinetic Model

The dependence of H_2 production rate on limiting substrate (starch) concentration in the presence of mixed anaerobic microflora was simulated by Lee et al. (2008) using a Monod-type kinetic model:

$$V_{H_2} = \frac{v_{\max H_2,} \cdot c_{\text{starch}}}{K_s + c_{\text{starch}}} \tag{9.42}$$

where $v_{\max H_2}$s is the maximum volumetric H_2 production rate (m³/m³ s), K_s is the saturation constant (kg COD/m³), and C_{starch} is the starch concentration (kg COD/m³). But many of the authors reported that classical Monod's model does not hold well for several instances such as inhibition due to high-substrate concentration, pH, presence of certain salt concentrations, substrate diffusion, maintenance, and so on.

9.7.4 Modification of Andrew's Model

Wang and Wan (2008) fitted Andrew's model to describe the biohydrogen production from glucose by mixed microflora using digested sludge with estimated correlation coefficient (R^2) 0.902.

$$V_{H_2} = \frac{67.1S}{47.7 + (s^2/13.5)} \tag{9.43}$$

where V_{H_2} is the volumetric H_2 production rate (m³/m³ s) and S is the substrate (glucose) concentration.

9.7.5 Generalized Monod-Type Model

A generalized Monod type originally proposed by Han and Levenspiel (1988) could probably better be used to account for substrate stimulation at low concentration and substrate inhibition at high concentration.

$$r = k(1 - (S/S_{\max}))^n \times \frac{S}{S + K_S - (1 - (S/S_{\max}))^m} \tag{9.44}$$

where r (m³/s) is the hydrogen production rate; k (m³/s) is the hydrogen production rate constant; S (kg/m³) is the substrate concentration; $S_{\max}$ (kg/m³) is the maximum substrate concentration above which the fermentative hydrogen production stops; K_S (kg/m³) is the saturation constant; m and n are the exponent constants the specific values of which determine the type of substrate inhibition such as noncompetitive, competitive, uncompetitive, and mixed inhibition.

9.8 Conclusion

Substrate consumption, increase in biomass, and product (biohydrogen) formation in the dark as well as photofermentation process can be predicted, with a significant level of accuracy using mathematical models. For the simulation of a process using a mathematical model, kinetic parameters must be estimated. Kinetic parameters K_S, μ_{max}, and K_i were estimated from the substrate inhibition model using the method of least squares. Regression analysis was carried out on experimental data and simulated values for Monod's and substrate inhibition model. R^2 values obtained for Monod's model were 0.987 and 0.979, and for the substrate inhibition model were 0.99 and 0.97 (substrate concentration profile and biomass concentration profile, respectively). Most of the models discussed in this chapter have reported values of $R^2 > 0.95$ indicating that these models are very reliable tools for process simulation. While studying the Luedeking–Piret model for product formation kinetics, it has been established that biohydrogen production is a growth-associated process indicating that the rate of biohydrogen production will increase with the increase in the growth rate of microorganisms. The modified Gompertz's equation was found to be a very accurate mathematical model for estimating the cumulative amount of biohydrogen produced. Several researchers have used it to model their processes and reported R^2 values around 0.99.

Nomenclature

μ	Specific growth rate of microorganism (s^{-1})
μ_{max}	Maximum specific growth rate of microorganism (s^{-1})
$H(t)$	Cumulative volume of hydrogen production (m^3)
I_{pH}	pH inhibition coefficient
K	Hydrogen production rate constant (m^3/s)
k_c	Apparent specific growth (s^{-1})
k_d	Biomass decay constant (s^{-1})
K_i	Substrate inhibition coefficient (kg/m^3)
K_s	Saturation constant (kg/m^3)
P	Gas production potential (m^3)
R	Hydrogen production rate (m^3/s)
R^2	Coefficient of determination
R_m	Maximum production rate (m^3/s)
S	Concentration of substrate (kg/m^3)
S_0	Initial substrate concentration (kg/m^3)
X	Cell mass concentration (kg/m^3)
X_0	Initial cell mass concentration (kg/m^3)

α_i Growth-associated formation coefficient of product i

β_i Nongrowth-associated formation coefficient of product i (s^{-1})

Λ Lag time (s)

N Specific product formation rate (s^{-1})

References

Andrews, J. F. 1968. A mathematical model for continuous culture of microorganisms using inhibitory substrates. *Biotechnology and Bioengineering, 10,* 707–723

Chen, W. H., Chen, S. Y., Khanal, S. K., and Sung, S. 2006. Kinetic study of biological hydrogen production by anaerobic fermentation. *International Journal of Hydrogen Energy, 31,* 2170–2178.

Das, D., Khanna, N., and Veziroglu, T. N. 2008. Recent developments in biological hydrogen production processes. *Chemical Industry & Chemical Engineering Quarterly* (CI & CEQ), *14,* 57–67.

Davila-Vazquez, G., Alatriste-Mondragón, F., de León-Rodríguez, A., and Razo-Flores, E. 2008. Fermentative hydrogen production in batch experiments using lactose, cheese whey and glucose: Influence of initial substrate concentration and pH. *International Journal of Hydrogen Energy, 33,* 4989–4997.

Han, K. and Levenspiel, O. 1988. Extended Monod kinetics for substrate, product, and cell inhibition. *Biotechnology and Bioengineering, 5,* 430–447.

Jung, K.-W., Kim, D.-H., and Shin, H.-S. 2010. Continuous fermentative hydrogen production from coffee drink manufacturing wastewater by applying UASB reactor. *International Journal of Hydrogen Energy, 35,* 13370–13378.

Kumar, N. and Das, D. 2000. Enhancement of hydrogen production by Enterobacter cloacae IIT-BT 08. *Process Biochemistry, 35,* 589–593.

Kumar, N., Monga, P. S., Biswas, A. K., and Das, D. 2000. Modeling and simulation of clean fuel production by *Enterobacter cloacae* IIT-BT 08. *International Journal of Hydrogen Energy, 25,* 945–952.

Lee, K.-S., Hsu, Y.-F., Lo, Y.-C., Lin, P.-J., Lin, C.-Y., and Chang, J.-S. 2008. Exploring optimal environmental factors for fermentative hydrogen production from starch using mixed anaerobic microflora. *International Journal of Hydrogen Energy, 33,* 1565–1572.

Lin, C.-Y, Chang, C.-C., and Hung, C.-H. 2008. Fermentative hydrogen production from starch using natural mixed cultures. *International Journal of Hydrogen Energy, 33,* 2445–2453.

Lo, Y.C., Chen, W.M., Hung, C.H., Chen, S.D., and Chang, J.S. 2008. Dark H$_2$ fermentation from sucrose and xylose using H$_2$-producing indigenous bacteria: Feasibility and kinetic studies. *Water Research, 42,* 827–842.

Monod, J. 1949. The growth of bacterial cultures. *Annual Review Microbiology, 3,* 371.

Mu, Y., Wang, G., and Yu, H.-Q. 2006. Kinetic modeling of batch hydrogen production process by mixed anaerobic cultures. *Bioresource Technology, 97,* 1302–1307.

Nath, K., Muthukumar, M., Kumar, A., and Das, D. 2008. Kinetics of two-stage fermentation process for the production of hydrogen. *International Journal of Hydrogen Energy, 33,* 1195–11203.

Nath, K. and Das, D. 2011. Modeling and optimization of fermentative hydrogen production. *Bioresource Technology, 102,* 8569–8581.

Ntaikou, I., Gavala, H. N., Kornaros, M., and Lyberatos, G. 2008 Hydrogen production from sugars and sweet sorghum biomass using Ruminococcus albus. *International Journal of Hydrogen Energy, 33,* 1153–1163.

Ntaikou, I., Gavala, H. N., and Kornaros, M. 2009. Modeling of fermentative hydrogen production from the bacterium Ruminococcus albus: Definition of metabolism and kinetics during growth on glucose. *International Journal of Hydrogen Energy, 34,* 3697–3709.

Obeid, J., Magnin, J. P., Flaus, J. M., Adrot, O., Willison, J. C., and Zlatev, R. 2009. Modeling of hydrogen production in batch cultures of the photosynthetic bacterium *Rhodobacter capsulatus*. *International Journal of Hydrogen Energy, 34,* 180–185.

Wang, J.-L. and Wan, W. 2008. The effect of substrate concentration on biohydrogen production by using kinetic models. *Scientific China Series B: Chemistry, 51,* 1110–1117.

Wang, J. L. and Wan, W. 2009. Factors influencing fermentative hydrogen production: A review. *International Journal of Hydrogen Energy, 34,* 799–811.

10

Scale-Up and Energy Analysis of Biohydrogen Production Processes

10.1 Introduction

Hydrogen is considered as a novel fuel for the 21st century, mainly due to its environmentally benign character. Production of hydrogen from renewable biomass has several advantages compared to that of fossil fuels. A number of processes are being practiced for efficient and economic conversion and utilization of biomass to hydrogen. Biohydrogen production from organic wastes is attractive due to two reasons: as a gaseous fuel production; and bioremediation. A limited number of pilot plant studies are reported and also little information on the economic analyses of biohydrogen production processes is available.

Biologically produced hydrogen is currently more expensive than other fuel options. Thus, if technology improvements succeed in bringing down costs, it is likely to play a major role in the economy in the future. Although rigorous techno-economic analyses are necessary to draw a cost-effective comparison between biologically produced hydrogen and the various other conventional fossil fuels, an economic survey, based on fuel cost estimation, turns out to be somewhat complicated when applied in practical terms. This is because of the intervening large number of other techno-economic parameters. The socially relevant costs of bringing any fuel to market must also include such factors as pollution and other short- and long-term environmental costs, as well as direct and indirect health costs. When these factors are taken into consideration, together with the initial cost competitiveness, hydrogen is surely the most logical choice as a worldwide energy source.

Hydrogen, the simplest and most abundant element, can be produced as a gas (H_2) from various resources and, when combusted, generates water as the only by-product. Its enabling technology, the H_2-based fuel cell, is twice as efficient as the traditional internal combustion engine currently used in automobiles. However, nearly all hydrogen gas being produced industrially comes from steam-reforming of natural gas, a fossil fuel. Thus, the development of alternative and renewable pathways for producing hydrogen fuels

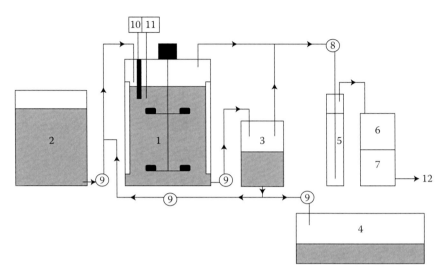

FIGURE 10.1
Schematic flow diagram of an industrial-scale biohydrogen production in continuous stirred tank reactor (CSTR): (1) reactor, (2) feed tank, (3) recycle tank, (4) drain tank, (5) CO_2 absorber, (6) flow meter, (7) sensor, (8) pressure control pump, (9) peristaltic pump, (10) pH-monitoring system, (11) thermostat, and (12) hydrogen storage section.

is of utmost importance. In the biological H_2 production process, the microorganisms can convert the high water-containing organic resources such as industrial effluents and sludge into useful hydrogen. So, biohydrogen production in a sense is an entropy-reducing process, which could not be realized by mechanical or chemical systems. Moreover, renewable biomass can be used as substrates for biological H_2 production facilitating both bioremediation and energy recovery.

Biohydrogen can be produced continuously both by using suspended and immobilized cells. Schematic diagram of the continuous biohydrogen production process using dark fermentation is shown in Figure 10.1. Figure 10.1 deals with the suspended cells and Figure 10.2 comprises the immobilized whole cell system.

10.2 Determination of Scale-Up Parameters

10.2.1 Geometric Similarity in Scale-Up

The geometry of the bioreactor is often the first consideration when scaling up a process, especially when fabrication is required. Maintenance of geometric similarity in a stirred vessel typically requires the conservation of a

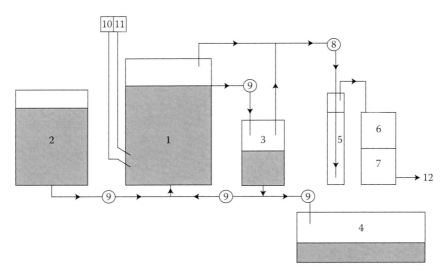

FIGURE 10.2
Schematic flow diagram of an industrial-scale biohydrogen production in pack bed reactor (PBR): (1) reactor, (2) feed tank, (3) recycle tank, (4) drain tank, (5) CO_2 absorber, (6) flow meter, (7) sensor, (8) pressure control pump, (9) peristaltic pump, (10) pH-monitoring system, (11) thermostat, and (12) hydrogen storage section.

set of ratios. This parameter, while being the first to be considered, is also unlikely to be completely conserved. If the process involves moving from a shaken flask (as is the case in the process defined in Section 10.2.3) to a bioreactor (stirred tank), geometric similarity cannot be conserved. Also if a large change in scale occurs, the aforementioned ratios are likely to change. Examples of the ratios concerned with maintaining geometric similarity within reactors are (Junker, 2004)

- Vessel height to diameter (H_v/D_v)
- Impeller diameter to vessel diameter (D_i/D_v)
- Vessel diameter to vessel volume (D_v/V)

The change in the above ratios, along with a possible change in the number, and design of the impellers within the vessel is designed to improve air utilization (in case of aerobic fermentation process) and decrease energy input as bioreactors increase in scale.

10.2.2 Scale-Up Based on Volumetric Power Consumption

Power consumption is an engineering characterization parameter and is a key parameter in biochemical engineering. Subsequently, it is a strong candidate for use as a criterion for bioreactor design and process scale-up. It is also referred to as volumetric power consumption (P/V) and is defined as the amount of energy required to generate movement of a fluid within a vessel in a given period of time. Apart from power drawn by the fluid relevant to the outcome of a given process, excess power is required to account for energy losses, mainly due to friction and power consumption by motors and gearboxes. This excess power is relevant in terms of overall power consumption but is not considered for design or scale-up of the process. The volumetric power consumption is representative of the turbulence degree and media circulation in vessels, and influences heat and mass transfer, mixing and circulation times. Constant volumetric power input was applied successfully as scale-up parameter for the early industrial penicillin fermentations (1 hp/gallon, equivalent to 1.8 kW/m³), and in fermentations with low-energy inputs, but it is limited in fermentations requiring high energy inputs, such as recombinant *E. coli* cultures, possibly due to the high associated costs and to high shear stress in the larger-scale stirred vessels.

The power input for the agitation of a nonaerated mixture, P_0 can be expressed in terms of the dimensionless variable power number (N_p) as

$$P_0 = \frac{N_p \rho N^3 D^5}{g} \tag{10.1}$$

where N is the stirrer speed, ρ is the density of the fluid, g is the conversion factor, and D is the stirrer diameter (Rushton et al., 1950a,b). The power number itself is dependent on other dimensionless groups such as the Reynolds number and the Froude number, as well as on the number of agitator turbines.

The following relationship for estimating the volumetric power consumption (P) has been suggested by Hughmark (1980):

$$\frac{P}{P_0} = 0.1 \left(\frac{N^2 D^4}{g W V^{\frac{2}{3}}} \right)^{-\frac{1}{6}} \left(\frac{Q}{NV} \right)^{-\frac{1}{4}} \tag{10.2}$$

where P_0 is the power consumption of nonaerated system, W is the width of turbine blades, and Q is the volumetric gas flow rate. Modifications to take into account the number and type of impellers and reactor dimensions for the improvement of this model were proposed over the years.

10.2.3 Volumetric Power Consumption in Agitated System

Büchs and Zoels (2011) measured the power consumption in shaken flasks at high- and low-medium viscosities. The experimental assembly was a simple rotary shaking machine fixed to a frame, combined with a torque sensor attached to the powering drive. Torque and shaking speed were monitored and correlated by the following relation:

$$\frac{P}{V} = Ne'\rho \frac{N^3 d^4}{V^{2/3}} = C_3\rho \frac{N^3 d^4}{V^{2/3}} Re^{-0.2} \qquad (10.3)$$

where Ne' is the modified Newton number of shake flasks, d is the impeller diameter, and C_3 is a fitting parameter. The correlation implies that the specific power consumption is dependent on the shaking frequency according to $P \approx N^{2.8}$. Such a correlation is typical of that found in unbaffled-agitated tank reactors. The model includes C_3 as the only fitting parameter using least-squares nonlinear fitting for the description of all the experimental results obtained in their study, having a value of 1.94.

The values for volumetric power consumption calculated by this empirical correlation and the experimentally measured values fit within a deviation range that does not exceed 30%. In the range 0.01–0.2 kW/m³, with shaking frequency from 1.33 to 6.33 s⁻¹, filling volumes of 4% to 20% of nominal flask volume and shaking diameter from 0.025 to 0.05 m, larger discrepancies are found between calculated and measured values. The corresponding data were gathered at a low-shaking frequency (1.33–2 s⁻¹) and, therefore, at the low Reynolds numbers (Re ≈ 500 to 5000), possibly within the transition from laminar to turbulent flow, a feature that could account for the increased deviation between predicted and experimental values.

10.2.4 Constant Impeller Tip Speed

The impeller tip speed, v_{tip}, is expressed as

$$v_{tip} = \pi ND \qquad (10.4)$$

where N is the number of revolutions per second, and D is the impeller diameter. Tip speed is an essential parameter to consider for scale-up since the relationship between shear and morphology has not been understood completely. It has been established through several observations that cell damage can occur at tip speeds above 3.2 m/s, but the exact value is influenced by many factors such as broth rheology. Calculated tip speeds are usually >3 m/s for production scale reactors. Although useful for estimating the potential for hyphae breakage and thus alteration of broth morphology when filamentous bacterial fermentations are involved, tip speed is less useful for single-cell bacterial fermentations. If scale-up is carried out using constant tip speed (with geometric similarity), then the value of P/V is often lowered. This drawback can be overcome by using more impellers in the larger vessel in such a way that both tip speed and P/V are kept constant. Tip speed influences impeller shear, which is proportional to the product of impeller tip speed and impeller diameter, NDi, for turbulent flow conditions.

10.2.5 Reynolds Number

The Reynolds number is expressed as

$$Re = \frac{\rho ND^2}{\mu} \tag{10.5}$$

where N is the number of revolutions per second, D is the impeller diameter, μ is the viscosity, and ρ is the density of the fluid. The Reynolds number is a dimensionless number that gives a numerical value to the amount of turbulence within the flow regime of a system (Table 10.1). Turbulence in this case is used as a measure of homogeneity within the bioreactor system. This parameter is not generally used during scale-up of bacterial fermentations as it does not account for the effect of aeration on the process and the deviation in geometric similarity during the process scale-up also affects the Reynolds number (Ju and Chase, 1992; Junker, 2004). The use of dimensionless groups in scale-up can lead to infeasible operating conditions when used in conjunction with other parameters.

10.2.6 Constant Mixing Time

The mixing time t_m denotes the time required for the reactor composition to achieve a specified level of homogeneity following the addition of a pulse at a single point in the vessel. Mixing time gives an idea about the flow and mixing within the reactor and can be useful for biosynthesis process scale-up. The mixing time t_m can be expressed as given below:

$$t_m = \frac{V}{N_f ND^3} \tag{10.6}$$

where N is the number of revolutions per second, D is the impeller diameter, N_f is the pumping number, and V is the volume of the reactor. The mixing time is

TABLE 10.1

Examples of the Reynolds Number of Different Scales and Agitation Rates

Bioreactor Geometry	N (rps)	Re
Microwell (10^{-6} m^3)	8.33	700
	12.5	1060
	16.67	1400
Shake flask (10^{-4} m^3)	5	106,800
Stirred tank (1.4×10^{-3} m^3)	11.67	24,720
	16.67	35,320

Source: Adapted from Micheletti, M. et al., 2006. *Chemical Engineering Science*, 61, 2939–2949.

typically measured in stirred vessels. Gerson et al. (2001) used a mixing probe to determine fluid mixing in a 10^{-3} m^3 shaken flask with filling volume 5.4×10^{-4} m^3 at different shaking frequencies. The results demonstrated that it is possible to make such measurements, and that the mixing intensity rises monotonically with shaking frequency in both stirred reactors and shake flasks.

Nealon et al. (2006) developed a high-speed video technique for the accurate quantification of jet macromixing times in static microwell plates. Three microwell geometries were investigated: a single well from a standard 96-round well plate and dimension modified 96-well plate. A general correlation was found for the time to reach 95% homogeneity:

$$t_{95} = \frac{2.6D^{1.5}h^{0.5}}{u_0 d_n} \tag{10.7}$$

where d_n is the nozzle diameter, D is the well or vessel diameter, h is the liquid height, and u_0 is the nozzle velocity.

10.2.7 Superficial Gas Velocity and Volumetric Gas Flow per Unit of Liquid

Superficial gas velocity is described as the ratio of the volumetric flow of the gas into the vessel and the vessel cross-sectional area. Maintaining the superficial gas velocity maintains gas hold-up and prevents gas flooding within the vessel. When scaling up using this criterion, the superficial gas velocity must also be addressed as increasing the VVS with increasing scale can lead to flooding, impeller overloading, and excessive foaming in industrial size equipment. This parameter is effectively used to scale-up the processes that do not have mechanical agitation.

10.3 Scale-Up Methods

10.3.1 Significance of Scale-Up

The purpose of scaling up is to obtain conditions similar to those inside a laboratory scale reactor, in an industrial scale reactor. The conditions are pH, temperature, homogeneity, and agitation among others. To do this, some of the parameters associated with the process or the fermentor need to be altered while other parameters may be kept constant. It is obvious that the volume of the vessel will be magnified since the process is a scale-up. This magnification in volume has to be met with appropriate changes in the other parameters.

The following case can serve as an example as to how some parameters vary with scale-up of the process. The process was carried out initially at laboratory scale (5×10^{-4} m^3) and later scaled up to a 0.02 m^3 pilot-scale version.

Both batch and continuous operation was performed at the lab scale, while the pilot scale was a continuous operation.

10.3.2 Laboratory Scale Study

10.3.2.1 Batch Fermentation

The batch experiments were carried out in a double-jacketed reactor made up of glass with working volume of 5×10^{-4} m³ (Figure 10.3). The maximum yield of biohydrogen achieved in batch process was 8.23 mol H_2/kg $COD_{removed}$ with 10 kg COD/m³ of molasses, 1% (w/v) of malt extract, 0.4% (w/v) of yeast extract at a temperature of 310 K and pH of 6.5 ± 0.2. Cane molasses concentration of 10 kg COD/m³ was supplemented with 1% (w/v) of malt extract and 0.4% (w/v) of yeast extract in order to supply the bacteria with adequate nitrogen, phosphorus, and growth factors. After supplementation, the production media contained 22.4 kg COD/m³. The maximum cumulative hydrogen production and yield were 1.1×10^{-3} m³ and 8.23 mol H_2/kg $COD_{removed}$, respectively. The total COD and carbohydrates reduction were 53% and 69%, respectively. The hydrogen yield was increased 1.5 times higher than production media without supplementation.

10.3.2.2 Continuous Fermentation

The continuous experiments were carried out in a double-jacketed reactor made up of glass with working volume of 5×10^{-4} m³. The bioreactor was

FIGURE 10.3
(**See color insert.**) Batch experimental set up with double-jacketed reactor for the biohydrogen production.

packed with coconut coir as carrier material to a packing density 50 kg/m³. After packing, the void volume of the reactor was 4.5×10^{-4} m³. To immobilize the whole cells, 30 kg/m³ of cells suspension in 0.1 M phosphate buffer pH 6 was passed through the packed column at a dilution rate of 5.55×10^{-5} s⁻¹ with the help of a peristaltic pump. The bioreactor was thereafter kept alone for 12 h for cells to stabilize. The production media was passed inside the reactor after stabilization of cells on the matrix. The immobilized whole cells were immersed in production media till the rate of gas production reached its maximum and became steady. This was followed by passing the production medium at different dilution rates. The maximum rate of hydrogen production and yield were 3.47×10^{-4} m³/m³ s and 11.6 mol H_2/kg $COD_{removed}$, respectively, with COD reduction of 47% at a dilution rate of 1.67×10^{-4} s⁻¹. The hydrogen yield was increased 1.4 times higher than the batch study.

10.3.3 Scale-Up Study

10.3.3.1 Experimental Setup for Continuous Hydrogen Production

Organism: Immobilized *Enterobacter cloacae* IIT-BT 08. The bioreactor was packed with pretreated lignocellulosic solid matrices (Figure 10.4).

FIGURE 10.4
Continuous hydrogen production using immobilized whole cell (*E. cloacae* IIT BT 8) in 20 L bioreactor.

TABLE 10.2

Yield of H_2 at Different Dilution Rates in the 0.020 m³ Bioreactor Using 1% w/v Cane Molasses

Dilution Rate (s⁻¹)	Rate of Hydrogen Production (mol H_2/m³ s)	Amount of $COD_{reduced}$ (%)	Yield (mol H_2/ kg $COD_{reduced}$)
5.5×10^{-5}	0.0091	53.7	2.54
8.3×10^{-5}	0.0228	49.3	7
1.38×10^{-4}	0.0331	46.3	10.66
1.66×10^{-4}	0.0342	39.4	12.7
1.94×10^{-4}	0.0305	39	11.5
2.22×10^{-4}	0.0272	37	10.8
2.5×10^{-4}	0.0244	35	10.2

Production media: Malt extract, 1% (w/v), yeast extract, 0.4% (w/v), and cane molasses, 1% (w/v).

Vessel: Total volume −0.020 m³, Void volume −0.014 m³.
Temperature: 310 K.
pH: 6.5 ± 0.2.

To immobilize the whole cells, 50 kg/m³ of cell suspension in 0.1 M phosphate buffer (pH 6) was passed through the packed column at a dilution rate of 5.55×10^{-5} s⁻¹ with the help of a peristaltic pump for 8 h. Thereafter, the reactor was kept at rest for 12 h in 4°C. The immobilized cells were immersed in production media (for about 8–10 h) till the gas production rate reached its maximum and became steady. In the scale-up studies, hydrogen yield was increased to 12.7 mol H_2/kg $COD_{removed}$ (about 9% increase) (Table 10.2 and Figure 10.5).

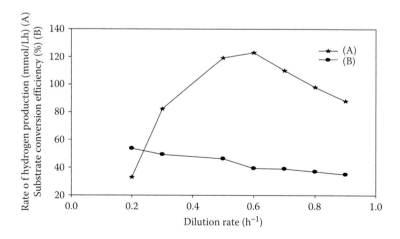

FIGURE 10.5
Plot of hydrogen production versus dilution rate.

10.4 Case Studies on Pilot-Scale Plants

10.4.1 Case I: Pilot-Scale Plant Using Mixed Microflora at Feng Chia University, Taiwan

Features of pilot plant system (Lin et al., 2011):

- Feedstock storage tanks (2): Capacity = 0.75 m^3
- Feedstock: Synthetic wastewater containing pure carbon sources (glucose, xylose, and sucrose)
- Nutrient storage tank: Capacity = 0.75 m^3
- Mixing tank: Volume = 0.6 m^3
- Agitated granular sludge bed fermentor: Working volume = 0.4 m^3
- Gas–liquid–solid separator: Volume = 0.4 m^3

Operational conditions:

- Duration = 67 d
- Temperature = 308 K
- Organic-loading rate (OLR) = 46-278 × 10^{-5} kg COD/m^3 s
- Influent sucrose concentration = 20 and 40 kg COD/m^3
- Fermentor agitation speed = 0.17–0.25 rps

10.4.1.1 Microbial Community

The seed mixed microflora was obtained from a lab-scale agitated granular sludge bed bioreactor. *Clostridium pasteurianum*, *Bifidobacteria* sp., and *Clostridium tyrobutyricum* were the significant bacterial strains present in the fermentors.

10.4.1.2 Operation Strategy and Hydrogen Production in the Fermentor

At the beginning, 0.06 m^3 of cultivated seed sludge and 0.3 m^3 of substrate and nutrient solution were mixed in the fermentor using a batch strategy to grow the biomass. The mixed liquid in the fermentor was first purged with argon gas for 600 s to ensure an anaerobic condition prior to H$_2$ fermentation experiments. The system was then switched to a continuous-feeding mode (HRT 4.3 × 10^4 s) after 48 h in batch operation. After reaching steady-state condition (based on a constant volumetric H$_2$ production rate with a variation of within 10% for 2.6–4.3 × 10^5 s), the HRT was reduced progressively from 4.3 to 1.4 × 10^4 s.

This pilot-scale fermentor was fed with sucrose (organic loading rate (OLR) from 46–278 × 10^{-5} kg COD/m^3 s and concentrations of 20 and 40 kg COD/m^3) and operated at 308°C and 0.25 rps agitation. The nutrient solution contained the same nutrients (Endo formulation) used for the seed sludge. A varied NaHCO$_3$ concentration (2.62–5.24 kg/m^3) was used for maintaining the pH to 6.0.

During the experiments, the quantity and composition of gas products (mainly, CO_2 and H_2) and the soluble microbial products (SMP) were monitored at a designated time interval. The hydrogen yield was calculated based on the moles of H_2 produced per kg of sucrose consumed.

Hydrogen production rate and yield:

- Yield of hydrogen production: 1.04 mol H_2/mol sucrose
- Hydrogen production rate (volume/volume): 1.8×10^{-4} m³/m³ s

Energy efficiency of the pilot plant:
Energy consumption (or energy input) occurred due to the following:

- Agitator motors for substrate tank, nutrient tank, and fermentor
- Feeding pump
- Temperature control system
- Compressor power
- Electricity needed for the control center panel

The heating value of hydrogen is 285.8 kJ/mol, at the normal condition ($T = 293$ K and $P = 1.01 \times 10^5$ Pa). The energy output was obtained by detecting the hydrogen production rate and using the following relation:

$$\text{Energy output} = \text{HPR} \times \frac{\text{Fermenter volume}}{\text{Volume of 1 mole of } H_2} \times \text{Heating value of } H_2 \text{ (per mole)} \tag{10.8}$$

The energy factor (E_f) was calculated using

$$E_f = \frac{\text{Energy output}}{\text{Energy input}} \tag{10.9}$$

and was found to lie in the range 13.65–28.68.

10.4.2 Case II: Pilot-Scale Plant Using Distillery Effluent to Produce Biohydrogen

Features of pilot plant system (Vatsala et al., 2008):

- Volume: 100 m³.
- The lab-scale process of H_2 production was then scaled up from 10^{-3} m³ to 100 m³.

- A sequence of bioreactors was constructed for inoculum preparation for the 100 m^3 reactor. The total volumes of the reactors were 0.125, 1.25, 12.5 and 125 m^3, and the working volumes were set from 0.1 to 100 m^3.
- The height to diameter (H/D) ratio of each reactor was set to 1.28.
- Feedstock: Distillery effluent.

Operational conditions:

- Duration: 40 h
- Temperature: Not controlled
- pH: Not controlled
- Initial effluent temperature: 333 K
- Initial COD: 101.2 kg/m^3
- Initial BOD: 58.8 kg/m^3

10.4.2.1 Microbial Community

Heterotrophic bacterial strains (HTB) were isolated from the effluent treatment plant, EID Parry (I) Ltd., Nellikuppam, Tamil Nadu, India. Photoheterotrophic bacterial strains (PTB) were isolated from pond soil, Chengalpet, Tamil Nadu, India. Based on the H$_2$-producing ability, only three strains were subjected to phenotypic and genotypic characterization. The isolated strains HTB01, HTB10, and PTB2 were identified and denoted as *C. freundii* C01, *E. aerogenes* E10, and *R. palustris* P2, respectively.

   Hydrogen production and yield:

- 21.38 kg of H$_2$, corresponding to 10,692.6 mol, which was obtained through batch method from reducing sugar (3862.3 mol) as glucose.
- The average yield of H$_2$ was 2.76 mol/mol$_{glucose}$.
- The rate of H$_2$ production was estimated to be 1.47×10^{-6} kg/m^3 s.

10.4.3 Case III: Biohydrogen Production from Molasses by Anaerobic Fermentation with a Pilot-Scale Bioreactor System

Features of pilot plant (Ren et al., 2006):

- Volume of the reactor: 2 m^3 anaerobic hydrogen bioproducing reactor (HBR) with an effective volume of 1.48 m^3.
- Feedstock: normal molasses, containing about 53% sugars, diluted by water to certain loading rate (3.6–99×10^{-5} kg$_{COD}$/m^3 s).

Operational conditions:

- Duration: Over 200 days.
- Temperature: The temperature was automatically maintained at the level of 308 $K \pm 1$.
- pH: Influent pH was kept above 7 by adding $CaCO_3$ in order to prevent low pH (<4).
- OLR: the reactor was operated at a high organic loading rate (OLR > 3.5×10^{-5} $kg_{COD}/m^3_{reactor}$ s).
- Start-up OLR: 7.31×10^{-5} $kg_{COD}/m^3_{reactor}$ s with a HRT of 4.1×10^4 s substrate concentration of 3 kg_{COD}/m^3.

10.4.3.1 Microbial Community

The seed sludge in the HBR was collected from a local municipal wastewater treatment plant. During the operation of the reactor, the microbial profile was not analyzed. However, *Clostridium* spp., *Enterobacter aerogenes* and *Ectothiorhodospira vacuolata*, were expected to be the major contributors to the hydrogen production process.

Hydrogen production rate (HPR):

- A maximum hydrogen production rate of $6.45 \times 10^{-5} m^3$ $H_2/m^3_{reactor}$ s, with a specific hydrogen production rate of 8.68×10^{-6} m^3 H_2/kg_{MLVSS} s, was obtained in the reactor.
- The hydrogen yield reached 26.13 mol/$kgCOD_{removed}$ within OLR range $40–64 \times 10^4$ $kgCOD/m^3_{reactor}$ s.

Important observations during operation of the reactor:

- Both hydrogen yield and specific hydrogen production rate increased when OLR ranged from 6.32 kg COD/m^3 day to 79×10^{-5} kg COD/m^3 s, during which the maximum hydrogen yield was 9.5×10^{-5} m^3/s (at OLR of 79×10^{-5} kg COD/m^3 s).
- Hydrogen production decreased when the OLR was more over 79×10^{-5} kg COD/m^3 s, due to the accumulation of VFAs in the HBR system.
- The hydrogen production rate was affected by the characteristics of liquid fermentation products (such as ethanol and acetate), and the maximum hydrogen yield was obtained while the ratio of ethanol to acetate was about 1, due to the adjustment of $NAD^+/NADH + H^+$ through fermentation pathways.

TABLE 10.3

Comparison of Features of Pilot-Scale Plants for Biohydrogen Production

Features/Parameters	Case I	Case II	Case III
Reactor volume (m³)	0.4	100	1.48
Feed	Synthetic wastewater containing pure carbon sources	Distillery effluent	Normal molasses, containing about 53% sugars
Duration of process	5.8×10^6 s	1.4×10^5 s	$>1.7 \times 10^7$ s
Mode of operation	Continuous	Batch	Continuous
Temperature	308 K	Not controlled	308 ± 1 K
Initial COD	20 and 40 kg COD/m³	Initial COD: 101.2 kg/m³	3.11–85.57 kgCOD/m³
OLR	46-278×10^{-5} kg COD/m³ s	N/A	$>3.5 \times 10^{-5}$ kgCOD/$m^3_{reactor}$ s
Microbial community	*Clostridium pasteurianum, Bifidobacteria* sp. and *Clostridium tyrobutyricum*	*C. freundii E. aerogenes* and *R. palustris*	*Clostridium* spp., *Enterobacter aerogenes* and *Ectothiorhodospira vacuolata*
Yield	1.04 mol H_2/mol sucrose	2.76 mol H_2/mol glucose	26.13 mol H_2/kgCOD$_{removed}$
HPR	1.8×10^{-4} m3/m³s	1.47×10^{-6} kg/m³ s.	8.68×10^{-6} m³ H_2/kg$_{MLVSS}$ s

- Ethanol-type fermentation was favorable for hydrogen production.
- High concentration of biomass in the pilot-scale HBR system improved hydrogen yield to a considerable degree, due to a good settleability and compactness of anaerobic-activated sludge.
- The hydrogen production rate of maturated anaerobic-activated sludge was as high as 8.68×10^{-6} m³ H_2/kg$_{MLVSS}$ s.

10.4.4 Comparative Study among Different Pilot-Scale Plants

Comparative studies on the above three pilot plants are given in Table 10.3.

10.5 Mass and Energy Analysis

Biohydrogen is a gaseous fuel with great potential that is highest gravematric calorific value. In this section, mass and energy calculations have been performed on the overall biohydrogen production process.

10.5.1 Material Balance of the Biohydrogen Production Process

In a hybrid reactor system, that is, a reactor where dark fermentation and photofermentation, the following set of chemical reactions, can be used to depict the hydrogen production process:

Stage 1 (Dark fermentation)

$$C_6H_{12}O_6 + 2H_2O \rightarrow 4H_2 + 2CH_3COOH + 2CO_2$$

Stage 2 (Photofermentation)

$$CH_3COOH + 2H_2O + light \rightarrow 4H_2 + 2CO_2$$

The above is a two-stage process performed in order to improve the overall yield of hydrogen production. Stage 1 is the anaerobic fermentation of glucose by fermentative bacteria, which is followed by stage 2, the photofermentation of acetic acid by anoxygenic phototrophic bacteria (Nath and Das, 2004). Theoretically, 1 mol of glucose should yield 4 mol of H_2. However, the highest H_2 yield reported by Morimoto et al. (2004) and Kumar et al. (2001) are 2.0–2.4 and 3.85 mol H_2/mol glucose, respectively. Production of butyrate rather than acetate may partially explain the deviations from the theoretical yield. Utilization of substrate as an energy source for bacterial growth is another main reason for obtaining the yields lower than theoretically predicted. The following is the material balance of a pilot-scale biohydrogen production plant (Section 10.4.2, Case II).

More than 98% of the carbon input was recovered as various metabolites as shown in Table 10.4. Succinate and ethanol were the major end products of 22% of the carbon spent by *E. aerogenes* E10.

10.5.2 Energy Analysis of Biohydrogen Production Process

The energy data from the fermentation process of wastewater containing cane molasses for biohydrogen in a batch process are presented in Table 10.5, and the GER and NUEP for the process have been calculated.

Net utilizable energy product (NUEP) is expressed as

$$NUEP = \frac{\text{Externally utilizable energy} - \text{Secondary energy input}}{\text{Primary energy available (raw material)}} \times 100$$

(10.10)

From Table 10.5, it has been found that the gross energy requirement (GER) is found equal to 15.7% of energy recovered as hydrogen. This comprises energy required for mixing, heating, and pumping. The NUEP was estimated to be 82.84%.

TABLE 10.4

Anaerobic Fermentation of Glucose and the Yield of H_2 and Other Carbon Metabolites by *E. aerogenes* E10 and Carbon Material Balance of the Process

Reactant/Product	Concentration (mol/m³)	Carbon (%)
Glucose	55.50	100
H_2	83.00	0
CO_2	64.83	19.03
Ethanol	38.5	22.60
2,3-Butanediol	2.85	3.35
Formate	25.80	7.57
Succinate	18.70	21.95
Lactate	4.45	3.92
Acetate	23.00	13.50
Biomass	—	6.32
Total		98.24

Source: Adapted from Vatsala, T.M., Mohan Raj, S., and Manimaran, A. 2008. *International Journal of Hydrogen Energy*, 33, 5404–5415.

TABLE 10.5

Energy Production and Requirement for Biohydrogen Production

Parameters	Energy Calculations
Gross energy of product as biohydrogen (GJ)	1.277
Energy content in substrate (GJ)	1.3
Total process energy required (GJ)	0.20
GER	0.157
NUEP (%)	82.84

Source: Adapted from Nayak, B.K., Pandit, S., and Das, D. 2013. Biohydrogen. In *Air Pollution Prevention and Control*, eds. C. Kennes, and R. C. Veiga, pp. 345–381. John Wiley & Sons Ltd.
Basis: 100 m³ biohydrogen production/day.

10.5.3 Biological Route versus Chemical Route

The energy requirements are computed for the production of hydrogen using the biological route and the chemical route (steam reforming). Steam reforming is a method for hydrogen production or other useful products from hydrocarbon fuel. This is achieved in a processing device called a reformer which reacts steam at high temperature with the fossil fuel. The following chemical reaction represents the steam-reforming of methane:

$$CH_4 + H_2O \xrightarrow{1000°K} CO + 3H_2$$

$$CO + 3H_2 + H_2O \xrightarrow{700°K} 4H_2 + CO_2$$

TABLE 10.6

Comparison of Energy Production and Requirement during Production of
Hydrogen Via the Chemical and Biological Route

Parameter	Chemical Route	Biological Route
Total energy content of product (GJ)	1.277	1.277
Energy content in substrate (GJ)	0.99	1.3
Total process energy (GJ)	0.4645	0.20
GER	0.3637	0.157
NUEP (%)	82.07	82.84

Source: Adapted from Das, D. 1985. Biomethanation of mixed agricultural residues, Ph.D.
Thesis, IIT, Delhi; Blok, K. et al., 1997. *Energy,* 22, 161–168.
Basis: 100 m³ hydrogen production/day.

The biological hydrogen production process is more promising as compared to the chemical process (Table 10.6).

10.5.4 Electrolysis of Water versus Biological Route for Hydrogen Production

Electrolysis of water is the decomposition of water (H_2O) into oxygen (O_2)
and hydrogen gas (H_2) due to an electric current being passed through the
water. Decomposition of pure water into hydrogen and oxygen at standard
temperature and pressure is not favorable in thermodynamic terms.

$$\text{Anode (oxidation): } 2\ H_2O(l) \rightarrow O_2(g) + 4\ H^+(aq) + 4e^-\ Eo_{ox} = -1.23\ V$$
$$\text{Cathode (reduction): } 2\ H^+(aq) + 2e^- \rightarrow H_2(g)\ Eo_{red} = 0.00\ V$$

Electrolysis will not generally proceed at these voltages, as the electrical input must provide the full amount of enthalpy of the $H_2 - O_2$ products
(286 kJ per mol). If a basis of 100 m³ of hydrogen gas is considered, the electrical energy required to perform electrolysis comes out to be around 2.5 GJ
while the biological route requires only 0.20 GJ of energy. This indicates that
the process energy requirement for electrolysis is about 10 times higher than
that required via the biological route. Also, it should be noted that energy
efficiency of the electrolysis process is lower since half the electrical energy
supplied to the cell is dissipated as heat.

10.6 Cost Analysis of the Process

10.6.1 Hydrogen as a Commercial Fuel

The suitability of any process lies in its capacity to be scaled up to the industrial level. Several large-scale processes exist for the production of hydrogen.
However, hydrogen does not find its place as a conventional fuel mainly due

TABLE 10.7

Various Methods of Hydrogen Production and Their Cost

Source and Process (Large-Scale Technology)	Cost of Hydrogen
Natural gas (produced via steam reforming at fueling station)	$4–5/kg
Wind (via electrolysis)	$8–10/kg
Nuclear (via electrolysis)	$7.50–9.50/kg
Nuclear (via thermochemical cycles)	$6.50–8.50/kg
Solar (via electrolysis)	$10–12/kg
Solar (thermochemical cycles)	$7.50–9.50/kg
Wastewater (dark fermentation)	$1.3/MBTU
Gasoline	$23.5/MBTU
Natural gas	$7/MBTU

Source: Adapted from Das, D., Khanna, N., and Veziroğlu, N.T. 2008. *Chemical Industry & Chemical Engineering Quarterly* (CI & CEQ), 14, 57–67.

to two reasons: (a) expensive production and (b) insufficient technology to harness the energy evolved in the combustion of hydrogen. Table 10.7 provides a list of technologies being used for large-scale production of hydrogen and the cost per kilogram of the hydrogen produced.

10.6.2 Cost Calculation of Continuous Biohydrogen Production Process Using Cane Molasses

The scale-up studies were conducted to explore the suitability of cane molasses for continuous bioH$_2$ production using a 0.020 m^3-immobilized anaerobic bioreactor (Das, 2009). The system was equipped with an indigenously developed "automated-logic-control-system" to operate at reduced partial pressure of H$_2$ ranging from 0.5×10^5 to 1.01×10^5 Pa. Organic loading rates (OLR) were varied from 1.38×10^{-3} to 8.33×10^{-3} kg COD/m^3 s and corresponding rate of hydrogen production was recorded between 3.33×10^{-5} and 2.3×10^{-3} m^3 H$_2$/m^3 s. The H$_2$ content of the gas fluctuated from 58% to 66% v/v. Maximum-specific hydrogen production rate of 2.58×10^{-5} m^3 H$_2$/kg VSS s was achieved at a dilution rate of 1.89×10^{-5} s^{-1} and recirculation ratio of 6.4. This rate, however, was observed to be inversely proportional to specific OLR when dilution rate was more than 1.94×10^{-5} s^{-1}. The maximum hydrogen yield was 17.94 mol H$_2$/kg COD$_{removed}$ at an OLR of 5.55×10^{-3} kg COD/m^3 s, which was comparatively higher than earlier reported values. A comparative study on different scale-up processes is shown in Table 10.8.

10.6.2.1 Cost Analysis

Maximum yield of H$_2$ in the process = 0.0454 m^3/kg COD reduced (Kotay, 2008)

TABLE 10.8

Comparative Analysis on the Bench Scale or Pilot Plant Operations Reported for Biohydrogen Production

Substrate	Organism	Reactor Capacity	Days of Operation	Yield/HPR	References
Wastewater from citric acid	*Clostridium pasteurianum*	50 m³ UASB	10–15	8.1×10^{-6} m³ $H_2/m^3_{reactor}$ s	Yanga et al. (2006)
Molasses	Anaerobic-activated sludge	1.48 m³	200	26.1 mol/ $kgCOD_{removed}$	Ren et al. (1997)
Cane molasses	*Enterobacter cloacae* DM 11	0.02 m³ immobilized packed bed reactor	20	17.9 mol H_2/ $kgCOD_{removed}$	Kotay (2008)

Basis: Considering sewage sludge as raw material for the biohydrogen production (Das, 2009).

Average COD value of the sludge = 88 kg/m³

COD reduction = 46%

Heating value of H_2 = 142,000 kJ/kg of H_2

Heating value of sewage sludge = 18,640 kJ/kg

H_2 yield = 0.0454 m³/kg COD reduced = 45.4460 mL/kg sludge = 0.0019 kg H_2/kg sludge

Energy recovery from the substrate = {(heating value of H_2 × yield of H_2)/(heating value of sewage sludge)} × 100 = {(142,000 × 0.0019)/18,640} × 100 = 1.44%

Amount of gaseous energy generated as hydrogen = 142,000 × 0.0019 = 270 kJ/kg sludge

Total volume of sludge produced in India = ~3 × 10⁻³ million m³/day

Total amount of sludge produced per day = 3 × 10⁶ × 0.088 kg

Therefore, total amount of energy produced as H_2 per day = 3 × 10⁶ × 0.088 × 270 kJ = 1200 MBTU (provided all the activated sludge is devoted for hydrogen biogeneration).

This value can become further promising, considering the amount of sludge that would be generated if the % of sewage treated increases, in other words with the increase in number of STPs there would be exponential increase in the amount of sludge produced and subsequently the hydrogen that can be produced from it.

Dilution rate = 1 h⁻¹

HRT of the process = 1 h

Volume of the reactor = 150 m³ (considering 20% head space)

Cost of the reactor and accessories = 150×1000 (at the rate of \$1000/m³) = \$150,000

Life of the reactor = 10 year

Assuming the cost of sludge = 0

Money to be returned/day = \$45 (capital + interest)

Labor/recurring expenditure = \$1500/day

Total expenditure = \$1545/day

Cost of hydrogen = \$1545/1200 MBTU = \$1.3/MBTU

Cost of natural gas (2013, India) = \$8.4/MBTU

Cost of gasoline 2013 = \$3.57/gallon

Heat of combustion of gasoline = 47,000 kJ/kg = 44,407 BTU/kg = 0.17 MBTU/gallon

Cost of gasoline = \$20.9/MBTU

There are scopes to use sugarcane juice, molasses, biomass, or distillery effluent as substrates, because they contain carbohydrates in significant quantities. Therefore, production as well as the unit energy cost of bioH$_2$ is to be determined. However, a rigorous techno-economic analysis is necessary to draw a cost-effective comparison between biologically produced hydrogen and the various other conventional fossil fuels. But economic survey, based on fuel cost estimation, turns out to be somewhat complicated when applied in practical terms. This is because of the intervening large number of other techno-economic parameters.

The calculation indicates that the unit cost of biologically produced hydrogen energy is more or less three times higher as compared to gasoline, when we consider pure sugar as substrate. As a matter of fact, this difference would appear less in magnitude if social and environmental costs as a result of using fossil fuels are contemplated. The socially relevant costs of bringing any fuel to market must also include such factors as pollution and other short- and long-term environmental costs, as well as direct and indirect health costs. When these factors are taken into consideration, together with its initial cost competitiveness, H$_2$ is surely the most logical choice for a worldwide energy medium. Table 10.9 gives an estimate of the revenues that can be generated from a large-scale (100 m³) wastewater to biohydrogen plant.

10.7 Conclusion

The purpose of scientific research is to realize the application of the concepts discovered and understood from it. Even though numerous experiments on

TABLE 10.9

Revenue from Biohydrogen Production Process Using Wastewater in a 100 m³ Reactor

Sales	Revenue (1000 $)
Revenue from H_2 sales	160
Revenue from CO_2 sales	24
Annual total revenue	184
Annual total cost for the system	100
Annual administration costs	3
Annual total profit	81
Return on investment (%)	81

Source: Adapted from Li, Y.-C. et al. 2012. *International Journal of Hydrogen Energy, 37,* 15704–15710.

a lab scale have been performed in biohydrogen production, industrial scale production of hydrogen is not established yet. The results of this method on a small scale are very promising and an industrial version of the process will become the much-needed alternative energy source in the future. The quest for scalingup of laboratory set-ups to large-scale reactor will continue until an economically efficient system is developed that remediates both the energy and organic waste treatment concerns.

Nomenclature

C_3 Fitting parameter (m⁻³)

D Stirrer diameter (m)

D_v Diameter of vessel/reactor (m)

H_v Height of vessel/reactor (m)

N Stirrer speed (m/s)

Ne' Modified Newton number for shake flasks (m⁻³)

N_f Pumping number

N_p Power number

P_0 Power input for agitation of a non-aerated mixture (W)

Q Volumetric gas flow rate (m³/s)

S_{max} Maximum substrate concentration above which hydrogen production stops (kg/m³)

t_m Mixing time (s)

v_{tip} Impeller tip speed (m/s)

W Width of turbine blades (m)

v_s Superficial gas velocity (m/s)

Glossary

COD	Chemical oxygen demand
GER	Gross energy requirement
HBR	Hydrogen bioreactor
HPR	Hydrogen production rate
HRT	Hydraulic retention time
HTB	Heterotrophic bacteria
NUEP	Net utilizable energy product
OLR	Organic-loading rate
PTB	Photoheterotrophic bacteria

References

Blok, K., Williams, R.H., Katofsky, R.E., and Hendriks, C.A. 1997. Hydrogen production from natural gas, sequestration of recovered CO_2 in depleted gas wells and enhanced natural gas recovery. *Energy*, 22, 161–168

Büchs, L. and Zoels, B. 2011. Evaluation of maximum to specific power consumption ratio in shaking bioreactors. *Journal Chemical Engineering of Japan*, 34, 647–653.

Das, D. 1985. Biomethanation of mixed agricultural residues, Ph.D. Thesis, IIT, Delhi.

Das, D. 2009. Advances in biohydrogen production processes: An approach towards commercialization. *International Journal of Hydrogen Energy*, 34, 7349–7357.

Das, D., Khanna, N., and Veziroğlu, N.T. 2008. Recent developments in biological hydrogen production processes. *Chemical Industry & Chemical Engineering Quarterly* (CI & CEQ), 14, 57–67.

Gerson, D. F., Bontius, M. L., Wojtyk, M. A., Lee, J. F., Johnson, J. J., Lee, D. C,. Kahn, R., and Gerson A. R. 2001. Measuring bioreactor mixing with mixmeter. *Genetic Engineering News*, 21, 68–69.

Hughmark, G.A. 1980. Power requirements and interfacial area in gas–liquid turbine agitated systems. *Industrial & Engineering Chemistry Process Designand Development*, 19, 638–641.

Ju, L.K. and Chase, G.G. 1992. Improved scale up strategies of bioreactors. *Bioprocess Engineering*, 8, 49–53.

Junker, B. 2004. Scale up methodologies for *Escherichia coli* and yeast fermentation processes. *Journal of Bioscience and Bioengineering*, 97, 347– 364.

Kotay, S.M. 2008. Microbial hydrogen production from sewage sludge, Ph.D. Thesis, IIT, Kharagpur.

Kumar, N., Ghosh, A., and Das, D. 2001. Redirection of biochemical pathways for the enhancement of H_2 production by Enterobacter cloacae. Biotechnology Letters, 23(7), 537–541.

Li, Y.-C., Liu, Y.-F., Chu, C.-Y., Chang, P.-L., Hsu, C.-W., Lin, P.-J., and Wu, S.-Y. 2012. Techno-economic evaluation of biohydrogen production from wastewater and agricultural waste. *International Journal of Hydrogen Energy*, 37, 15704–15710.

Lin, C.-Y., Wu, S.-Y., Lin, P.-J., Chang, J.-S., Hung, C.-H., Lee, K.-S. et al. 2011. A pilot-scale high-rate biohydrogen production system with mixed microflora. *International Journal of Hydrogen Energy*, *36*, 8758–8764.

Micheletti, M., Barrett, T., Doig, S.D., Baganz, F., Levy, M.S., Woodley, J.M., and Lye, G.J. 2006. Fluid mixing in shaken bioreactors: Implications for scale-up predictions from microlitre-scale microbial and mammalian cell cultures. *Chemical Engineering Science*, *61*, 2939–2949.

Morimoto, M., Atsuko, M., Atif, A.A.Y., Ngan, M.A., Fakhru'l-Razi, A., and Iyuke, S.E. 2004. Biological production of hydrogen from glucose by natural anaerobic microflora. *International Journal of Hydrogen Energy*, *29*, 709–713.

Nath, K. and Das, D. 2004. Biohydrogen production as a potential energy resource—Present state-of-art. *Journal of Scientific Industrial Research*, *63*, 729–738.

Nayak, B.K., Pandit, S., and Das, D. 2013. Biohydrogen. In *Air Pollution Prevention and Control*, eds. C. Kennes, and R. C. Veiga, pp. 345–381. John Wiley & Sons Ltd., West Sussex, UK.

Nealon, A.J. O'Kennedy R. D., Titchener-Hooker, N. J.and Lye, G. J. 2006. Quantification and prediction of jet macro-mixing times in static microwell plates. *Chemical Engineering Science*, *61*, 4860–4870.

Ren, N., Li, J., Li, B., Wang, Y., and Liu, S. 2006. Biohydrogen production from molasses by anaerobic fermentation with a pilot-scale bioreactor system. *International Journal of Hydrogen Energy*, *31*, 2147–2157.

Ren, N., Wang, B., and Huang, J.C. 1997. Ethanol-type fermentation from carbohydrate in high rate acidogenic reactor. *Biotechnology and Bioengineering*, *54*, 429–433.

Rushton, J.H., Costich, E.W., and Everett, H.J. 1950a. Power characteristics of mixing impellers: Part I. *Chemical Engineering Progress*, *46*, 395–404.

Rushton, J.H., Costich, E.W., and Everett, H.J. 1950b. Power characteristics of mixing impellers: Part II. *Chemical Engineering Progress*, *46*, 467–476.

Vatsala, T.M., Mohan Raj, S., and Manimaran, A. 2008. A pilot-scale study of biohydrogen production from distillery effluent using defined bacterial co-culture. *International Journal of Hydrogen Energy*, *33*, 5404–5415.

Yanga, H., Shaoc, P., Lub, T., Shena, J., Wangb, D., and Xub, Z. 2006. Continuous biohydrogen Production from citric acid wastewater via facultative anaerobic bacteria *International Journal of Hydrogen Energy*, *31*, 1306–1313.

11

Biohydrogen Production Process Economics, Policy, and Environmental Impact

11.1 Introduction

In his State of the Union message in January 2003, President Bush announced a major new initiative: powering the world through hydrogen. He proposed a huge research funding amassing $1.2 billion that he said would enable the United States to "lead the world in developing clean, hydrogen-powered automobiles." Spurred by this new federal support, Bush said, "our scientists and engineers will overcome obstacles to taking these cars from laboratory to showroom, so that the first car driven by a child born today could be powered by hydrogen and be pollution-free." However, it is not as simple as it sounds. The major consideration of regarding the hydrogen as a proposed fuel is the fuel efficiency and the economics of the process.

This chapter examines the present status of hydrogen energy and looks at process economics, environmental impact, and special system applications of hydrogen energy around the world at the end of the 20th century. It must be noted that since microbial hydrogen production contributes a very small percentage of the current total hydrogen production, therefore all the policies, environmental and political impacts, have been stated in view of the same.

11.2 Process Economy

The economics of a bioprocess is dependent on the performance of the technologies, ease of operation, maintenance, and longevity. Collectively these factors impact the owner's cost to operate a bioprocess manufacturing operation. The preliminary economic evaluation of a process for manufacturing a biological product usually involves the estimation of capital investments, estimation of operating costs, and analysis of profitability. A critical hurdle on the

IS HYDROGEN THE IDEAL FUEL?

Hydrogen has a low-energy density by volume at a standard temperature and atmospheric pressure. One gram of hydrogen gas at room temperature occupies about 11 L of space. Storing the gas under pressure or at temperatures below −253°C, at which point it turns into a liquid, raises its volumetric density. In liquid form, hydrogen has only one-third as much energy per liter as gasoline. If stored as compressed gas at 300 atmospheres (a more practical option), it delivers less than one-fifth the energy per volume as gasoline. Such low-energy density means that fuel storage would take up lots of room in a hydrogen-powered car—or, alternatively, a modest-sized fuel tank would severely restrict the vehicle's range between fill-ups (Mudler, 2006). However, these considerations are technological hurdles, which may be overcome. Presently, technology is being developed to allow higher pressures that would make hydrogen cars more attractive.

road to the hydrogen economy is the efficient and clean production of electricity. As an energy carrier, hydrogen has to be manufactured from a primary energy source. There are many industrial methods currently available for the production of hydrogen, but all of them are expensive compared to the cost of supplying the same amount of energy with conventional forms of energy.

THE HYDROGEN CALCULATIONS

Bossel et al. (2006) gave calculations regarding the economy of production of hydrogen. One U.S. gallon of gasoline can be replaced by 1 kg of hydrogen. Considering the water electrolysis process, about 200 MJ (55 kWh) of dc electricity is needed to generate 1 kg of hydrogen from 9 kg of water by electrolysis. A quantity of 4.5 kg of water is required to produce 1 kg of hydrogen by steam reforming of methane (natural gas). Concomitant to hydrogen production, this process also releases 5.5 kg of CO_2. Similarly, considering the thermal process, 3 kg of coal and 9 kg of water can produce 1 kg of hydrogen and simultaneously also evolve 11 kg of CO_2. Moreover, presently even the most efficient fuel cell is known to convert no more than 50% of the hydrogen higher-heating value into electricity.

Considering this, the hydrogen economy looks bleak! However, when the world started out with petroleum from the wood, did the economy look any brighter? The larger question looming is probably to economize the hydrogen production from renewable resources employing nature's catalyst: microorganism.

THE HYDROGEN ECONOMY

The "hydrogen economy" is being promoted as one of the solutions to the world's energy problems. The term "hydrogen economy" was first used during the oil crisis of the 1970s to describe a national (or international) energy infrastructure based on hydrogen produced from nonfossil energy sources (NY, State Energy Research and Development Authority, report). The cost of hydrogen production from different sources varies widely given unique capital equipment costs, feedstock cost, availability, transport, and technology maturity. Presently, 48% of hydrogen is currently produced from natural gas, 30% from oil and 18% from coal. The remaining 4% is produced through water electrolysis (US DoE Report, 2003). Conventionally produced hydrogen gas costs about twice that of natural gas or oil and about three times more than coal. At present, only the space industry seems to be willing to pay the high cost of hydrogen energy. These industrial methods mainly consume fossil fuels as energy source and are considered to be energy intensive and not always environmental friendly.

11.2.1 Technical and Cost Challenges

Cost analysis of hydrogen is based on a wide range of factors. These include large-scale infrastructure to transport, distribute, store, and dispense it as a fuel for vehicles or for stationary uses (Table 11.1). It should be noted that the direction and importance of these factors may change over time and hence the costs of the product may vary accordingly.

TABLE 11.1

Factors Affecting the Cost of Hydrogen Production

Drivers of Hydrogen Economics	Rationale
Production	Since hydrogen is an energy carrier, it needs to be extracted from other sources such as water, biomass, fossil fuel, etc. The process involves thermal, electrolytic, and photolytic processes.
Storage	Hydrogen needs to be stored like other forms of energy. Conventional sources involve liquefied and gaseous energy storage. Recent advancements include reversible and irreversible metal hydride systems and carbon nanotubes.
Distribution	It involves the distribution of hydrogen from production and storage sites. Involves pipelines, trucks, rails, etc.
Supply cost and demand	Involves the end use of hydrogen such as in vehicular transport and fuel additives and the cost of such a supply.
Conversion	The making of energy and/or thermal energy. May involve fuel cells and the likes.

11.2.1.1 Production

The cost of production of hydrogen depends on the mode of generation of the product and the feedstock/raw material used. Generally, the higher costs of commercial rates for feedstock and utilities coupled with lower-operating rates lead to higher hydrogen costs. It is well known that the technology which competitively enters the market early will be facilitated by further development of infrastructure to accommodate it. Under such circumstances, further understanding of the process will be realized by practical application of the technology, which may add to further lower the cost. Alternatively, railroads, gas pipelines, biomass collection systems, nuclear thermochemical, and solar thermochemical could be developed in regional markets, where these energy sources are economically advantageous. It must be realized that costs will vary immensely between centralized and decentralized plants for generation of energy.

11.2.1.2 Storage

It is well known that hydrogen on one hand has the highest gravimetric density of any fuel, while on the other hand it is also the lightest element on the earth. As mentioned elsewhere in this book, the density of hydrogen is so low that all free available hydrogen escaped the earth's surface millions of years ago, such that the hydrogen available on the earth's surface is only that which is bound to other elements. However, this property of hydrogen has negative implications on hydrogen storage. Table 11.2 highlights the different types of hydrogen storage that are currently under use or investigation. Due to the low-volumetric energy density of hydrogen, it must be compressed and stored as a gas in a pressurized container or chilled and stored in a cryogenic

TABLE 11.2

Different Storage Containers Used to Store Hydrogen in the Three Different Physical States (Solid, Liquid, Gas)

State of Storage of Hydrogen	Types of Storage Containers
Gaseous State	
Compressed fuel storage as gas	• Cylindrical tanks
	• Quasi tanks
Liquid State	
Liquid hydrogen storage	• Cylindrical tanks
	• Elliptical tanks
Solid State	
Solid state conformable storage	• Hydride storage material
	• Carbon adsorption
	• Glass microspheres

liquid tank. Both the techniques are commercially used today. Compact storage of hydrogen gas in tanks is the most mature storage technology, but is difficult because of low density of hydrogen at standard temperature and pressure. This is addressed through compression to higher pressures or interaction with other compounds. In addition, storage tank materials are advancing—they are getting lighter and are able to provide improved containment. Some have a protective outside layer to improve impact resistance and safety. However, the techniques fall short of meeting the vehicle manufacturers' goals for safe and efficient energy storage. On the other hand, liquid hydrogen is stored in cryogenic containers, which requires less volume than gas storage. However, the liquefaction of hydrogen consumes large quantities of electric power, equivalent to about one-third the energy value of the stored hydrogen. Also, extremely efficient tank insulation is required to maintain this low temperature. Sizes of the tank can range from several liters, though smaller tanks are much less efficient.

Previous attempts were made to store hydrogen as chemically bound to hydrocarbon materials, including gasoline, methane, and methanol. However, it was further realized that applying such materials to vehicular transportation machinery required an onboard fuel processor to release and separate the hydrogen. Some of the impurities remaining in the separated hydrogen stream from this onboard processing have been found to adversely affect the performance of fuel cells. This has led the U.S. Department of Energy (DOE) and the automotive industry to re-address the onboard hydrogen storage issue. Presently, the potential for storing hydrogen safely and efficiently in a solid state is under investigation. In another technology, hydrogen can be stored "reversibly" and "irreversibly" in metal hydrides. In reversible storage, metals are generally alloyed to optimize both the system weight and the temperature at which the hydrogen can be recovered. When the hydrogen needs to be used, it is released from the hydride under certain temperature and pressure conditions and the alloy is restored to its previous state. In irreversible storage, the material undergoes a chemical reaction with another substance, such as water, that releases the hydrogen from the hydride. The by-product is not reconverted to a hydride.

CENTRALIZED VERSUS DECENTRALIZED PRODUCTION OF HYDROGEN

Renewable fuel can be mostly produced in two ways depending on their end usage. For bulk consumption, it is produced in centralized plants, however for smaller requirements it may be produced in small-scale distributed facilities. The choice has important implications on infrastructure needs. The benefit of a centralized system is that it is more economical if higher end usage is predicted. However, the drawback of such a system is the transportation of hydrogen gas possibly over large distances. This requires the development of infrastructure for the transport of gas with very low density. Presently, the pipelines used to transport natural gas may be used with modifications in its valve, compressors, etc.

Decentralized plants are set up to cater smaller needs where plenty of feedstock is readily available. It may drastically help reduce transportation costs. However, the drawback of this system is that maintenance and carbon capture of smaller units are more economically cumbersome as compared to centralized systems.

11.2.1.3 Distribution Cost

A hydrogen energy infrastructure would include production and storage facilities, structures and methods for transporting hydrogen, fueling stations for hydrogen-powered applications, and technologies that convert the fuel into energy through end-use systems that power buildings, vehicles, and portable applications. Hydrogen can be distributed through various means depending on the load to be distributed. Mostly, till date, gas such as the natural gas has been distributed through pipelines. However, it remains to be seen whether the same can be used for distribution of low-density hydrogen. Distribution through such pipelines may incur heavy diffusion loses, brittleness of materials and seals, incompatibility of pump lubrication with hydrogen and other technical issues. However, calculations show that pipeline transport may offer a cost-effective option for short transport distances with high-flow rates. For example, the cost for a 10 km pipeline transmitting about 400,000 Nm^3/day could be of the order of £1/GJ (Ogden et al., 1999). Hydrogen is, however, expensive to transport over long distances by pipeline (it requires about four to five times more energy to move than is needed for natural gas). It remains to be seen how successful research will help in bringing down the cost of hydrogen transportation over long distances. This will determine the construction and operation of centralized or decentralized hydrogen production plants.

11.2.1.4 Supply Cost and Demand

Cost of hydrogen will also depend on the cost of supply of raw material and demand of the product. The cost of supply of raw material will vary from time to time depending on inflation and the demand of raw material. Perse, as the fossil fuels are declining, the choice of these as raw materials for hydrogen production would weigh heavily on the economics of the process. As demand for natural gas increases, prices will rise and alternative technologies may become competitive. Furthermore, the availability of natural gas could be problematic if overall demand is not matched with an adequate production, delivery, and storage system. For coal, the capacity of freight rail may be the limiting factor. With a greater demand for hydrogen from coal, shipments of coal from the mine-mouth to the terminal where hydrogen is produced will rise, and a lack of freight capacity could limit the production of hydrogen (Ogden and Kaijuka, 2003).

11.2.1.5 Conversion

As mentioned, hydrogen is an energy carrier that requires production by an energy source (e.g., fossil, renewable, or nuclear) using a feedstock (e.g., fossil, biomass, or water) followed by consumption of the hydrogen by a particular end-use device to produce heat or electricity. Hydrogen can be converted to energy via traditional combustion methods and through electrochemical processes in fuel cells. Therefore, economics of fuel cell and conversion of hydrogen into energy becomes a crucial process of this economics.

11.2.1.5.1 Fuel Cells for Mobile and Stationary Use

The concept of fuel cell is not new, it dates back to the 18th century. An Englishman, William Groove, invented the cells in 1839. In 1889, Ludwig Mond and Charles Langer coined the term "Fuel cells." As part of commercial applications, the fuel cells were used in the United States and Soviet space programs in the 1960s and 1970s. Over the years, there is a general consensus that fuel cells appear to be the most promising technology to exploit hydrogen because of their high efficiency. Since electrochemical reactions generate energy more efficiently than combustion, fuel cells can achieve higher efficiencies than internal combustion engines. Current fuel cell efficiencies are in the 40–50% range, with up to 80% efficiency reported when used in combined heat and power applications.

A fuel cell is a device that uses a hydrogen-rich fuel and oxygen to produce electrical energy by means of an electrochemical reaction. The cell consists of two electrodes—an anode (negative) and a cathode (positive)—sandwiched around an electrolyte. Hydrogen is fed to the anode and oxygen to the cathode as shown in the schematic illustration of the fuel cell. The electron goes through an external circuit, creating an electrical charge. The proton migrates directly through the electrolyte to the cathode, where it reunites with the electron and reacts with oxygen to produce water and heat.

Thus, fuel cells are similar to batteries in that they are composed of positive and negative electrodes with an electrolyte or membrane. The difference between fuel cells and batteries is that energy is not recharged and stored in fuel cells as it is in batteries. Fuel cells receive their energy from the hydrogen or similar fuel that is supplied to them. No charge is thereby necessary (Figure 11.1).

Water vapor is released as the only end product. The other advantage associated with fuel cell is that it has no moving part, is small, compact, and

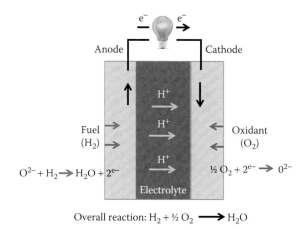

FIGURE 11.1
Schematic representation of the fuel cell.

mobile. This makes it a promising technology especially for the transport industry. Besides, stationary fuel cell for onsite production of heat and electricity is also beginning to be commercialized. Their main use is expected to be in auxiliary and distributed power generation.

The main disadvantage of the fuel cell technology is the present high costs. It is estimated that presently the fuel cells cost up to 50 times more than the traditional gasoline-filled internal combustion engine, though fuel cell efficiency is twice as high. Moreover, presently, the sources used to produce hydrogen include steam methane reformation and gasification. These processes produce CO_2, which lead to secondary issues of carbon dioxide capture and storage. However, research on microorganism-dependent microbial fuel cell for electricity production and simultaneous hydrogen production is also gaining much momentum.

11.2.1.5.2 Microbial Fuel Cell Technology

Microbial fuel cells are not new—the concept of using microorganisms as catalysts in fuel cells was explored from the 1970s (Suzuki, 1976) and microbial fuel cells treating domestic wastewater were presented in 1991 (Habbermann and Pommer, 1991). However, it is only recently that microbial fuel cells with an enhanced power output have been developed providing possible opportunities for practical applications. An MFC converts energy, available in a bio-convertible substrate, directly into electricity. This can be achieved when bacteria switch from the natural electron acceptor, such as oxygen or nitrate, to an insoluble acceptor, such as the MFC anode (Figure 11.2). This transfer can occur either via membrane-associated components or soluble electron shuttles. The electrons then flow through a resistor to a cathode, at which the

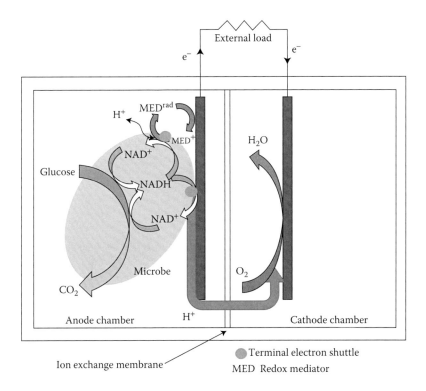

FIGURE 11.2
Design and working model of an MFC.

electron acceptor is reduced. In contrast to anaerobic digestion, an MFC creates electrical current and an off-gas containing mainly carbon dioxide. MFCs have operational and functional advantages over the technologies currently used for generating energy from organic matter. Firstly, the direct conversion of substrate energy into electricity enables high-conversion efficiency. Secondly, MFCs operate efficiently at ambient, and even at low, temperatures distinguishing them from all current bioenergy processes. Thirdly, an MFC does not require gas treatment because the off-gases of MFCs are enriched in carbon dioxide and normally have no useful energy content. Fourthly, MFCs do not need energy input for aeration provided the cathode is passively aerated. Fifth, MFCs have potential for widespread application in locations lacking electrical infrastructures and also to expand the diversity of fuels we use to satisfy our energy requirements.

11.2.2 Economics of a Hydrogen Infrastructure

Presently, there is no clear picture of what the infrastructural organization of a hydrogen economy will look like in detail. Cost analysis of hydrogen

TABLE 11.3

Cost Analysis of Hydrogen Produced from Various Sources

Hydrogen Production Method	Efficiency of CO_2 Capture	Capital Cost (£/GJ)	Hydrogen Cost (£/GJ)
Natural gas steam reforming	No	9.06	3.27
Partial oxidation of heavy hydrocarbons	No	13.59	4.57
Partial oxidation of coal	No	20.38	6.46
Partial oxidation of biomass (gasification)	No		5.67
Biomass pyrolysis	—		5.79
Methane pyrolysis	—		3.79
Electrolysis	—	7.88	6.53

reveals that costs vary significantly with the mode of production (Table 11.3). Moreover, with the present methods of production of large-scale hydrogen (conventional means), it is unclear whether introducing hydrogen can culminate into CO_2 reduction. This primarily depends on the source of production of hydrogen (Table 11.4).

Besides, it needs to be studied in detail about the cost analysis of implementation of a radically new technology. It must be understood that the introduction of hydrogen on a large scale would require a radical transformation of the energy-supply system. New hydrogen-related infrastructure (distribution facility, modified engines) will be required to replace the existing fossil fuel-based system. A vast infrastructure to produce, transport, store, and deliver hydrogen, as well as to manufacture fuel cells would need to be built. However, end consumers will be required to buy the technology. Moreover, large-scale investment would be required to capture the carbon released during its production. The total investment required would depend on the time of implementation of the technology, subsidies levied by the government, reduction of costs by research efforts, and the extent

TABLE 11.4

CO_2 Generation from Various Sources Used to Produce Hydrogen

Fuel Type	Chemical Formula	Value kJ/kg	CO_2 Generated/kg Fuel
H_2 from non-fossil energy	H_2	141,825	0
H_2 from coal liquid	H_2	141,825	36.37
Bituminous coal	$C_{135}H_{96}O_9NS$	29,527	5.71
Fuel oil/gasoline	$(CH_2)_n$	45,570	6.92
Natural gas (NG)	CH_4	55,800	6.06
H_2 from natural gas	H_2	141,825	15.43

to which the technology is applied. Cost may not be the only deterrent. As with any new technology, use of hydrogen in the transport sector could face the classical chicken and hen conundrum. Big companies hesitate in investing in this area since there is no distribution network available. On the other hand, governments are not investing in developing distribution networks due to lack of available end users. Considering these aspects, it may be a wise move to launch hybrid cars. The prospect looks lucrative on several grounds:

- The small I.C.E. in a hybrid can be configured to run on both gasoline and hydrogen interchangeably—possibly with a flip of a switch on the dashboard. This would ensure that vehicle performance largely remained unaffected (acceleration comes from the battery).
- The design of the hybrid car would be such that compressed gas storage for hydrogen would occupy only some but not a large space as traditionally predicted, with the current storage techniques.
- Lastly, the hybrid cars would have the advantage to run on hydrogen or gasoline, depending on the availability of the fuels. When hydrogen refueling is available *per se* in city center, the hybrid vehicle would run on hydrogen; outside the city center, it could run on gasoline.

Biomass, especially organic waste, offers an economical, environment-friendly way for renewable hydrogen production. The list of some biomass material used for hydrogen production is given in Table 11.5.

Biologically produced hydrogen is currently more expensive than other fuel options using carbohydrate-rich synthetic substrate. However, with a suitable renewable biomass/wastewater and ideal microbial consortia can convert this biomass efficiently into hydrogen. However, with a suitable renewable biomass/wastewater as substrate and ideal microbial consortia that can convert this biomass efficiently into hydrogen it is possible to bring down the overall production cost. Das (2009) showed that by efficient utilization of, municipal sewage/wastewater the cost of hydrogen could be as low as $1.3/GJ. Thus, if technology improvements succeed in lowering the costs, it is likely to play a major role in the economy in the future. The socially relevant costs of bringing any fuel to market must also include such factors as air pollution and other short- and long-term environmental costs as well as direct and indirect health costs. When these factors are taken into consideration, together with the initial cost competitiveness, hydrogen is surely the most logical choice as a worldwide energy medium (Das et al., 2008).

Every biohydrogen process has its own merits and demerits in terms of technology and productivity, none of them has been evaluated rigorously in terms of the cost for commercialization (Hallenbeck, 2009; Sinha and

TABLE 11.5

List of Biomass Used for Hydrogen Production by Various
Conversion Methods

Biomass Species	Main Conversion Process
Biomass species	Main conversion process
Bionut shell	Steam gasification
Olive husk	Pyrolysis
Tea waste	Pyrolysis
Crop straw	Pyrolysis
Black liquor	Steam gasification
Municipal solid waste	Supercritical water extraction
Crop grain residue	Supercritical fluid extraction
Pulp and paper waste	Microbial fermentation
Petroleum basis plastic waste	Supercritical fluid extraction
Manure slurry	Microbial fermentation
Palm oil waste	Microbial fermentation
Distillery waste	Microbial fermentation
Whey	Microbial fermentation
Waste from oil refinery	Microbial fermentation
Rice slurry waste	Microbial fermentation
Food waste	Microbial fermentation
Industrial waste water	Microbial fermentation
Sugar industry waste	Microbial fermentation
Paper industry waste	Microbial fermentation

Pandey, 2011). Only a limited number of economic analyses of biohydrogenation processes are available. However, it is to be noted that the biobased
hydrogen production is the most cost-effective way of producing hydrogen
and is also the most environmentally cooperative process. As discussed earlier, it is based on photolysis of naturally available abundant water including
water from waste streams and other biobased feedstocks. In case of biophotolysis of water, water is used as a substrate and thus the operating costs of
such processes are low as compared to dark fermentation, which requires
carbohydrates like glucose increasing the overall costs of the process. In
the biobased processes, the production costs are based on the reactor type
used for the process. It has been reported that hydrogen yield is directly
proportional to the operating costs while rate is directly proportional to
the reactor costs or the installation costs. In case of photosynthesis though
operating costs are low (due to the low cost of the raw materials used in the
process), total yield is also low. To compensate for such low yields, larger
reactor sizes need to be considered to compensate the low production rates
(Hallenbeck, 2012).

Benemann (2000) provided a preliminary cost estimate for an indirect microalgal biophotolysis system. The assumed plant capacity was 280,000 Nm^3 H_2/day, equivalent to 3600 GJ/day or 1.2 million GJ/year (at 90% plant capacity). The total capital costs for the system were estimated at US$ 43 million, the annual operating costs at US$ 12 million/year, and the total H_2 production costs at US$ 10 GJ^{-1}. In this analysis, the capital costs were almost 90% of total costs at a 25% annual capital charge. The costs of the algal ponds were estimated at US$ $6/m^2$. The photobioreactors, with assumed costs of US$ 100 m^{-2}, were the major capital and operating cost factors, while the costs of gas handling were also significant. Tredici et al. (1998) carried out a preliminary cost analysis for a large-scale single-stage algal or cyanobacterial biophotolysis process in a near-horizontal tubular reactor (NHTR) system. The analysis was based on favorable assumptions, including 10% solar energy conversion efficiency. The costs of the NHTR were projected at U.S. $ 50 m^{-2}, as a maximum allowable cost target. However, the analysis did not include costs for gas handling and assumed a relatively low annual capital charge (17%). The capital-fixed costs amounted to approximately 80% of total costs of which the tubular material for the NHTR was the largest single expense. Accordingly, the H_2 production costs were estimated at US $ 15/GJ (Dasgupta et al., 2010).

For dark fermentation the costs can be further reduced by using cheaper raw substances such as sewage sludge, distillery waste, and so on. de Vrije and Claasen (2003) reported the cost of hydrogen production using lignocellulosic feedstock available locally. The plant was operated at a capacity of 10,200 Nm^3 H_2/day and consisted of a 95 m^3 thermobioreactor for hydrogen fermentation followed by photofermentation in a 300 m^3 photobioreactor for conversion of the organic metabolites into hydrogen and CO_2. They estimated an overall cost of €2.74/kg H_2 produced based on zero feedstock value and zero hydrolysis costs. The cost of hydrogen production from various processes is depicted in Table 11.6 (Blencoe, 2009).

Economic assessments of solar and biologically based hydrogen production processes have focused on the steps that need to be taken to improve the competitive position of these technologies. Within the United States Hydrogen Program, a number of technologies are used to evaluate and pare the hydrogen production technologies. Technical and economic analyses are conducted to determine the application feasibility of the process. Analyses have been conducted on the process being studied at the National Renewable Energy Laboratory to produce hydrogen from renewable resources. These processes include gasification and pyrolysis of biomass followed by steam reformation, photoelectrolysis, hydrogen production by green alga, and water–gas shift by immobilized bacteria. Additionally, a novel storage medium and a hydrogen leak detector have been investigated. Results have determined at which points these various research tasks will become feasible in the near-, mid-, and long term. Further, the outcome of each analysis has been used to plan future research.

TABLE 11.6

Unit Cost of Energy Obtained from Different Processes against the Reference of Well-Established Bioethanol

Hydrogen Source	Unit Cost of Energy of Fuel (€/GJ)	Conversion Efficiency (%)	Comments
Considering fermentative bioethanol as a standard reference	~24.22	15–30	In use in various countries including Brazil where nearly two-thirds of the cars are fuelled with ethanol.
Fermentative hydrogen	~30.76	~10	Not in practical use. Though the costs are high, the rate of hydrogen production is much higher as compared to other biological routes. Moreover, use of waste as substrate can considerably lower costs.
Photobiological hydrogen	~7.69	~10	Most economical process but is not in practical use. However, the low rate of the process is a major drawback.

Source: Adapted from Demirbas, A. 2008. *Energy Convers Management*, 49, 2106–2116.

11.3 Environmental Impact

Air quality is a major global concern. It has been estimated that 60% of the Americans live in areas where levels of one or more air pollutants are high enough to affect the public health and/or the environment. Mostly personal

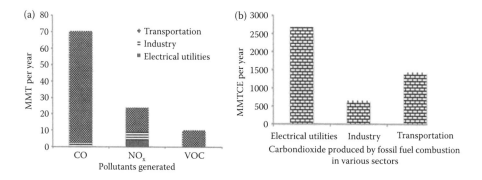

FIGURE 11.3

Emissions from fossil fuel combustion in various sectors including transportation, industry and electrical utilities. (a) Shows release of CO, NO_x, and CO. (b) Release of CO_2. (Adapted from U.S. Department of Energy (USDoE), National Energy Technology Laboratory. 2004. *Hydrogen Infrastructure Delivery Reliability R&D Needs*. Pittsburgh, Pennsylvania.)

vehicles and electric power plants are significant contributors to the nation's air quality problems (Figure 11.3). Most nations are now developing strategies for reaching national ambient air quality goals and bringing their major metropolitan areas into attainment with the requirements of the clean air act. The introduction of hydrogen-based commercial bus fleets is one of the approaches that states are considering to improve air quality.

THE NEGATIVE SIDE OF HYDROGEN!

On the negative side of the hydrogen production technology, some scientists are of the view that hydrogen is not as eco-friendly as it is proposed to be! In fact, it interferes with the environment and causes global warming thought at a lower level, as compared to fossil fuels.

British scientists have recently reviewed current understanding of the fate and behavior of hydrogen in the atmosphere and characterized its major sources and sinks. They have showed that contrary to the common belief that hydrogen is an environmentally benign fuel, it has an indirect potential for global warming. The scientists are of the opinion that with the manufacture of hydrogen its large-scale leakage into the atmosphere is inevitable. They showed that the released hydrogen would react with tropospheric OH–radicals, which would disrupt the distribution of methane and ozone, the second and third most important greenhouse gasses. Limitations to the distribution of methane and ozone would further increase their burden on the earth leading to further global warming. Therefore, hydrogen can be considered as an indirect greenhouse gas with the potential to increase global warming. Further, the group compared the potency of hydrogen as a contributor to global warming to that of CO_2 emanating from fossil-based fuel to that of hydrogen. The group found that the potency of hydrogen as a contributor to global warming was far less as compared to the present-day fossil fuel-based energy systems. However, such impacts will depend on the rate of hydrogen leakage during its synthesis, storage, and use. The researchers have calculated that a global hydrogen economy with a leakage rate of 1% of the produced hydrogen would produce a climate impact of 0.6% of the fossil fuel system it replaces. If the leakage rate was 10%, then the climate impact would be 6% that of the fossil fuel system.

Thus, the study suggests that the future hydrogen-based economy would not be completely free from climate disturbance, although this may be considerably less pronounced than that caused by the current fossil fuel energy systems. Thus, it is important to control hydrogen leakage at all stages from production to storage to conversion to minimize the effect of such indirect climatic disturbances.

Recent research regarding air pollution effects on human health describes serious lung damage sustained from fossil fuel combustion. Substituting hydrogen for fossil fuel will result in improved physical health (Zweig, 1995). The combustion of hydrogen does not produce CO_2, CO, SO_2, volatile organic carbon (VOC), and particles. However, it entails the emission of vapor and NO_x. The formation of NO_x is a function of flame temperature and duration. Considering the wide flammability range of hydrogen, its combustion can be influenced by the design of the engine so that the NO_x emission can be reduced (Momirlan and Veziroglu, 2002). Considering the health impacts, some scientists are of the opinion that introduction of hydrogen as fuel seems almost inevitable. Hydrogen introduction dramatically affects carbon dioxide in the atmosphere, which is estimated to reach the maximum before 2050 at 520 ppm. Figure 11.4 illustrates what would happen if transition to the solar hydrogen system is delayed 25 years. Energy consumption and economic activity would be higher than in "no hydrogen" scenario, but much lower than the case when hydrogen is introduced in the year 2000. Carbon dioxide would continue to increase until approximately 2070 reaching 620 ppm. If transition starts at 2050, there would be almost no positive effects. This suggests that an early transition to the hydrogen energy system would benefit the economy and the environment in the long run (Barbir et al., 1995).

In view of this, in recent years, there is a new important development in automotive technology which is aimed toward more efficient and less polluting vehicles. In view of excess pollution, several countries are promoting hydrogen-based vehicular transport to control the pollution levels. In-depth analysis shows that the fuel cell drive for city buses offers significant environmental improvements compared to diesel internal combustion engines. This refers to emissions of greenhouse gases as well as to local emissions of trace gases. The main improvement with regard to the global warming problem can nonetheless only be achieved if renewable fuels are introduced (Wurster et al., 1998). A major advantage of the fuel cell vehicles is that they represent an inherently clean and efficient technology that can optimize the use of fuels from environmentally benign energy sources and feed stocks such as solar, wind, geothermal, and biomass.

In today's world where the primary concern is energy security and a degrading environment, hydrogen seems to be the need of the hour. It makes sense to use the forms of energy that are abundant, clean, and renewable.

11.4 Hydrogen Policy

In recognition of the reduction targets for greenhouse gas set by the Kyoto Protocol, studies have focused on hydrogen as a means of meeting the demand for clean energy. In view of this, due to the wide-scale advantages

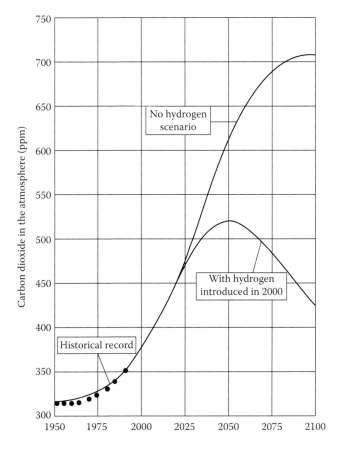

FIGURE 11.4

Carbon dioxide in the atmosphere (base case scenario). (Reprinted from Momirlan, M. and Veziroglu, T. N. 2002. Current status of hydrogen energy. *Renewable and Sustainable Energy Reviews*, 6, 141–179. With permission.)

of hydrogen, hydrogen policies are today under the energy policy for most countries.

For the smooth transition to a hydrogen-based economy, energy policy matters need to be discussed. It calls for a strong co-operation between the government and industry. Energy policy attributes to production, consumption (efficiency and emission standards), taxation and other public policy techniques, energy-related research and development, energy economy, general international trade agreements and marketing, energy diversity, and risk factors contrary to possible energy crisis. There are two key areas to be focused: (1) research, development, and demonstration of hydrogen technologies by industries and (2) incentives to encourage investment in hydrogen infrastructure by the government. In a well-planned program, the entire

CURRENT STATUS OF HYDROGEN TECHNOLOGY

Presently, it is a great challenge to get an internal combustion engine running well on hydrogen because of significantly different properties of hydrogen as compared to gasoline, particularly the density and the self-ignition energy, among other things. Still there were several studies undertaken for use of hydrogen in IC engines. The U.S. Department of Energy (DOE) undertook to test whether existing gasoline or natural gas engines could work on either pure hydrogen or hydrogen blended with other fuels. Accordingly, they tested four internal combustion vehicles using hydrogen: a Dodge Ram van and a Ford F-150 with engines designed for compressed natural gas, a Ford F-150 with a gasoline engine that was modified to run on a hydrogen-natural gas blend and a Mercedes van with a gasoline engine modified to run on pure hydrogen. The tests showed that engines driven on pure hydrogen or their blends were more economical and efficient as compared to the engine on natural gas or gasoline alone (Morrison et al., 2012). In another study, engine emissions from two modified passenger buses in Northern California powered by 20–80 volumetric H-CNG blends were studied (Burnham et al., 2004). It was found that constant power could be achieved while reducing NO_x emissions between 85% and 91% and increasing fuel economy by 15–25% as compared to pure CNG buses.

Similarly, in India, Ministry of New and Renewable Energy (MNRE), Government of India has taken extensive steps and developed a hydrogen road map to initiate hydrogen research and application. Hydrogen Energy Centre at BHU carried out test drives to demonstrate hydrogen-fuelled road transport. Their vehicles of choice were those that were used most commonly by the local population, two-wheelers, three-wheelers, and small cars. In view of their success, the International Cars and Motors Ltd. (ICML) undertook to manufacture 10 three-wheelers. These were planned to run between the Central Secretariat and Lodhi Road, New Delhi, India. Similar efforts are also being made for two-wheelers with the help of the Society for Indian Auto Manufacturers (SIAM), which has access to various two-wheeler manufacturers in India (Leo et al., 2009).

transition can be divided into four phases: the technology development phase, initial market establishment phase, infrastructure investment phase, and the last phase of a well-developed open market (Figure 11.5). In accordance, Department of Energy, the USA has developed a four-stage road-map for implementation of the hydrogen technology. According to the plan, in phase one, private organizations will research, develop, and demonstrate technologies prior to major investment in infrastructure. During this phase,

ADVANTAGES AND DISADVANTAGES OF HYDROGEN AS TRANSPORT FUEL

Hydrogen as a future fuel has a number of advantages, provided the technical hurdles can be overcome regarding its implementation. One of hydrogen's primary advantage is that it can be produced from a variety of primary resources such as biomass and water. Another important advantage of hydrogen over other fuels is that on combustion it produces water vapor as the only by-product. This can significantly limit the greenhouse gas emissions. Another property that makes it advantageous is that hydrogen can be used as a transportation fuel whereas neither nuclear nor solar energy can be used directly. It has good properties as a fuel for internal combustion (IC) engines in automobiles including a rapid burning speed, a high effective octane number, and no toxicity or ozone-forming potential. It has much wider limits of flammability in air (4–75% by volume) than methane (5.3–15% by volume), and gasoline (1–7.6% by volume). Moreover, in countries that face extreme cold winter, hydrogen may prove to be the ideal fuel as it remains in a gaseous state until it reaches a low temperature such as 20 K. This may be important to maintain the longevity of the engine.

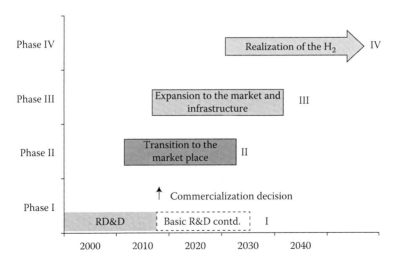

FIGURE 11.5
Roadmap for hydrogen research. (Adapted from U.S. Department of Energy (USDoE), National Energy Technology Laboratory. 2004. *Hydrogen Infrastructure Delivery Reliability R&D Needs.* Pittsburgh, Pennsylvania.)

public awareness and education would also be focused in concurrence to the ongoing research. In the next Phase II of the technology mission, the technology will be made available on wide-scale basis in the market. However, it remains to be seen which technology would penetrate the market. However, by this time, sufficient public awareness is required to purchase these products as end consumers. Here, the role of the government will be required to provide incentives to the end consumers. Such incentives could be justified by the long-term social economic and environmental benefits. They could be in the form of favorable taxation, carbon penalties, and so on. all of which would favor hydrogen. Governments will also need to work with fuel providers, equipment manufacturers, car makers, and standard setting bodies for designing, building testing, and ultimately marketing hydrogen-related equipment. As the markets become established, government can continue to foster their further growth in Phase III by patronizing the technology to stimulate the market. There are plenty of precedents in which proactive government action has caused a shift in the pattern of energy use. For example, in central Europe, government used preferential taxes to encourage the development of natural gas transmission and distribution network. In the fourth phase (Phase IV) of the roadmap, with the realization of a hydrogen market, the product can be integrated into the national infrastructure.

Further, to the benefit of the industry and stake-holders, policy interest in moving toward a hydrogen-based economy is rising. However, presently, due to some technological and economic consequences, practical experiences of hydrogen energy do not have wide applications either in the richest countries or in the poorest countries. In most countries, research is still in the R&D phase. The developed countries are ahead and are working through collaborative international programs, to facilitate the introduction of new hydrogen technologies as they become competitive. Investment in research has already led to significant developments in hydrogen-related technologies in countries such as member states of the European Union (EU), the USA, Canada, and Japan. These countries account for about two-thirds of total public hydrogen R&D spending. However, international organizations should extend their support to the developing countries for the transition to a hydrogen economy as well as hydrogen production and distribution.

In practical advancement and proto-type demonstration of the technology, British Petroleum is providing the hydrogen delivery infrastructure for transport demonstration projects in 10 cities around the world, including the CUTE (Clean Urban Transport for Europe) bus project in London (Hughes, 2007). The aim of demonstration of the projects is to thrust hydrogen technologies from the stage of research and development to the commercialization level. Current EU policies on alternative motor fuels focus on the promotion of biofuels. The definition of the marginal producer depends on the policy stance on biofuels. It should be applied in such a way that it does not create cross-subsidies between classes of consumers (Balat, 2007). In a proposed biofuels' directive from 2009, a mandatory to minimum blending shares of the biofuel has been

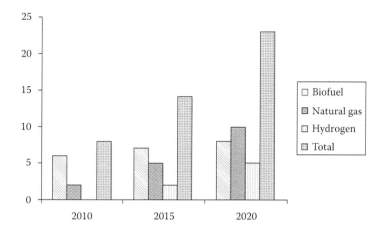

FIGURE 11.6
Share of alternative fuels in total automotive fuel consumptionin the EU under the optimistic development scenario of the EU Commison.

proposed (Figure 11.6) (Demirbas, 2008). However, to make hydrogen a part of the blend, concerted steps need to be taken to promote the end use.

A possible development path for hydrogen infrastructure includes a step-wise transition to avert high costs and address the chicken-and-egg problem. Initially when demand for hydrogen energy is low, instead of laying extensive pipeline, hydrogen can be delivered by trucks from centralized plants. Mobile refuel tanks might be used or alternatively hydrogen could be produced onsite. The advantage of onsite production is that it avoids the cost of hydrogen distribution and allows the supply to grow incrementally with demand. In the later phase, as demand increases, installation of pipelines could be considered, which would be more economical for a wider consumer. The existing infrastructure is bound to affect the way hydrogen evolves in the future. Therefore, the policy makers, academia, and industry need to act now to have a sustainable hydrogen future.

Moreover, public understanding of hydrogen will have a tremendous impact on current as well as future policy initiatives for vehicle as well as portable and stationary applications. Numerous studies have been performed to analyze the public's current perception and understanding of hydrogen. Most energy-producing technologies have an attached combination of positive and negative stigmas and means of understanding by the general public. For example, individuals who are aware of the environmental effects of a possible nuclear meltdown may deem nuclear power as a negative entity. On the contrary, individuals who are aware of the quantities of carbon emissions being reduced by using one less fossil fuel-driven power plant may find nuclear power as a positive entity. A combination of scientific understanding with common associated social themes anchored by pre-existing knowledge will have a significant impact on the future hydrogen policy.

11.5 Issues and Barriers

- The current yield and rates of production especially from the microbial process may not be economically feasible.
- For hydrogen renewables, the issue at hand is primarily the high cost of the project. In particular meeting, matching supply and demand during a transition at low cost is a key issue.
- Strong government support to help penetrate the technology into the market.
- Development of a consistent energy policy to address societal problems of climate change, air pollution, and national security. A strong consensus stand is required to cut carbon emissions.
- Public awareness must be developed about the safety and the use of hydrogen. To this end, it is important to develop safety procedures and codes for use of hydrogen in energy applications.
- The major issues facing the fast development of a hydrogen-based economy is the lack of interaction between the developers of the technology and the large-scale buyers who can put it to end use. This causes lack of hydrogen infrastructure for introduction of fuel cell vehicles.
- Moreover, the rates of hydrogen produced by the various biohydrogen systems are expressed in different units, making it difficult to assess and compare the rates and amounts of hydrogen synthesized by different biohydrogen technologies.
- Other crucial drawbacks of using hydrogen as a transportation fuel are huge on-board storage tanks, which are required because of hydrogen's extremely low density as already discussed before. However, the low-ignition temperature is one of the major advantages for hydrogen to be used directly as a fuel. Hence, it can be used as fuel indirectly by making fuel cells for producing electricity.
- Even under a scenario of technical success and strong policy, it may still be probably 10–15 years before hydrogen energy technologies start to enter the market. Therefore, they may have no immediate effect on the current oil usage and or carbon dioxide emissions.

11.6 Status of Hydrogen in the Developed and the Developing Countries

Biofuels are attracting growing interest throughout the world, with some governments announcing commitments to the biofuel programs as a way to reduce greenhouse gas emissions and dependence on petroleum-based fuels. Among all the bio-based fuel, hydrogen gains popularity. Over the next 10–20 years, vigorous government-supported RD&D programs on hydrogen and fuel cell technology will be pursued. However, the extent to which hydrogen is considered to play a role in the global energy system in the future ranges widely across the world depending on the policy adopted, the cost analysis, and the technology breakthroughs. Under the most favorable conditions, hydrogen vehicles are projected to reach shares of 30–70% of the global vehicle stock by 2050, resulting in a hydrogen demand of 7 EJ to 16 EJ. As per the current situation, it is predicted that most of this demand will be from central Europe, North America, and China. The resulting reduction of oil consumption would be in the range of 7–16 million barrels per day (Ball and Wietschel, 2009).

Although most hydrogen research is taking place in the industrialized countries, developing economies must also decide to invest in this venture, since their economies are more liable to suffer due to the political instability of the oil-producing countries. However, due to the lack of necessary funds, the developing countries may not be able to afford the cost of participating in R&D. However, engaging the developing countries early in the process may help to speed up the transition to hydrogen. Statistics indicate that in the near future the developing countries are expected to account for the bulk of the increase in global energy consumed in the coming decade. This would mean more utilization of fossil fuels, if breakthrough in energy policy is not endorsed. Therefore, the earlier these countries begin the transition to hydrogen, the quicker the world could achieve energy stability.

By far the largest ongoing projects related with hydrogen are being carried by the United States, Japan, and the EU. The challenge is to link the international and regional activities using common methodologies and tools, augmenting analysis for all countries and supporting development efforts. Some countries have integrated R&D programs that cover all elements of hydrogen supply and end uses. Primarily, the approach the governments adopt to implement the hydrogen program will reflect their own needs and resources. For example, Australia focuses on hydrogen production from coal, because it has large coal reserves. However, Germany focuses on fuel cell for vehicles as they are leaders in vehicle manufacturing. Accordingly, the amount of budget distributed and the resources used by different countries varies within their framework.

11.6.1 United States

Hydrogen received a great boost in the United States during the second term of the Bush administration. The then government showed much interest in developing hydrogen fuel cell technologies within the transportation sector. This interest was mainly driven by the desire to decrease the dependence of the United States on foreign oil and reduce the environmental impact caused due to the burning of fossil fuels. The initiative to promote fuel cell technology was announced by former President Bush in his State of the Union Address in 2003. In accordance to this, a new national commitment, a Hydrogen Posture Plan, was created in order to begin to map the future of hydrogen technology research, development, and demonstration. In order to accelerate research, development, and demonstration, former President Bush announced plans to appropriate $1.2 billion to hydrogen research.

Most of the hydrogen and fuel cell research in the United States is funded by the Department of Energy (DOE). The government's strategy is to concentrate funding on high-risk applied research on technologies in the early stages of development and leverage private sector fundings through partnerships. The United States Federal Government has created several programs to promote alternative fuels including hydrogen. These include programs such as "clean cities and clean construction the United States." The clean city program mostly focuses on practices to reduce the greenhouse gas emissions by the transport sector. Clean cities performs these duties through a network of more than 80 offices that develop public/private partnerships to promote alternative fuels, advanced vehicles, fuel blends, and hybrid vehicles. They also provide information about financial opportunities, coordinate technical assistance projects, update and maintain energy databases, and publish fact sheets, newsletters, and related technical and informational material. Clean Ports USA is an incentive-based program designed to help reduce emissions by encouraging port authorities to redesign and replace older diesel engines with new technologies and cleaner fuels. The U.S. Environmental Protection Agency's National Clean Diesel Campaign offers funding to port authorities to help them overcome the obstacles that prevent the adoption of cleaner diesel technologies. These are just two examples. There are numerous such ongoing programs run by the DoE to effectively reduce the greenhouse gas emissions and dependence on fossil fuel in the coming decade.

11.6.2 Europe

Since 1986, the EU has funded some 200 projects on hydrogen and fuel cell energy technologies with a total contribution of over EUR 550 million. These projects focused on advancements in research in all the basic areas of hydrogen research including production, storage, delivery, and use of hydrogen in cost-effective fuel cell, hydrogen-fuelled vehicles, and other related policies aimed at transition of hydrogen. Further, these projects foster long-term collaborations among different organizations that are active in the same field. By working together in projects, they exchange experience and create links that might continue cooperation even after the project has finished. Importantly, research is also channeled toward marketable solutions, as businesses and universities co-operate and partners are found to create supply chains. Further, to accelerate development and deployment of hydrogen as a fuel in the most efficient way, the EU has joint forces with European industry and research institutes in a public–private partnership of the Fuel Cells and Hydrogen (FCH), and Joint Technology Initiative (JTI).

Together, the partners will implement a programe of research, technological development, and demonstration to accelerate the commercialization of FCH technologies in a number of application areas. Additionally, recently, the Danish Government has announced a new Energy Plan 2020 that includes establishment of a range of initiatives for hydrogen infrastructure and FCEVs with the overall aim to reach 100% fossil independence by 2050. The government initiatives follow the recommendations from a recent Danish industry coalition analysis and roadmap on "Hydrogen for transport in Denmark onwards 2050."

The German government has planned to take the hydrogen research to all new levels. The Ministry of Transport, Building and Urban Development has taken initiatives to solve the classic "chicken-and-egg" problem. It has announced to develop 50 new public hydrogen fuel stations. The ministry is estimated to spend more than $50 million in the endeavor within the next year. Germany is home to 15 hydrogen fuel stations, enough to power the 5000 hydrogen-powered vehicles that are currently operating in the country. The 50 additional stations will pave the way for the rapid adoption of new hydrogen vehicles being released between the years of 2013 and 2015.

11.6.3 Asia–Pacific

It has been suggested that the pace of funding for the design and rollout of hydrogen refuelling stations will pick up in 2013, especially in Asia–Pacific and Europe. Strong interest in hydrogen refuelling stations was apparent in 2012. Japan, for example, released subsidies in 2013 to kick-start the building program of hydrogen-refuelling stations called "Subsidy for Hydrogen Supply Facility Preparation." The Japanese government has set aside a war chest of $0.5 billion for 2013. The program will provide for half a station construction cost. Japan was among the first countries to invest in hydrogen research in a 10-year project, which was completed in 2002. Soon after the finish of the first project the, Japanese government initiated a second project in 2003. The Japanese government is confident that with continued funding in this area, economical hydrogen-based fuel may soon be a reality.

11.7 Future Outlook

The role of hydrogen in today's society is inevitable if we want to realize both the energy security and the control of the pollution. In fact, globally the demand for energy is increasing in concurrence with socioeconomic standard of living. According to survey of the International Energy Agency, world energy demand will increase by half around the year 2030, with more than two-thirds of this increase will come from developing and emerging countries. For socioeconomic development, the alternative energy plays an important role. The majority of the experts consider that hydrogen has a great role to play as an important energy carrier in the future energy sector.

Biological production of hydrogen may play a key role though its contribution in today's time may occur to be insignificant. For the role of biological hydrogen to be realized, yields will have to be improved and further the economics of the process needs to be looked into greater depths by analyzing the reactor sizes and efficiencies of production. Presently, the size of the bioreactors required is too large for any practical application of the process. Concerted efforts would no doubt bring about a greater contribution of this technology to the existing hydrogen production technologies.

Although research on hydrogen production has come a long way, still concerted efforts are required for an industrial scale production. For realistic applications that are economically feasible, the hydrogen yields and production rates must surpass considerably the present achievements. More research on pilot scale productions is warranted.

Already significant work has been reported by various groups in terms of biohydrogen production and yields. There is still much work to be done before this translates into any kind of commercial application. Thus, the transition from a fossil-based economy to a hydrogen economy is a daunting task and the following points need immediate attention: A common platform such as an organization or an institute which can globally monitor the hydrogen research. Moreover, initiatives should be taken to enable researchers working with H_2 production, storage, and application research such as fuel cell to work under a common umbrella.

Glossary

CUTE	Clean urban transport for Europe
DOE	Department of energy
EU	European Union
GJ	Gigajoules
IC-engine	Internal combustion engine
PPM	Parts per million

References

Balat, M. 2007. Hydrogen in fueled systems and the significance of hydrogen in vehicular transportation. *Energy Sources Part B, 22,* 49.

Ball, M. and Wietschel, M. 2009. The future of hydrogen—Opportunities and challenges. *International Journal of Hydrogen Energy*, 34, 615–627.

Barbir, F., Plass, H. J., and Veziroglu, T. N. 1995. *Hydrogen Energy System Production and Utilization of Hydrogen and Future Aspects*, ed. Yurum, Y. *NATO ASI Series*. New York: Springer.

Benemann, J. R. 2000. Hydrogen production by microalgae. *Journal of Applied Phycology*, 12, 291–300.

Blencoe G. 2009. Cost of hydrogen from different sources. Hydrogen Car Revolution. http://www.h2carblog.com/?p. 461

Bossel, U. 2006. Does a hydrogen economy make sense? *Proceedings of the IEEE*, 94, 1826–1837.

Burnham, A., Burke, A. Collier, K., Forrest, M., McCaffrey, Z., and M. Miller. 2004. Hydrogen bus technology validation program: Analysis and update. *Proceedings, Annual Meeting of the National Hydrogen Association*, Los Angeles, CA.

Dasgupta, C. N., Jose Gilbert, J., Lindblad, P. et al. 2010. Recent trends on the development of photobiological processes and photobioreactors for the improvement of hydrogen production. *International Journal of Hydrogen Energy*, 35, 10218–10238.

Das, D., Khanna, N., and Veziroğlu, T. N. 2008. Recent developments in biological hydrogen production processes. *Chemical Industry and Chemical Engineering Quarterly*, 14, 57–67.

Das, D. 2009. Advances in biohydrogen production processes: An approach towards commercialization. *International Journal of Hydrogen Energy*, 34, 7349–7357.

Demirbas, A. 2008. Biofuels sources, biofuel policy, biofuel economy and global biofuel projections. *Energy Convers Management*, 49, 2106–2116.

Habermann, W. and Pommer, E. H. 1991. Biological fuel cells with sulphide storage capacity. *Applied and Microbial Biotechnology*, 35, 128–133.

Hallenbeck, P. C. 2009. Fermentative hydrogen production: Principles, progress. & prognosis. *International Journal of Hydrogen Energy*, 34, 7379–7389.

Hallenbeck, P. C. 2012. Hydrogen production by cyanobacteria. In *Microbial Technologies In Advanced Biofuels Production*, ed. P. C. Hallenbeck, pp. 15–28. New York: Springer.

Hughes, A. N. 2007. Organizations and institutions relating to the development of hydrogen and fuel cell activities in the UK. *UKSHEC social science working paper no. 34*. London: Policy Studies Institute.

Leo, H. M. S., Dubey, P. K., Pukazhselvan, D. et al. 2009. Hydrogen energy in changing environmental scenario: Indian context. *International Journal of Hydrogen Energy*, 34, 7358–7367.

Momirlan, M. and Veziroglu, T. N. 2002. Current status of hydrogen energy. *Renewable and Sustainable Energy Reviews*, 6, 141–179.

Morrison, G. M., Kumar, R., Chugh, S., Puri, S. K., Tuli, D. K., and Malhotra, R. K. 2012. Hydrogen transportation in Delhi? Investigating the hydrogen-compressed natural gas (H-CNG) option. *International Journal of Hydrogen Energy*, 37, 644–654.

Mudler, R. A. 2006. A pollution-free hydrogen economy? Not so soon. *Technology Review Online*. http://muller.lbl.gov/tressays/18_hydrogen.html

Ogden, J. M., Steinbugler, M. M. and Kreutz, T. G. 1999. A comparison of hydrogen, methanol and gasoline as fuels for fuel cell vehicles. *Journal of Power Sources*, 79, 143–168.

Ogden, J. and Kaijuka, E. 2003. New methods for modeling regional hydrogen infrastructure development. *Presented at the 14th National Hydrogen Association Meeting*, Washington, DC.

Sinha, P. and Pandey, A. 2011. An evaluative report and challenges for fermentative biohydrogen production. *International Journal of Hydrogen Energy*, 36, 7460–7478.

Suzuki, S. 1976. Fuel cells with hydrogen-forming bacteria. Hospital hygiene. *Gesundheitswesen und desinfektion, 1*, 159.

Tredici, M. R., Zittelli, G. C., and Benemann J. R. 1998. A tubular internal gas exchange hydrogen production: Preliminary cost analysis. In *BioHydrogen*, ed. O. Zaborsky, pp. 391–402. New York: Plenum Press.

US DoE Energy Efficiency and Renewable Energy. 2003. Hydrogen, fuel cells and infrastructure technologies programme, production and delivery.

U.S. Department of Energy (USDoE), National Energy Technology Laboratory. 2004. *Hydrogen Infrastructure Delivery Reliability R&D Needs*. Pittsburgh, Pennsylvania.

de Vrije, T. and Claasen P. A. M. 2003. Dark hydrogen fermentation. In *Biomethane and biohydrogen*, eds. J. H. Reith, R. H. Wijffels, and H. Barten, pp. 103–123. The Hague, The Netherlands: Dutch Biological Hydrogen Foundation.

Wurster, R., Altmann, M., Sillat, D. et al. 1998. Hydrogen energy progress XII. *Proceedings of the 12th World Hydrogen Energy Conference*, Buenos Aires, Argentina: 3.

Zweig, R. M. 1995. The hydrogen economy—Phase 1. *Proceedings of the Ninth World Hydrogen Energy Conference*, PARIS, France, 1995.

Index

Printed and bound by CPI Group (UK) Ltd, Croydon, CR0 4YY

18/10/2024

01776271-0009